Exercises and Experiments in Physics

JOHN E. WILLIAMS
FREDERICK E. TRINKLEIN
H. CLARK METCALFE

HOLT, RINEHART and WINSTON PUBLISHERS
New York London Sydney Toronto

John E. Williams 25702 Via Viento, Mission Viejo, California 92675; formerly teacher of chemistry and physics at Newport Harbor High School, Newport Beach, California, and Head of the Science Department, Broad Ripple High School, Indianapolis, Indiana.

Frederick E. Trinklein Dean of Faculty and Physics Instructor, Long Island Lutheran High School, Brookville, New York, and Adjunct Professor, Physical Science Department, Nassau Community College, Garden City, New York.

H. Clark Metcalfe P.O. Box V2, Wickenburg, Arizona, 85358; formerly teacher of chemistry at Winchester-Thurston School, Pittsburgh, Pennsylvania, and Head of the Science Department, Wilkinsburg Senior High School, Wilkinsburg, Pennsylvania.

Acknowledgment of source appears with each photograph in the book.

Printed in the United States of America

ISBN: 0-03-089796-3

23 111 9

Preface

Exercises and Experiments in Physics is designed primarily as a supplement to the 1976 edition of MODERN PHYSICS by Williams, Trinklein, and Metcalfe, but may be used with any standard text for secondary schools. Exercises, each dealing with a major topic of physics according to the order of presentation in MODERN PHYSICS, are grouped together at the front of the book; experiments, in similar order, are found in the latter part of the book.

The exercises, as well as the experiments, are in semi-looseleaf form and may be detached or left intact at the option of the instructor. The authors recommend that the exercises be left in this book. It is easy to locate and check the exercise after it has been completed, and the completed exercises then become a source of readily available review material for the student. Many types of questioning techniques have been used, including multiple choice, matching, direct questions, completion, and diagram types. The exercises have been designed to promote good study habits. Many require clear thinking before writing in complete statement answers. Others promote understanding of basic physical principles by means of a sequential development of related items within the exercise. The problem work in the exercises has been designed to check the student's understanding of the mathematical principles involved. Each exercise begins on a right-hand page, with the spaces for answers to the right of each page. This makes for easy and rapid checking. The exercises can be used either as study assignments or as reviews to precede tests.

The experiments have meaningful laboratory activities and a great variety of experimental procedures. We have included sufficient experiments to permit the teacher to make selection of ones which will best fit in with the local plan of work. Physics students should have as much laboratory experience as can be provided. The experiments have been chosen as basic and desirable for an adequate program of physics experimentation. They are adjusted for successful use in one 50-minute laboratory period.

The Introduction contains many helpful suggestions for working out both the exercises and experiments. Careful attention has been given to providing clear and detailed instructions for basic laboratory procedures. A discussion of problem-solving techniques, including the methods of performing calculations with significant figures, has been included.

The Appendix contains specific instructions for the use of electric instruments, a Mathematics Refresher, and tables of physical quantities which are useful not only in working out the experiments, but also in verifying the results.

A table is included on pages xii and xiii which correlates the Exercises and the Experiments with the sections and chapters of the MODERN PHYSICS textbook.

John E. Williams
Frederick E. Trinklein
H. Clark Metcalfe

Introduction: General Suggestions to Students

Exercises and Experiments in Physics is intended to serve you as a guide for study and as a laboratory manual. It is planned to make your study time more profitable, and to give you a better understanding of physics. Some general directions are given here for both exercises and experiments. You should study these procedures before you start to use this book. As necessary, you will be referred back to various sections of this Introduction throughout the course of your work. Other special instructions for some of the exercises and experiments are included within these exercises and experiments.

EXERCISES. The exercises are based on the fundamental principles of physics found in any good textbook written for secondary school students. There are several ways in which the exercises may be used. For example:

(1) The student studies the textbook assignment thoroughly. Then, from memory, he works out as much of the exercise as he can. These responses are checked against the textbook. Imperfectly learned and omitted items are reviewed, corrected, and completed after further textbook study.

(2) The student reads fairly rapidly the assignment in the textbook. Then he works out the exercise, referring to the textbook as frequently as necessary. If any questions are answered from memory, their correctness is verified by checking against the textbook when the exercise is completed.

(3) The student begins his work with the exercise itself, and looks up each answer in the textbook as he proceeds. When the exercise is finished, the entire section being studied is read through in sequence for a more complete understanding of the topic.

Unless directed by your instructor to make omissions, you should answer all the exercise items. Some exercises have blank spaces to be filled in or questions to be answered by complete statements. Other exercise items are of the multiple choice or matching types. Still others consist of diagrams which you are to draw or complete.

PROBLEMS. It is absolutely necessary to follow an orderly procedure when solving problems. Such a procedure is discussed in Section 2.9 of MODERN PHYSICS, and it is repeated below for your convenience. You should refer to this procedure frequently to solve the problems in this book.

1. Read the problem carefully and make sure that you understand all the terms and symbols that are used in it. Write down all given data.
2. Write down the physical quantity or quantities called for in the problem, together with the appropriate units.
3. Write down the equation relating the known and unknown quantities of the problem. This is called the *basic equation*. In this step you will have to draw upon your understanding of the physical principles involved in the problem. It is usually helpful to draw a sketch of the problem and to label it with the given data.
4. Solve the basic equation for the unknown quantity in the problem, expressing this quantity in terms of those given in the problem. This is called the *working equation*.
5. Substitute the given data into the working equation. Be sure to use the proper units and carefully check the significant figures in this step.
6. Perform the indicated mathematical operations with the units alone to make sure that the answer will be given in the units called for in the problem. This is called *dimensional analysis*.
7. Estimate the order of magnitude of the answer.

8. Perform the indicated mathematical operations with the numbers. Be sure to observe the rules of significant figures.
9. Review the entire solution.

Solving a problem in physics is not merely a matter of "plugging in" a set of numbers into the appropriate formulas. Understanding the underlying principles of the problem is much more important than "getting the right answer." Carefully study the following example to see how the steps of problem solving are applied in an actual example.

Example

The volume of the moon has been calculated at $2.20 \times 10^{19}\,\text{m}^3$. A rectangular specimen of rock on the earth has the following dimensions: 1.52 cm × 2.63 cm × 2.15 cm. The rock has a mass of 28.8 g. Assuming that the rock has the same mass density as the moon, find the moon's mass.

Solution

We are given the volume of the moon together with both mass and volume information about the earthbound rock. We are asked to determine the mass of the moon.

This is basically a density problem and we have a known mass density relationship as a *basic equation* to start with. Assuming that the density of the moon is the same as the density of the rock, we have the means of deriving a *working equation* into which the problem information can be directly substituted.

Basic equation: $D = \dfrac{m}{V}$

For the moon:

$$m_\text{m} = D_\text{m} V_\text{m}$$
$$D_\text{m} = D_\text{r}$$
$$m_\text{m} = D_\text{r} V_\text{m}$$
$$D_\text{r} = \frac{m_\text{r}}{V_\text{r}} = \frac{m_\text{r}}{(l \times w \times h)_\text{r}}$$

Working equation: $m_\text{m} = \dfrac{m_\text{r} V_\text{m}}{(l \times w \times h)_\text{r}}$

We now have an equation giving the mass of the moon in terms of the volume of the moon, the mass of the rock, and the dimensions of the rock—the information given in the problem. Since our answer should be in kilograms rather than grams, and the different volume units must be reconciled, the necessary factors are inserted in the substitution step which follows. Such conversion factors are not a part of the working equation.

$$m_\text{m} = \frac{(28.8\text{ g})(2.20 \times 10^{19}\text{ m}^3)}{(2.63\text{ cm})(1.52\text{ cm})(2.15\text{ cm})}\left(\frac{10^6\text{ cm}^3}{\text{m}^3}\right)\left(\frac{\text{kg}}{10^3\text{ g}}\right)$$

By solving the units, we recognize that our answer will be in kilograms.

$$\frac{\cancel{\text{g}} \times \cancel{\text{m}^3}}{\cancel{\text{cm}} \times \cancel{\text{cm}} \times \cancel{\text{cm}}} \times \frac{\cancel{\text{cm}^3}}{\text{m}^3} \times \frac{\text{kg}}{\cancel{\text{g}}} = \text{kg}$$

By estimating the order of magnitude of the answer, we would expect to arrive at a numerical answer of the order of 10^{23}. Solving for m_m,

$$m_\text{m} = 7.35 \times 10^{22}\text{ kg}$$

You will notice that the rules of significant figures have been observed in this solution, so that the answer has the same precision as the data in the problem. Orders of magnitude were used to check on the reasonableness of the answer. In addition, dimensional analysis was used to check the unit of the answer.

EXPERIMENTS. It will save you much time, and you will be more successful in your laboratory work if you read each experiment through before coming to the laboratory. The *Purpose* of the experiment tells you what you are expected to learn, to do, and to observe. The *Introduction* and the *Suggestions* are intended to explain the apparatus used, to describe the problem presented, or to show a practical application of the experiment.

When you enter the laboratory, you should procure the necessary apparatus, set it up according to the directions in the experiment or furnished by your instructor, have your setup checked by the instructor if necessary, and proceed to obtain the necessary data. Make sure the apparatus is working properly during the course of the experiment. You should strive for as high a degree of accuracy as possible, but a less accurate result honestly obtained is always better than the worthless perfect results obtained by "doctoring" readings and other data. High school students with high school apparatus are not expected to obtain results which are as accurate as those obtained by experienced physicists with highly refined apparatus.

During the course of the experiment, the printed sheet should be used only for the directions it contains. No data should be entered on this sheet during the laboratory period. All original data should be recorded on tablet paper or in a special notebook. Only after you have completed your calculations and had them checked by the instructor, should you enter them on the printed report sheet.

MEASUREMENTS. Obtaining data in most physics experiments consists of making measurements. For making measurements we use such instruments as meter sticks, spring balances, platform balances, burets, thermometers, watches, and electric meters. Generally the smallest divisions on the scale of any of these instruments are not numbered. With a Celsius thermometer, for example, there are no numbers between the 0° C mark and the 10° C mark. Since the space between these divisions is subdivided into ten equal parts, each subdivision must represent 1° C.

In reading thermometers, voltmeter and ammeter scales, etc., you should count the number of divisions between consecutive numbers and then divide the difference between such numbers by the number of divisions. The quotient equals the value of the smallest scale division. The reading should then be made by estimating to the nearest tenth of this smallest subdivision.

When we look at the level of water or other liquid which wets glass in a buret or a graduated cylinder, we see that the liquid surface in contact with the glass is lifted slightly above the level of the water in the center of the container. This lifting of the edges of the surface produces a curved liquid surface called a *meniscus.* To read the level of the liquid correctly, you must have your eye level with the top of the column of liquid. You should look along the line AC, Fig. A. Read the graduations on the scale at the *bottom* of the meniscus. The lower edge of the meniscus appears slightly darker than the remainder of the liquid. If you look along the line BC or the line DC, you will not get a correct scale reading. If the liquid in the tube is mercury, or some other liquid which does not wet the glass, the liquid surface is convex. The level of such a surface is read at the *top* of the meniscus.

Fig. A

The numbers which represent the value of a measurement are *significant figures. The numbers obtained directly from the scale graduations and one number (the rightmost) obtained as an estimate between two successive scale graduations are significant figures.* For example, with a meter stick, which is graduated in tenths of a centimeter, we can measure the length of a block with certainty to the nearest tenth of a centimeter and estimate the length to the nearest hundredth of a centimeter. The figures which represent the value of this measurement are significant figures. The measurements in the experiments in this book are to be made in significant figures.

CALCULATIONS. In making calculations with measured quantities, we must follow certain precautions so that the calculated values obtained from the measurements are not expressed with greater precision than the precision of the original measurements warrants. Detailed directions for performing calculations with significant figures are given in MODERN PHYSICS, Section 2.5. A brief summary is given here.

(1) Addition or subtraction. When adding or subtracting similar measurements, each measurement should be rounded off so that the rightmost column of digits is the only one containing uncertain digits.

Example

The lengths of three wooden blocks are 5.3 cm, 4.57 cm, and 1.385 cm. What is their combined length?

Solution

The combined length of the three blocks cannot be expressed more precisely than the least precisely measured block. In the addition set up below, the final digit of each of the numbers is uncertain. Hence the first column containing an uncertain digit is the tenths column.

5.3 cm
4.57 cm
1.385 cm

The values are rounded off to tenths, as shown below, and the addition performed. The combined length of the three blocks is 11.3 cm.

$$\begin{array}{rl} 5.3 & \text{cm} \\ 4.6 & \text{cm} \\ 1.4 & \text{cm} \\ \hline 11.3 & \text{cm} \end{array}$$

(2) Multiplication or division. When multiplying or dividing measurements, the product or quotient should not usually contain more significant figures than are found in the term with the least number of significant figures used in the multiplication or division. In a series of multiplications or divisions, round off after each separate calculation.

Example

What is the area of the surface of a rectangular block which is 7.64 cm long and 5.45 cm wide?

Solution

The calculation of the surface area is shown at the right; the product is 41.6 cm^2. There is no justification in keeping more than three digits because an area cannot be determined precisely to the nearest ten thousandth of a square centimeter if the length and width are measured only to the nearest hundredth of a centimeter.

$$\begin{array}{r} 7.64\ \text{cm} \\ 5.45\ \text{cm} \\ \hline 3820\phantom{\ \text{cm}} \\ 3056\phantom{0\ \text{cm}} \\ 3820\phantom{00\ \text{cm}} \\ \hline 41.6\not{3}\not{8}\not{0}\ \text{cm}^2 \end{array}$$

Example

Multiply 64.7 cm × 89.3 cm.

Solution

64.7 cm × 89.3 cm = 5777.71 cm^2. But since there are only three significant figures in each of the numbers multiplied, we are permitted to keep only three digits in the product. 5777.71 cm^2 expressed to three significant figures becomes 5780 cm^2 or 5.78×10^3 cm^2. See Section 2.6 of the textbook for the method of writing numbers in exponential notation.

ERRORS AND DEVIATIONS. In most of the experiments, you will be asked to compare your results with certain accepted values or to compare two or more of your findings with each other. The difference between your experimental or calculated results and an accepted value is called "absolute error." When expressed as a percentage of the accepted value, the difference is called "relative error." In other words,

$$E_a = O - A$$

where E_a is the absolute error, O is the observed or calculated value, and A is the accepted value. Also,

$$E_r = \frac{E_a}{A} \times 100\%$$

where E_r is the relative error, E_a the absolute error, and A the accepted value.

When you compare two or more findings with each other, and do not compare them with any accepted values, then the differences are called "deviations." The equations for finding the absolute and relative deviations of your data are

$$D_a = O - M$$

where D_a is the absolute deviation, O is an observed or calculated value, and M is the mean or average of several readings. Also,

$$D_r = \frac{D_a\,(\text{ave})}{M} \times 100\%$$

where D_r is the relative deviation, D_a is the absolute deviation, and M is the mean or average of the set of readings.

Be sure to record the kind of error or deviation called for in an experiment and then calculate it with the help of the equations given above.

OBSERVATIONS. In certain experiments we are not primarily concerned with making measurements, but we do perform certain operations and observe the results. In such experiments you are called upon to record your observations. These are to be concise statements indicating clearly what you observed when the experimental procedure was carried out. Record your observations in the form of complete sentences insofar as possible.

CONCLUSIONS. At the end of several experiments, you will be called upon to state your conclusions. Faulty conclusions often arise from incorrect observations, or from failure to understand some part of the experiment. Read all the instruments with the greatest possible precision, and then compare the purpose of the experiment with your observations and data before you attempt to draw conclusions.

QUESTIONS. The questions which appear at the end of the experiments are designed to help you check your understanding of the experiment, to help you draw the proper conclusions from your experimental data, or to point out other possible applications of the experiment. Answers should be complete and independent statements.

GRAPHS. One of the most useful methods of discovering the relationship between two variable quantities is by graphing the results of an experiment in which one quantity is varied, the resultant variation in the second quantity is noted, and all other factors are held constant. The Boyle's law experiment is an experiment of this type. Boyle's law states that if the temperature is held constant, the volume of a gas varies inversely with the pressure to which it is subjected. An inverse relation of this sort means that if we start with 1 liter of gas at 1 atmosphere pressure, by increasing the pressure to 2 atmospheres, the volume is reduced to 1/2 liter; if the pressure is increased to 3 atmospheres, the volume is reduced to 1/3 liter, and so on. If the pressure is reduced to 1/2 atmosphere, the volume becomes 2 liters. A complete summary of the results of such an ideal experiment is shown in the table at the right.

PRESSURE-VOLUME RELATIONSHIP OF A GAS (Temperature Constant)

Pressure in Atmospheres	*Volume in Liters*
0.20	5.00
0.25	4.00
0.33	3.00
0.50	2.00
1.00	1.00
1.50	0.67
2.00	0.50
3.00	0.33
4.00	0.25

In constructing a graph to illustrate such data, the values of the controlled variable are usually plotted on the horizontal X axis as abscissas, and the values of the dependent variable are plotted on the vertical Y axis as ordinates. The ratio between the width and height of the grid, or rulings, should be about 1.5 to 1 for best results. In order to obtain this ratio, proper adjustment of the size of scale values must be made.

In the graph, Fig. B, each unit on the X axis represents 1 atmosphere pressure, while to maintain the proper ratio of width to height, each unit on the Y axis represents 1.50 liters. The scale should be so planned that the plotted data will require the use of almost all the scale with some space on the grid above and below the plotted values. While the scale values in the lower left hand corner of the graph should both be zero, it is permissible to "break" the scale both horizontally and vertically so as to center the curve and use as large a portion of the grid as possible.

The location of the plotted points is made in the customary algebraic manner. If the relationship is direct and should produce a straight line graph, a ruler is used and the graph drawn to give an average location of ten points, as if they did fall along a straight line. If the relationship is non-linear, a French curve is used to guide the plotting of the graph.

Scale captions are placed on both axes to clearly identify the scale values. Both the subject of the scale and the units used must be indicated. The caption for the horizontal scale is generally placed under the horizontal axis at the center. The caption for the vertical axis is best placed at the top left of the axis. The words or letters of the caption are printed parallel to the base of the graph so that the paper need not be turned to read the label.

The title of the graph must be complete, and should give the what, where, and when of the data in the order indicated. The position of the title is determined by the use of the graph. In general the title is centered at the top of the graph, and printed in larger letters than any other lettering on it.

It is usual practice to include a table giving the data from which the graph was constructed, and this should always be included in the laboratory report.

When several variables are included on the same graph, it is necessary to identify each by using a key or legend. A sample of each type of line, color, shading, or cross hatching used for distinguishing purposes is indicated and identified. This identifying key or legend is generally enclosed within a box, and where possible, is placed within the grid itself.

The proper placing of the elements of a graph is shown in Fig. B.

The following characteristics of the parts of the model graph should be noted:

1. The completeness of the title, explaining in order, what, where, when.
2. The relative size of the lettering in the title, scale captions, etc.
3. The location of the scale captions.
4. The "round" values on the scale.
5. The emphasized curve lines.
6. The order of the plotted data, from the smallest to the largest.

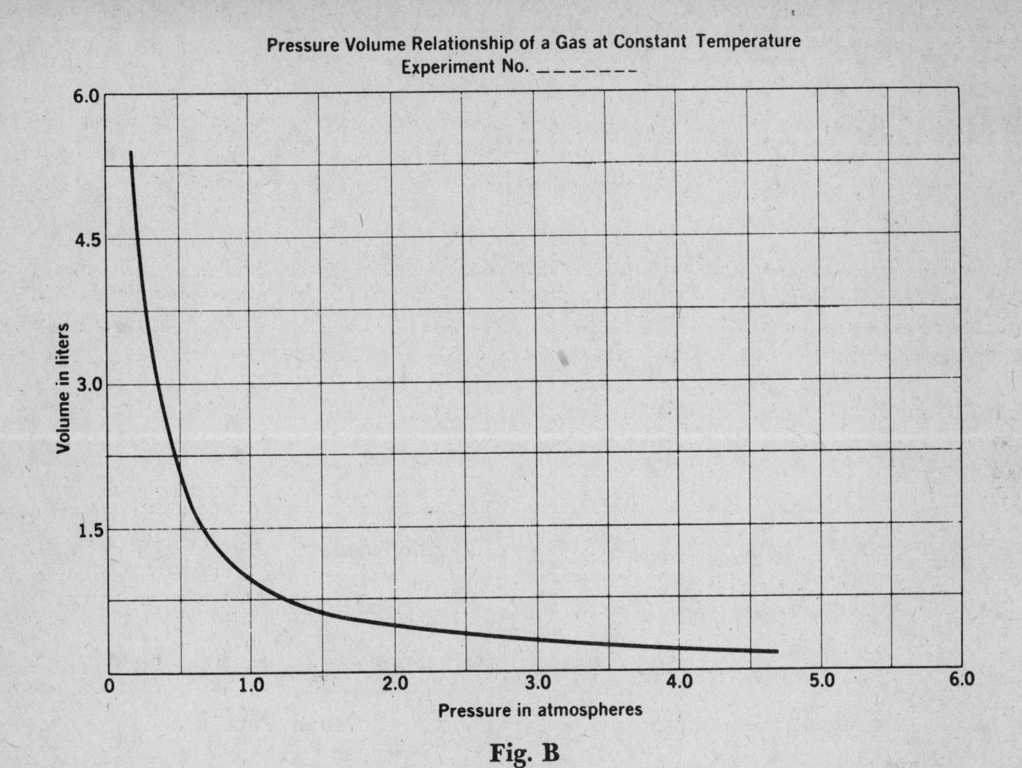

Fig. B

Contents

Exercises

Experiments

Correlation Table of Exercises and Experiments in Physics with Modern Physics 1976

Number	Exercise	Text Sections
1	Physics: The Science of Energy	1.1—1.15
2A	Units of Measurement	2.1—2.4
2B	Making and Recording Measurements	2.5—2.6
2C	Solving Problems	2.7—2.9
3A	Velocity	3.1—3.4
3B	Acceleration	3.5—3.7
3C	Newton's Laws of Motion—Gravitation	3.8—3.14
4A	Resolution of Forces	4.1—4.4
4B	Composition of Forces	4.5—4.6
4C	Friction	4.7—4.10
4D	Parallel Forces	4.11—4.14
5A	Circular Motion	5.1—5.4
5B	Rotary and Simple Harmonic Motions	5.5—5.12
6	Conservation of Energy and Momentum	6.1—6.16
7A	Molecules and Atoms—Solids	7.1—7.11
7B	Liquids and Gases	7.12—7.24
8A	Temperature and Heat Measurement	8.1—8.7
8B	Thermal Expansion	8.8—8.14
8C	Change of Phase	8.15—8.25
9	Heat and Work	9.1—9.7
10	The Nature of Waves	10.1—10.16
11	Sound Waves	11.1—11.16
12A	The Nature of Light	12.1—12.13
12B	Illumination	12.14—12.18
13	Reflection	13.1—13.12
14A	Optical Refraction	14.1—14.14
14B	Dispersion	14.15—14.22
15	Interference, Diffraction, and Polarization	15.1—15.12
16	Electrostatics	16.1—16.17
17A	Direct-Current Circuits	17.1—17.5
17B	Series and Parallel Circuits	17.6—17.13
18A	Heating Effects	18.1—18.6
18B	Electrolysis	18.7—18.9
19A	Magnetism	19.1—19.10
19B	Electromagnetism—d-c Meters	19.11—19.18
20A	Induced Currents—Generators	20.1—20.13
20B	Motors—Inductance	20.14—20.22
21A	Alternating Current	21.1—21.8
21B	Series and Parallel a-c Circuits	21.9—21.11
21C	Resonance	21.12—21.16
22	Electronic Devices	22.1—22.23
23	Atomic Structure	23.1—23.12
24	Nuclear Reactions	24.1—24.12
25A	Detection Instruments—Particle Accelerators	25.1—25.8
25B	Fundamental Particles—Atomic Models	25.9—25.15

Number	Experiment	Chapter in Modern Physics Textbook (1976)
1	Measuring Length	1, 2 The Science of Energy; Measurement and Problem Solving
2	Measuring Mass	
3	Measuring Time	
4	Mass Density	
5	Acceleration—I	3 The Nature and Cause of Motion
6	Acceleration—II	
7	Resolution of Forces	4 Resolution and Composition of Forces
8	Composition of Forces	
9	Coefficient of Friction	
10	Center of Gravity and Equilibrium	
11	Parallel Forces	
12	Centripetal Force—I	5 Curvilinear and Harmonic Motion
13	Centripetal Force—II	
14	The Pendulum	
15	Efficiency of Machines	6 Conservation of Energy and Momentum
16	Conservation of Momentum—I	
17	Conservation of Momentum—II	
18	Hooke's Law	7 Phases of Matter
19	Elastic Potential Energy	
20	Specific Heat	8, 9 Thermal Effects; Heat and Work
21	Coefficient of Linear Expansion	
22	Charles' Law	
23	Boyle's Law	
24	Heat of Fusion	
25	Heat of Vaporization	

Number	Experiment	Chapter in Modern Physics Textbook (1976)	
26	Pulses on a Coil Spring	10	Waves
27	Wave Properties		
28	Resonance; The Speed of Sound	11	Sound Waves
29	Frequency of a Tuning Fork		
30	Photometry	12	The Nature of Light
31	Plane Mirrors	13	Reflection
32	Concave Mirrors		
33	Index of Refraction of Glass	14	Refraction
34	Index of Refraction by a Microscope		
35	Converging Lenses		
36	Focal Length of a Lens		
37	Lens Magnification		
38	The Compound Microscope		
39	The Refracting Telescope		
40	Color		
41	Diffraction and Interference	15	Diffraction and Polarization
42	Wavelength by Diffraction		
43	The Polarization of Light		
44	Static Electricity	16	Electrostatics
45	Electrochemical Cells	17	Direct-Current Circuits
46	Combination of Cells—Internal Resistance		
47	Measurement of Resistance: Voltmeter-Ammeter Method		
48	Measurement of Resistance: Wheatstone Bridge Method		
49	Effect of Temperature on Resistance		
50	Resistances in Series and Parallel		
51	Simple Networks		
52	Electric Equivalent of Heat	18	Heating and Chemical Effects
53	Electrochemical Equivalent of Copper		
54	Magnetic Field About a Conductor	19	Magnetic Effects
55	Galvanometer Constants		
56	Electromagnetic Induction	20	Electromagnetic Induction
57	The Electric Motor		
58	Capacitance in a-c Circuits	21	Alternating-Current Circuits
59	Inductance in a-c Circuits		
60	Series Resonance		
61	Thermionic Emission	22	Electronic Devices
62	Diode Characteristics		
63	Triode Characteristics		
64	The Size and Shape of a Molecule	23	Atomic Structure
65	Radioactivity	24	Nuclear Reactions
66	Half-Life		
67	Simulated Nuclear Collisions	25	Particle Physics

NAME PERIOD DATE INSTRUCTOR'S COMMENTS

1 EXERCISE Physics: The Science of Energy

SECS. 1.1 - 1.15

DIRECTIONS: Write the answers to the following in the spaces provided. Where appropriate, make complete statements.

1. What is *science?* 1

2. What is *technology?* 2

3. Explain what is meant by a *scientific law.* 3

4. Explain what is meant by a *theory?* 4

5. What steps are taken by scientists in developing a scientific law or theory? 5

6. What limits the validity of a scientific conclusion? 6

7. What is the *mass* of an object? 7

8. What is *inertia?* 8

9. What is *energy?* 9

10. What is the relationship between *mass* and *energy?* 10

DIRECTIONS: In the blank space at the right of each statement, write the word or expression which BEST completes the meaning.

11. We are unable to convert the ample energy supplies in and around our planet into usable forms because of shortages of ..(11a).. and ..(11b)... .... 11a 11b

12. The complementary work of scientists and technologists is sometimes referred to as ..(12a).. and ..(12b)... .... 12a 12b

13. Civil laws restrict behavior while scientific laws ..(13).. behavior. 13

14. Scientific laws may be stated in words but are usually expressed by a(n) ..(14)... 14

15. The part of a theory which often helps scientists picture the way an observed phenomenon could be produced is an imaginary ..(15)... 15

16. A useful theory ..(16a).. past observations and ..(16b).. not yet observed behavior. 16a

.... 16b

17. A possible solution to a problem is called a(n) ..(17)... 17

18. Wherever a scientist must go to investigate a problem becomes a(n) ..(18)... 18

19. No amount of experimentation can ever ..(19a).. a scientific law absolutely, but a single crucial experiment can ..(19b).. one. 19a

.... 19b

20. Matter can be described in terms of its ..(20).. properties. 20

21. Mass measurements are based on the standard ..(21).. mass. 21

22. The mathematical equation for calculating m_2 when m_1, T_1, and T_2 are known is ..(22)... 22

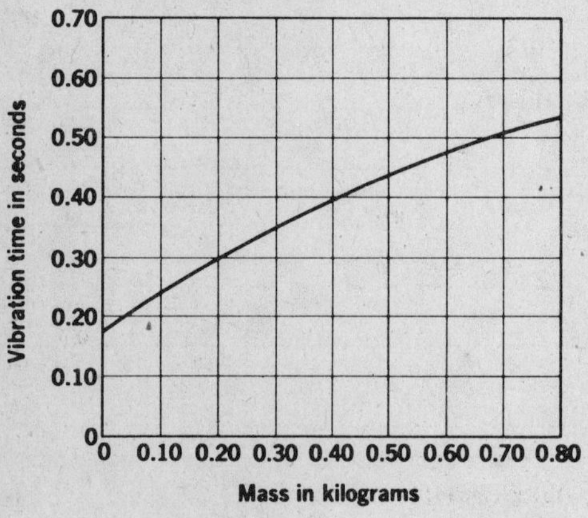

Vibration time in seconds: 0, 0.10, 0.20, 0.30, 0.40, 0.50, 0.60, 0.70

Mass in kilograms: 0, 0.10, 0.20, 0.30, 0.40, 0.50, 0.60, 0.70, 0.80

Figure 1-1

23. From Fig. 1-1, if the vibration time of the inertial balance is 0.45 sec, the combined mass of the object and the balance pan is ..(23)... 23

24. From Fig. 1-1, what mass was added to the pan of the inertial balance if the vibration time was increased from 0.30 sec to 0.50 sec? 24

25. In addition to the inertial balance, masses can be compared using a(n) ..(25a).. or a(n) ..(25b)... .... 25a

.... 25b

26. The mathematical equation for the volume of an object in terms of its mass and mass density is ..(26)... 26

27. A quantity which is descriptive of an object's environment, rather than of the object itself, is called a(n) ..(27)... 27

28. Energy of position is a form of ..(28a).. energy while energy of motion is a form of ..(28b).. energy. 28a

.... 28b

29. For each problem in gravitational potential energy, an arbitrary ..(29).. must be specified. 29

30. An object resting on the earth's surface is said to have ..(30).. kinetic energy. 30

31. The statement, "The total amount of energy of all kinds in a given situation is a constant," is known as the law of ..(28).. of energy. 31

32. The equivalence between heat and mechanical energy provided strong evidence that heat must be a form of ..(32)... 32

33. When an object is at rest with respect to an observer, the mass of the object is said to be the ..(33a).. mass; when an object is moving, its mass is called the ..(33b).. mass.

.... 33a

.... 33b

34. Einstein's equation means that matter and energy are different ..(34a).. of the same ..(34b)...

.... 34a

.... 34b

35. While we usually think of light energy in terms of ..(35a).., there are times when light acts as though it is made up of ..(35b)..; similarly ..(35c).. of matter show the properties of ..(35d)...

.... 35a

.... 35b

.... 35c

.... 35d

36. Physics is a(n) ..(36a).. science which is concerned with the relation between ..(36b).. and ..(36c)...

.... 36a

.... 36b

.... 36c

37. A distinguishing consideration in the study of physics is always the idea of ..(37)...

.... 37

38. The major divisions of physics are ..(38a).., ..(38b).., ..(38c).., ..(38d).., and ..(38e)...

.... 38a

.... 38b

.... 38c

.... 38d

.... 38e

DIRECTIONS: Formulate as many hypotheses as you can which may explain each observation.

39. A ball moves along a straight path, stops, and moves in the reverse direction along the same path.

....

....

....

....

.... 39

40. A steel ball rolls in a curved path on a plane.

....

....

....

....

.... 40

DIRECTIONS: In the parentheses at the right of each expression in the second column, write the letter of the expression in the first column which is MOST CLOSELY related.

a. Potential mechanical energy	a burning match	( )	41
b. Kinetic mechanical energy	stereo headphones	( )	42
c. Heat energy	television screen	( )	43
d. Sound energy	atomic-powered submarine	( )	44
e. Light energy	bowling ball rolling down the alley	( )	45
f. Electric energy	radio battery	( )	46
g. Nuclear energy	a hammer poised above a nail	( )	47
h. Chemical energy	a transistor	( )	48
	a charcoal grill	( )	49
	TV antenna	( )	50

NAME PERIOD DATE INSTRUCTOR'S COMMENTS

2A EXERCISE Units of Measurement

SECS. 2.1 - 2.4

DIRECTIONS: Write the answers to the following in the spaces provided. Where appropriate, make complete statements.

1. From which measurement system was the present International System of Units derived? 1

2. What is a *conceptual unit?* 2

3. What is a *standard unit?* 3

4. In what seven types of units may all physics measurements be expressed? 4

5. How was the meter originally defined? 5

6. How is the *meter* defined today? 6

7. How is the *liter* defined? 7

8. How is the *kilogram* defined? 8

9. What is a *newton?* 9

10. How is the *second* defined? 10

11. Is there a relationship between the second and the kilogram? Explain. 11

12. In the following, give the power of ten equivalent of the prefix, and give the prefix equivalent of the numerical value.

a. centi- = ...	e. atto- = ...	i. 10^{-15} = ...
b. kilo- = ...	f. pico- = ...	j. 10^{9} = ...
c. micro- = ...	g. 10^{-9} = ...	k. 0.001 = ...
d. mega- = ...	h. 10^{12} = ...	l. 0.1 = ...

DIRECTIONS: In the blank space at the right of each statement, write the word or expression which BEST completes the meaning.

13. The International System (SI) of Units is also called the ..(13).. system. 13

14. Three important fundamental SI units are a unit of length, the ..(14a)..; a unit of mass, the ..(14b)..; and a unit of time, the ..(14c)...

....14a

....14b

....14c

15. The use of prefixes with basic units gives units which are multiples or submultiples based on powers of ..(15)... 15

16. Atomic mass units are based on the mass of ..(16).. atoms. 16

17. To support a mass of one kilogram requires a force of about ..(17).. newtons. 17

Figure 2A-1 represents a cube with a volume of one cubic decimeter. Questions 18-20 refer to Fig. 2A-1.

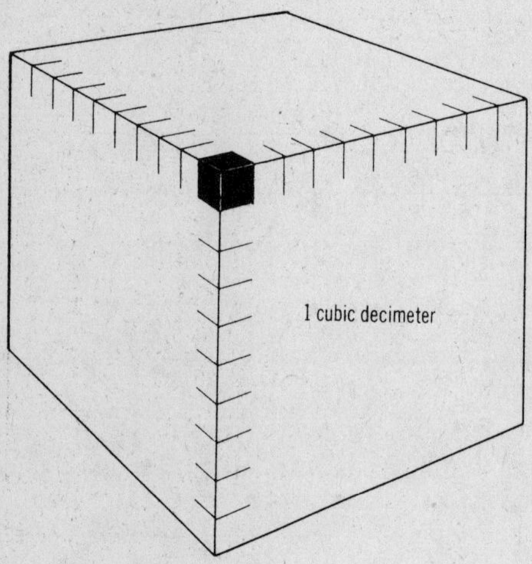

1 cubic decimeter

Figure 2A-1

18. The length of one side of this cube is ..(18a).. m or ..(18b).. cm.

....18a

....18b

19. The volume of this cube is ..(19a).. liter(s) or ..(19b).. cm^3.

....19a

....19b

20. The mass of such a volume of water at 4°C is ..(20a).. kg or ..(20b).. g.

....20a

....20b

21. An aspect of the universe such as mass is a(n) ..(21a).., while the gram in which mass is measured is a(n) ..(21b)...

....21a

....21b

22. The gravitational pull of the earth on an object is the ..(22).. of the object. 22

23. The ..(23a).. of an object does not depend on its location, but the ..(23b).. of an object does.

....23a

....23b

DIRECTIONS: Complete the following table of equivalents.

a. 369 mm	=	.... cm	g. 807 cm^3	=	.... ml
b. 0.427 m	=	.... mm	h. 14.5965 liters	=	.... cm^3
c. 704 cm	=	.... m	i. 108 g	=	.... kg
d. 615 cm	=	.... km	j. 0.823 g	=	.... mg
e. 8.973 km	=	.... m	k. 0.0009390 kg	=	.... mg
f. 2.46 ml	=	.... liter	l. 125 ml water	=	.... g water

NAME PERIOD DATE INSTRUCTOR'S COMMENTS

2B EXERCISE Making and Recording Measurements

SECS. 2.5 - 2.6

DIRECTIONS: Write the answers to the following in the spaces provided. Where appropriate, make complete statements.

1. Define *accuracy*. 1

2. Define *precision*. 2

3. Distinguish between *absolute error* and *relative error*. 3

4. Distinguish between *absolute deviation* and *relative deviation*. 4

5. What is meant by the *tolerance* of a measuring instrument? 5

6. What is the *order of magnitude* of a quantity? 6

7. How many significant figures are there in each of the following?

a.	273.16	....	e.	3,000,000	....	i.	$528\overline{0}$	....
b.	186,000	....	f.	$24\overline{0},000$	....	j.	0.070830	....
c.	707	....	g.	0.00623	....	k.	6.52×10^{-2}	....
d.	$10\overline{0}0$	....	h.	40.070	....	l.	9.040×10^{5}	....

8. In the following, convert numbers in common notation to exponential notation, and convert numbers in exponential notation to common notation.

a.	93,000,000	=	....	d.	2.997×10^{10}	=	....
b.	0.000019	=	....	e.	8.03×10^{-5}	=	....
c.	404.39	=	....	f.	4.5359×10^{2}	=	....

9. Give the order of magnitude of each of the following.

a.	8.3×10^{6}	....	d.	1.428×10^{-2}	....
b.	2.75×10^{4}	....	e.	7.793×10^{-8}	....
c.	6.023×10^{23}	....	f.	4.99×10^{-11}	....

10. Round off each of the following to the number of digits indicated.

	Number	5 digits	4 digits	3 digits	2 digits
a.	3.14159	....	....	....	....
b.	7.57530	....	....	....	....
c.	9.35550	....	....	....	....
d.	9.25450	....	....	....	....

11. In the following, which is the rightmost column to be retained in the result expressed in the proper number of significant figures?

a. 10.7 m + 6.01 m + 152.110 m = ... 11a

b. 4053 liters − 60 liters = ... 11b

c. 69.57 km + 348.01 km + 55 km = ... 11c

d. $100\overline{0}$ g − 70.0 g = ... 11d

e. 0.07050 cm + 9.60 cm + 6000.280 cm = ... 11e

f. 10.790 ml −8.0 ml = ... 11f

12. In the following, how many digits should appear in a result expressed in the proper number of significant figures?

a. 791.6 × 52 = ... 12a

b. 852 × 27.3 = ... 12b

c. 1025 × 35 = ... 12c

d. 98.4 ÷ 10.375 = ... 12d

e. $6.02 \times 10^{23} \times 12.00 \times 1.660 \times 10^{-24}$ = ... 12e

f. $453.6 \times 9.030 \times 10^{4} \times 269.1$ = ... 12f

DIRECTIONS: Place the answers to the following problems, expressed in the proper number of significant figures, in the spaces provided at the right.

13. Express the sum of 30.7 mm, 44.5 cm, and 7.01 m in meters. 13

14. If 32.5 liters of gasoline is drawn from a tank originally containing 85 liters of gasoline, what volume of gasoline remains in the tank? 14

15. What is the area of the bottom of a tank 30.0 cm long and 15.0 cm wide? 15

16. If the tank in Problem 15 has a volume of 2.25×10^{4} cm^3, what is its height in meters? 16

17. What is the capacity in liters of the tank of Problem 15? 17

18. How many kilograms of water at 4°C will the tank of Problem 15 hold? 18

19. What is the weight in newtons of an object which has a mass of 6.0 kg? 19

20. A graduated cylinder contains $4\overline{5}0$ ml of water. A lump of coal is lowered into the water and the liquid level rises to 625 ml. What is the volume of the lump of coal expressed in cm^3? 20

21-25. Suppose you obtain the following data for a measurement:

Trial	Mass	Accepted Value
1	13.26 g	13.20 g
2	13.18 g	
3	12.95 g	

Find the following:

21. E_a for Trial 3. 21

22. E_r for Trial 1. 22

23. D_a for Trial 2. 23

24. D_a (average) for the data. 24

25. D_r for the data. 25

NAME PERIOD DATE INSTRUCTOR'S COMMENTS

2C EXERCISE Solving Problems

SECS. 2.7 - 2.9

DIRECTIONS: Write the answers to the following in the spaces provided. Where appropriate, make complete statements.

1. Briefly summarize the nine steps in solving a physics problem.

   a. 1a

   b. 1b

   c. 1c

   d. 1d

   e. 1e

   f. 1f

   g. 1g

   h. 1h

   i. 1i

2. Distinguish between a *direct* and *inverse* proportion. 2

DIRECTIONS: Graph the data given in each table on the grid at the right of the table and give complete sentence answers to the corresponding questions. Instructions for preparing graphs are given in the Introduction.

3.

Elapsed Time (hr)	(min)	Distance Traveled (km)	Elapsed Time (hr)	(min)	Distance Traveled (km)
0	$2\overline{0}$	24	5	$3\overline{0}$	396
0	$5\overline{0}$	$6\overline{0}$	6	$0\overline{0}$	432
1	$4\overline{0}$	$12\overline{0}$	7	$1\overline{0}$	516
2	$3\overline{0}$	$18\overline{0}$	8	$4\overline{0}$	624
4	$1\overline{0}$	$30\overline{0}$	9	$5\overline{0}$	708

a. What is the shape of the graph? 3a

b. What is the relationship between the distance traveled and the elapsed time? ....

.... 3b

c. What name is given to the ratio of the distance traveled to the elapsed time? ....

.... 3c

d. What characteristic of the graph gives the value of the ratio in part c? ....

.... 3d

e. What is the value of the ratio in part c? ....

.... 3e

4.

Pressure (atm)	Volume (cm^3)	Pressure (atm)	Volume (cm^3)
1.0	$50\overline{0}$	6.5	77
1.5	333	7.2	69
2.5	$20\overline{0}$	8.8	57
3.0	167	9.4	53
5.0	$10\overline{0}$	10.0	$5\overline{0}$

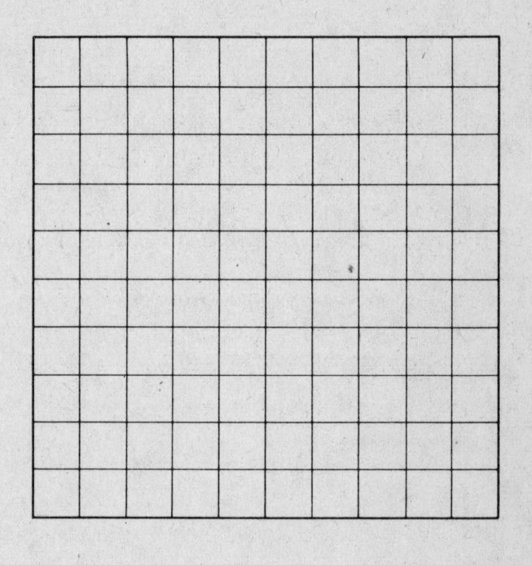

a. What is the shape of the graph? 4a

b. What is the relationship between the volume and the pressure? ....

.... 4b

c. What must be true of the product of the pressure and the volume? ....

.... 4c

d. From your graph, what is the volume when the pressure is 4.0 atm? ....

.... 4d

e. What is the calculated value of the volume in part d? ....

.... 4e

DIRECTIONS: Place the answers to the following problems, expressed in the proper number of significant figures, in the spaces provided at the right.

5. What is the average speed in km/hr of an automobile which travels 585 km in 7 hr $3\overline{0}$ min? 5

6. Find the mass of a piece of gold leaf 2.00×10^{-5} cm thick and $1.00\ m^2$ in area. The mass density of gold is $1.93 \times 10^4\ kg/m^3$. 6

7. The equation for calculating the relativistic mass of a moving object is

$$m = \frac{m_0}{\sqrt{1 - (v^2/c^2)}}$$

where m is the relativistic mass of the object, m_0 is the rest mass of the object, v is the velocity of the object, and c is the velocity of light. Calculate the relativistic mass to rest mass ratio, m/m_0, for an electron which has a velocity with respect to the observer that is 70.0% of the velocity of light. 7

DIRECTIONS: In the blank space at the right of each statement, write the word or expression which BEST completes the meaning.

8. The purpose of a scientific experiment is to discover ..(8).. between the quantities measured. 8

9. The estimation of values between the plotted points on a graph is ..(9a).., while that beyond the measurements made in an experiment is ..(9b)... .... 9a

.... 9b

10. A quantity such as mass, which can be expressed completely by a single number with an appropriate unit, is called a(n) ..(10).. quantity. 10

11. A quantity such as velocity, which requires a magnitude and a direction for its complete description, is called a(n) ..(11).. quantity. 11

12. A vector is usually represented by a(n) ..(12a).. whose ..(12b).. represents the magnitude of the vector and whose ..(12c).. shows the direction of the vector. 12a

.... 12b

.... 12c

13. The sum of two or more vectors is called the ..(13)... 13

14. A straight river with parallel banks flows at 5 km/hr. A boat moves directly downstream relative to the water at 12 km/hr. The velocity of the boat relative to the river bank is ..(14)... 14

15. If the boat moves directly upstream relative to the water of the river of No. 14 at 9 km/hr, its velocity relative to the river bank is now ..(15)... 15

DIRECTIONS: Construct the diagrams for the following problems to a suitable scale using a well-sharpened pencil, a straight edge, and protector. Label the diagrams to agree with the vector quantities described.

A

B

16. In Block A represent an eastward vector quantity OE of 45 units and a northward vector quantity ON of 35 units. By construction, find the resultant vector, OR. Indicate on the diagram the magnitude and direction of OR found graphically.

17. In Block B represent a southward vector quantity OS of 5.0 units and a vector quantity OV of 8.0 units 15° north of east. By construction, find the resultant vector, OR. Indicate on the diagram the magnitude and direction of OR found graphically.

NAME PERIOD DATE INSTRUCTOR'S COMMENTS

3A EXERCISE Velocity

SECS. 3.1 - 3.4

DIRECTIONS: In the parentheses at the right of each word or expression in the second column, write the letter of the expression in the first column which is MOST CLOSELY related.

a. Speed in a particular direction
b. Characterized by a changing velocity
c. Speed over all randomly chosen intervals has the same value
d. Change of position in a particular direction
e. Displacement of an object in relation to objects considered to be at rest
f. Motion along a curved path
g. Diagonal of a velocity parallelogram
h. Time rate of motion
i. Motion about an axis
j. Represents the magnitude and direction of a velocity
k. Represented by the slope of the tangent to a curve of a graph
l. Characterized by a constant velocity

displacement ( ) 1
motion ( ) 2
speed ( ) 3
constant speed ( ) 4
velocity ( ) 5
uniform motion ( ) 6
variable motion ( ) 7
instantaneous velocity ( ) 8
velocity vector ( ) 9
resultant velocity ( ) 10

DIRECTIONS: In the blank space at the right of each statement, write the word or expression which BEST completes the meaning.

11. A displacement is a vector quantity. This means it has ..(11a).., ..(11b).., and point of ..(11c)...

.... 11a

.... 11b

.... 11c

12. If we compare the displacement of one object with respect to another, we can detect ..(12).. motion.

.... 12

13. The slope of a line of a graph is the ratio of the change in the quantity plotted ..(13a).. divided by the corresponding change in the quantity plotted ..(13b)...

.... 13a

.... 13b

14. If path length in meters is plotted on the vertical axis of a graph, and time in seconds is plotted on the horizontal axis, the slope of a line on the graph represents ..(14a).. in the units ..(14b)...

.... 14a

.... 14b

15. If a graph representing speed is a straight line, the speed is ..(15)...

.... 15

16. Average speed is calculated by ..(16a).. the total ..(16b).. by the ..(16c)...

.... 16a

.... 16b

.... 16c

17. A graph representing variable speed will be a(n) ..(17).. line.

.... 17

18. Instantaneous speed on the graph of No. 17 is represented by the ..(18a).. of the ..(18b).. to the line at any point.

.... 18a

.... 18b

19. Because speed is completely described in terms of magnitude alone, it is a(n) ..(19).. quantity. 19

20. When a motion is at a particular speed in a particular direction, the term ..(20).. is used. 20

21. If *s* represents displacement during the time interval *t*, the equation for average velocity v_{av} is ..(21)... 21

22. The ratio *s/t* at a particular instant is the ..(22)... 22

23. If the ratio *s/t* is constant, the motion is said to be ..(23)... 23

24. If the ratio *s/t* changes, the motion is said to be ..(24)... 24

25. Since velocities require magnitude and direction for their description and they combine appropriately, velocities are ..(25).. quantities. 25

DIRECTIONS: Place the answers to the following problems in the spaces provided at the right.

26. An object moves along a straight path 3.0 m during 1.0 sec, 6.0 m during 2.0 sec, and 9.0 m during 3.0 sec. What is its velocity? 26

27. The distance from Cincinnati to Atlanta is $75\bar{0}$ km. **a.** If the driving time is 11.25 hr, what is the average speed? **b.** What is the average direction of this displacement? 27a

.... 27b

28. A motorboat travels 20.0 km/hr in still water. What will be the magnitude and direction of the resultant velocity if the boat is headed directly across a river flowing at the rate of 8.00 km/hr? 28

29. What is the magnitude and direction of the resultant of a velocity of $15\bar{0}$ km/hr northward and a velocity of 25.0 km/hr eastward? 29

30. An airplane flies southward at $80\bar{0}$ km/hr. The wind is from the west at 50.0 km/hr. What is the magnitude and direction of the resultant velocity of the airplane? 30

NAME PERIOD DATE INSTRUCTOR'S COMMENTS

3B EXERCISE Acceleration

SECS. 3.5 - 3.7

1. Complete the following table by obtaining velocity data from Fig. 3B-1 and calculating the average acceleration and displacement for each time interval. Then calculate the total displacement.

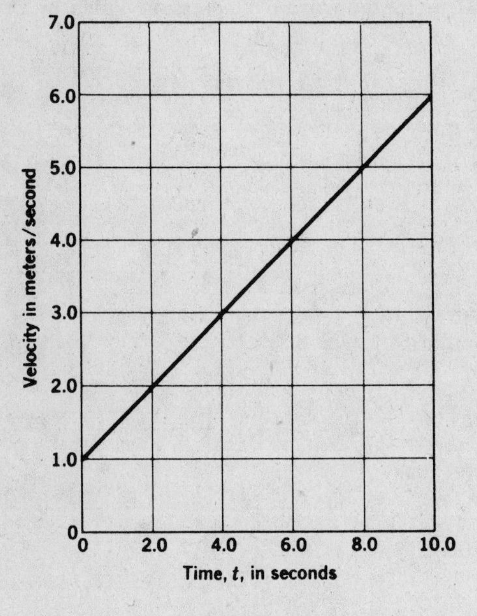

Fig. 3B-1

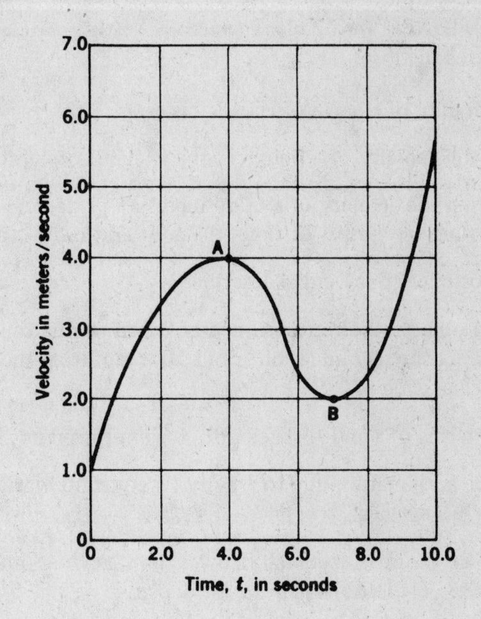

Fig. 3B-2

Time (sec)	Velocity (m/sec)	Acceleration (m/sec^2)	Displacement (m)
0.0			
2.0			
5.0			
10.0		Total displacement (m)	

2. Complete the following table by obtaining velocity data from Fig. 3B-2 and calculating the average acceleration and displacement for each time interval. Then calculate the total displacement.

Time (sec)	Velocity (m/sec)	Acceleration (m/sec^2)	Displacement (m)
0.0			
1.0			
2.0			
3.0			
4.0			
5.0			
6.0			
7.0			
8.0			
9.0			
10.0			
		Total displacement (m)	

DIRECTIONS: In the blank space at the right of each statement, write the word or expression which BEST completes the meaning.

3. The rate of change of velocity is called ..(3)... 3

4. If the velocity of a body is constant, its acceleration is ..(4)... 4

5. When the velocity of a body increases or decreases the same amount in successive units of time, the acceleration is ..(5)... 5

6. Motion of this kind is described as ..(6).. accelerated motion. 6

7. An object which moves linearly 2.0 m during the first second, 6.0 m during the second second, and 10.0 m during the third second has an acceleration of ..(7)... 7

8. In terms of *s* and *t*, acceleration is expressed as ..(8)... 8

9. The acceleration due to gravity (is constant, varies) ..(9).. over the earth's surface. 9

10. In air, dense objects fall (slower than, at the same speed as, faster than) ..(10).. objects of lower density. 10

DIRECTIONS: Write the answers to the following in the spaces provided. Where appropriate, make complete statements.

11. What algebraic relationship expresses the displacement of an object in terms of its average velocity and elapsed time of travel? 11

12. In cases of uniformly accelerated motion, how may the final velocity be expressed in terms of initial velocity, acceleration, and elapsed time? 12

13. During any interval of time in which the initial and final velocities are known, how is the average velocity expressed algebraically? 13

14. When we substitute the expression for final velocity from No. 12 in the equation of No. 13, what does the expression for average velocity become? 14

15. Substituting the expression for average velocity from No. 14 in the equation of No. 11, what does the displacement of the object now equal? 15

16. Solve the equation of No. 12 for t.

....16

17. If the equations of No. 15 and No. 16 relate to the same observation of the same system, t in No. 16 may be substituted for t in No. 15. Make this substitution and solve for v_f. 17

18. Do the equations developed above apply to cases of uniformly decelerated motion? ....

....

....18

19. Do the equations for accelerated motion apply to freely falling bodies? ....

....19

20. What is the effect of the force of gravity on an object thrown upward? ....

....20

NAME PERIOD DATE INSTRUCTOR'S COMMENTS

3c EXERCISE Newton's Laws of Motion-Gravitation

SECS. 3.8 - 3.13

DIRECTIONS: In the blank space at the right of each statement, write the word or expression which BEST completes the meaning.

1. A push or a pull is a(n) ..(1)... 1
2. Quantities which require magnitude, direction, and point of application for their description are ..(2).. quantities. 2
3. A physical quantity that can affect the motion of an object is a(n) ..(3)... 3
4. When no net force acts on a body, either ..(4a).. force acts on the body or the ..(4b).. of all forces acting on the body is ..(4c)... .... 4a
   .... 4b
   .... 4c
5. If there is no net force acting on a body, it will continue in its state of ..(5a).. or will continue moving along a(n) ..(5b).. line with ..(5c).. velocity. 5a
   .... 5b
   .... 5c
6. A common unbalanced force which makes it difficult to prove Newton's first law of motion experimentally is ..(6)... 6
7. The property of matter which is the concern of the first law of motion is ..(7)... 7
8. If an object is stationary, its inertia tends to keep it ..(8a)..; if an object is in motion, its inertia tends to keep it ..(8b)... .... 8a
   .... 8b
9. Uniform motion in a straight line is the only motion possible for a(n) ..(9).. object. 9
10. Non-uniform motion of an object is always caused by the ..(10).. of some other object. 10

DIRECTIONS: In the space at the right, place the letter or letters indicating EACH choice which forms a correct statement.

11. The acceleration of a body is (a) directly proportional to the force exerted, (b) inversely proportional to the mass of the body, (c) in the same direction as the applied force, (d) dependent on its initial velocity. 11
12. If a body is moving in a straight line and a force is applied in the direction of its motion, (a) the velocity is constant, (b) the acceleration is constant, (c) the body will increase in speed as long as the force continues, (d) the body will decrease in speed. 12
13. If a body is moving in a straight line and a force is applied in the direction opposite to its motion, (a) the velocity is changing, (b) the acceleration is changing, (c) the body may stop, (d) the body may move in the opposite direction. 13

14. The *newton* (a) is a unit of force in the MKS system, (b) is the force required to accelerate 1 kg of mass at the rate of 1 m/sec^2, (c) is a derived unit, (d) is 1 kg m/sec^2. 14

15. The force required to accelerate an object of known weight is (a) directly proportional to the weight of the object, (b) directly proportional to the acceleration desired, (c) directly proportional to the acceleration due to gravity, (d) not related in any way to the acceleration due to gravity. 15

16. The relationship betweenthe mass of a body and its weight is (a) weight = mass $\times$ 9.80 m/sec^2, (b) weight $\div$ mass = acceleration due to gravitational attraction, (c) $F_w = mg$, (d) $F = ma$. 16

17. Newton's third law of motion deals with (a) one object and two forces, (b) two objects and one force, (c) two objects and two forces, (d) action and reaction. 17

18. When an automobile is accelerated forward, (a) the tires exert a forward action force on the road, (b) the road exerts a forward reaction force on the tires, (c) the tires exert a rearward action force on the road, (d) the road exerts a rearward reaction force on the tires. 18

DIRECTIONS: Write the answers to the following in the spaces provided. Where appropriate, make complete statements.

19. Write the equation for the law of universal gravitation and explain each of the terms. ....

....

....

.... 19

20. If the masses of both objects are doubled, how is the force of gravitational attraction affected? ....

.... 20

21. If the distance between their centers of mass is doubled, how is the force of gravitational attraction affected? ....

.... 21

22. If a person's mass remains constant, what variation, if any, is there in weight descending from a mountain into an adjacent valley? 22

23. What is a gravitational field? ....

.... 23

24. In which areas of physics is the study of force fields useful? ....

.... 24

25. What force, in newtons, is required to accelerate a body which has a mass of 40.0 kg at the rate of 2.0 m/sec^2 in a westward direction? 25

26. What is the mass of a man whose weight on the earth's surface is $81\overline{0}$ n? 26

27. Calculate the acceleration due to gravity on the surface of Jupiter. 27

28. In imagination, what would be the mass and the weight of the man of No. 26 on the surface of Jupiter?

.... 28

NAME PERIOD DATE INSTRUCTOR'S COMMENTS

4A EXERCISE Resolution of Forces

SECS. 4.1 - 4.4

DIRECTIONS: In the space at the right, place the letter or letters indicating EACH choice which forms a correct statement.

1. A net force (a) changes the state of motion of an object, (b) must be a contact force, (c) may act over long distances, (d) produces a constant velocity. 1

2. A book rests on the horizontal surface of a desk. (a) The weight of the book on the desk is a force. (b) The desk pushes up on the book. (c) The force of gravitational attraction on the book and the force of the desk on the book are a pair of forces. (d) The force of the book on the desk and the force of the desk on the book act in the same direction. 2

3. A spring balance is a device (a) in which the force of gravitation on an object is equalized by the restoring force of the spring, (b) in which the amount of stretch equals the force on the spring, (c) used to measure weight, (d) which may be marked off in newtons. 3

4. The arrow designating a force vector (a) is represented as pushing on the point where the force acts, (b) represents the magnitude of the force by its length, (c) represents the direction in which the force acts by its position, (d) is directed away from the point where the force is applied. 4

5. In a resolution-of-force parallelogram (a) the given force vector is the diagonal, (b) the component force vectors are opposite sides, (c) the component force vectors act from the same point as the given force vector, (d) component force vectors may never be as long as the given force vector. 5

6. The usual angle between the components into which a single force is resolved is (a) 30°, (b) 45°, (c) 60°, (d) 90°. 6

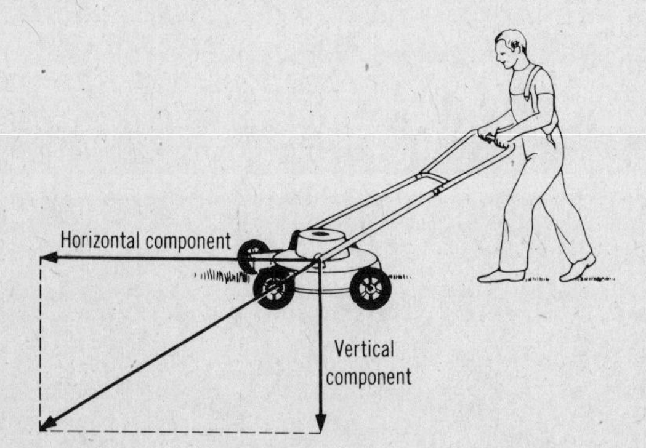

Figure 4A-1

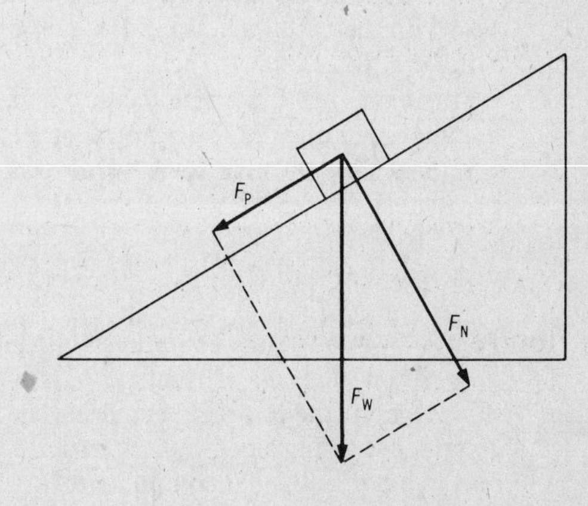

Figure 4A-2

7. When pushing a lawn mower, the force is applied at an angle to the ground as shown in Fig. 4A-1. **(a)** The desired motion is in the direction of the vertical component. **(b)** The effective force in the direction of the motion is less than the applied force. **(c)** If the applied force remains the same but the handle is raised, the vertical component becomes greater. **(d)** If the mower is pulled with the same force with which it was previously pushed, the magnitudes of the components remain the same but they act in opposite directions.

 7

8. A crate rests on an inclined plane as shown in Fig. 4A-2. **(a)** The force exerted by the crate perpendicular to the plane is less than the weight of the crate. **(b)** The plane exerts no force on the crate. **(c)** One component of the weight of the crate tends to cause the crate to slide down the plane. **(d)** The crate may weigh $5\overline{0}0$ kg.

 8

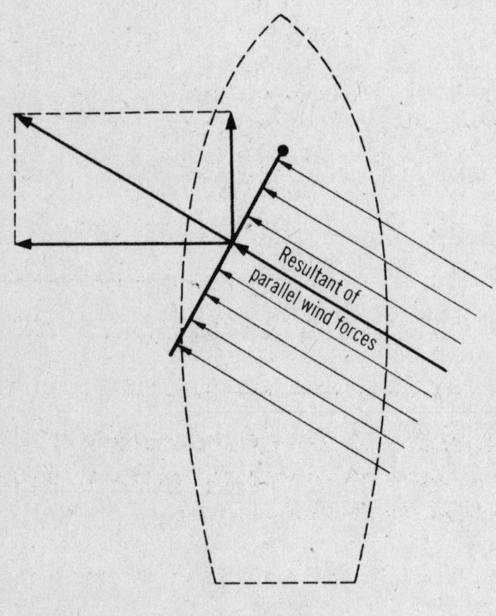

Figure 4A-3

O
X
Y
W
B
A
D
C

The weight of the roof OW may be resolved into the components OX and OY. OX applied at O equals AD applied at A.

Figure 4A-4

9. Figure 4A-3 shows the components of the parallel wind forces acting on the sail of a boat. **(a)** The forward component force is the desired one. **(b)** The sidewise component force tends to tip the boat. **(c)** The useful component force is greater than the undesired one. **(d)** More efficient use of the wind force would result if the boat turned and took a course $3\overline{0}^\circ$ to the right.

 9

10. Figure 4A-4 shows the components of the weight of a roof as they act on the walls of a building. **(a)** The components of the weight of the roof may act along the rafters. **(b)** Each component acting along the rafters may be further resolved into two perpendicular components. **(c)** The horizontal one of these components tends to spread the walls. **(d)** The vertical one of these components tends to push the walls into the ground.

....10

DIRECTIONS: (A) Construct the diagrams for the following problems to a suitable scale using a well-sharpened pencil, a straight edge, and protractor. Label the diagrams to agree with the forces described. (B) Place the answers to the following problems in the spaces provided at the right.

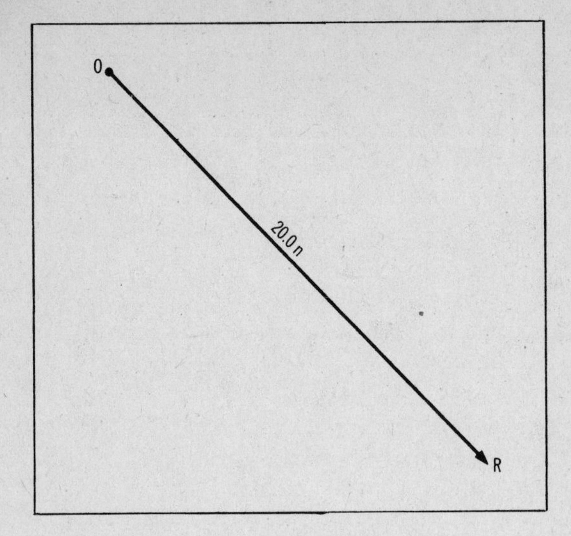

Figure 4A-5

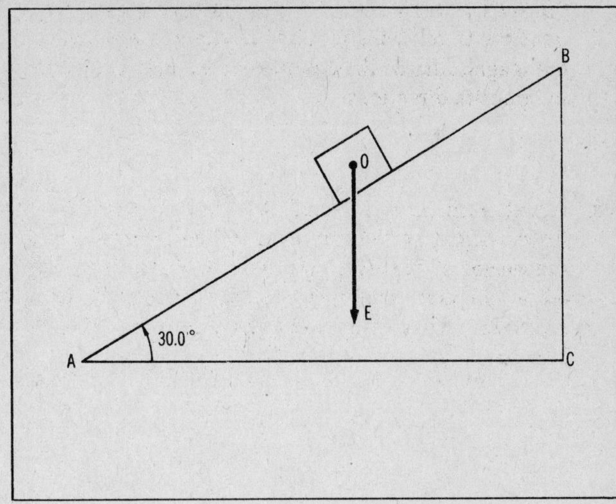

Figure 4A-6

11. In Fig. 4A-5, OR is a vector representing a force of 20.0 n acting on point O at an angle of 45° south of east. By construction, resolve OR into its horizontal and vertical components, OH and OV.

12. What is the angle HOR formed by the force vector OR and its horizontal component OH? 12

13. What is the magnitude of OH found graphically? 13

14. What is the calculated magnitude of OH? 14

15. What is the magnitude of OV? 15

16. In Fig. 4A-6, a 150.0-n box is placed on the incline AB which makes a 30.0° angle with the floor AC. The vector OE represents the weight of the box. Complete the diagram showing the component OF which acts to pull the box down the incline, and the component OD which tends to break the incline.

17. What is the magnitude of OF found graphically? 17

18. What is the magnitude of OD found graphically? 18

19. What is the computed magnitude of the force OF? 19

20. What is the computed magnitude of the force OD? 20

NAME PERIOD DATE INSTRUCTOR'S COMMENTS

4B EXERCISE Composition of Forces

SECS. 4.5 - 4.6

DIRECTIONS: In the blank space at the right of each statement, write the word or expression which BEST completes the meaning.

1. When two or more forces act concurrently at a point, the single force applied at the same point that would produce the same effect is the ..(1).. force. 1

2. The resultant of two forces acting in the same or in opposite directions upon the same point has a magnitude equal to the ..(2a).. of the forces, and acts in the direction of ..(2b)... .... 2a 2b

3. The resultant of a force of $40\bar{0}$ n northward and a force of $30\bar{0}$ n northward is ..(3)... 3

4. The resultant of a force of $40\bar{0}$ n northward and a force of $30\bar{0}$ n southward is ..(4)... 4

5. The resultant force vector of two forces acting concurrently at any angle upon a given point may be found by constructing a(n) ..(5a).. using the two force vectors as the ..(5b)... .... 5a 5b

6. The magnitude of this resultant force vector is represented by the ..(6a).. of the ..(6b).. drawn from the point on which the two force vectors act. .... 6a 6b

7. The line of action of the resultant force vector is represented by the ..(7a).. of the ..(7b)... .... 7a 7b

8. The state of a body in which there is no change in its motion is called ..(8)... 8

9. When no unbalanced forces act on a body, it is possible that the body has no ..(9)... 9

10. When there are no unbalanced forces acting on a body, the vector sum of all the forces acting on the body in any direction is ..(10)... 10

11. When two or more forces act concurrently at a point on an object, that single force applied at the same point that produces equilibrium is called the ..(11).. force. 11

12. The equilibrant force has the same ..(12a).. but opposite ..(12b).. to the resultant of two or more unbalanced concurrent forces. ....12a 12b

DIRECTIONS: (A) Construct the diagrams for the following problems to a suitable scale using a well-sharpened pencil, a straight edge, and protractor. Label the diagrams to agree with the forces described. (B) Place the answers to the following problems in the spaces provided at the right.

A

B

13. In block A, represent two forces acting concurrently on point O. One force, OW, acts westward with a magnitude of 24.0 n; the other force, OS, acts southward with a magnitude of 18.0 n. Compose these two forces into their resultant force, OR, by construction. Indicate on the diagram the magnitude and direction of OR found graphically.

14. What is the computed magnitude and direction of OR in No. 13? 14

15. What would be the magnitude and direction of the equilibrant in No. 13? 15

16. In block B, represent a force OW, 10.0 n west, and a force OA, 15.0 n 30.0° W of S. Construct the resultant force vector OR and indicate its magnitude and direction, found graphically, on the diagram.

17. What is the computed magnitude and direction of OR in No. 16? 17

18. What would be the magnitude and direction of the equilibrant in No. 16? 18

19. Find the resultant of the vectors in the figure at the right by the head-to-tail addition method.

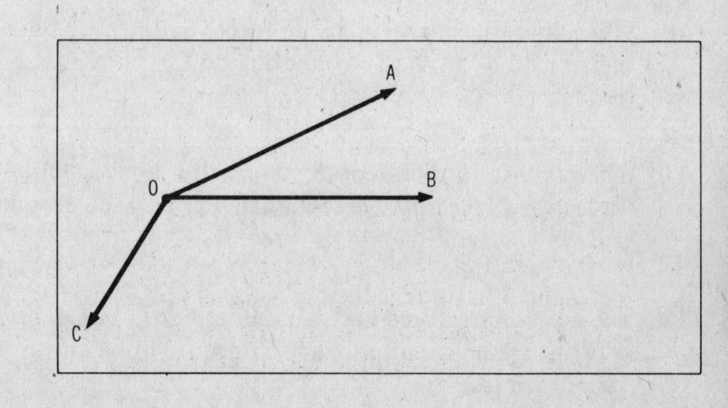

NAME PERIOD DATE INSTRUCTOR'S COMMENTS

4c EXERCISE Friction

SECS. 4.7 - 4.10

DIRECTIONS: In the blank space at the right of each statement, write the word or expression which BEST completes the meaning.

1. *Friction* is any force that ..(1a).. the relative sliding or rolling ..(1b).. of objects that are ..(1c).. each other. .... 1a .... 1b 1c
2. Some physicists believe that friction is caused by ..(2).. surfaces. 2
3. However, if surfaces are made very smooth, friction between them ..(3)... 3
4. Other causes of sliding friction may be the same ..(4a).. that hold ..(4b).. together. .... 4a 4b
5. Friction experiments are (easy, difficult) ..(5a).. to perform but the results (may always be, may not always be) ..(5b).. expressed simply. .... 5a 5b
6. Friction acts ..(6a).. to the surfaces which are sliding over one another, and in the (same, opposite) ..(6b).. direction as the motion. .... 6a 6b
7. Friction depends upon the ..(7a).. of the materials in contact and their ..(7b)... .... 7a 7b
8. Starting friction is (less than, greater than) ..(8).. sliding friction. 8
9. Within the range of medium speeds, sliding friction is nearly independent of both the ..(9a).. and the ..(9b).. of the surfaces. .... 9a 9b
10. Friction between a bullet and the gun barrel is (less than, the same as, greater than) ..(10).. the friction between slower-moving objects. 10
11. Friction is (directly proportional to, independent of, inversely proportional to) ..(11).. the force pressing the two surfaces together. 11
12. The ratio of the force of friction to the normal force pressing the surfaces together is called the ..(12)... 12
13. If an object moves along a horizontal surface under the influence of a horizontal force, the normal force pressing the surfaces together is the ..(13).. of the object. 13
14. If an object moves along a horizontal surface under the influence of a force acting upward at a $\overline{30}^{\circ}$ angle with the horizontal, the normal force pressing the surfaces together is the ..(14a).. of the object less ..(14b).. of the exerted force. .... 14a 14b

15. If an object moves along a slanted surface, the normal force pressing the surfaces together is only a(n) ..(15).. of the weight of the object. 15

16. The other ..(16a).. either ..(16b).. or ..(16c).. the force overcoming friction, depending on the direction of motion of the object. 16a

....................................16b

....................................16c

DIRECTIONS: Write the answers to the following in the spaces provided. Where appropriate, make complete statements.

17. List four situations in which friction is a help.

a. ....

b. ....

c. ....

d. 17

18. List four situations in which friction is a hindrance.

a. ....

b. ....

c. ....

d. 18

19. List four ways in which friction may be increased.

a. ....

b. ....

c. ....

d. 19

20. List four ways in which friction may be decreased.

a. ....

b. ....

c. ....

d. 20

DIRECTIONS: (A) Construct the diagrams for the following problems to a suitable scale using a well-sharpened pencil, a straight edge, and protractor. Label the diagrams to agree with the forces described. (B) Place the answers to the following problems in the spaces provided at the right.

A

B

21. A block weighing $30\bar{0}$ n is moved uniformly over a horizontal surface by a force of $6\bar{0}$ n. Construct the force diagram in block A on page 28.

22. What is the force needed to overcome friction? 22

23. What is the normal force pressing the surfaces together? 23

24. What is the coefficient of friction? 24

25. A block weighing 70.0 n is to be drawn up a 30.0° incline by a force parallel to the incline surface. Construct the force diagram in block B on page 28.

26. What is the normal force pressing the surfaces together? 26

27. If the coefficient of friction is 0.10, what is the force that is needed to overcome friction? 27

28. What is the component of the weight which must be overcome in moving the block up the incline? 28

29. What is the force needed to move the block up the incline? 29

NAME PERIOD DATE INSTRUCTOR'S COMMENTS

4D EXERCISE Parallel Forces

SECS. 4.11-4.14

DIRECTIONS: In the space at the right, place the letter or letters indicating EACH choice which forms a correct statement.

1. The *center of gravity* of an object (a) is that point at which all of its weight appears to be concentrated, (b) is the point of application of the resultant of the parallel forces due to the weight of the object, (c) is at the geometric center of a bar of uniform shape and density, (d) may be located outside the object. 1

2. *Torque* (a) equals the product of the magnitude of a force and the length of the torque arm on which it acts, (b) produces linear motion, (c) may be measured in newtons, (d) may act clockwise or counterclockwise. 2

3. The second condition of equilibrium (a) concerns the prevention of linear motion, (b) concerns the prevention of rotary motion, (c) requires that no unbalanced forces act on a body, (d) requires that no unbalanced torques act on a body. 3

4. The resultant of parallel forces (a) has a magnitude equal to their arithmetic sum, (b) has a magnitude equal to their algebraic sum, (c) acts in the same direction as the largest force, (d) acts in the same direction as the net force. 4

5. A force of $30\overline{0}$ n acting perpendicularly on the end of a lever arm 1.5 m long produces a torque of (a) $30\overline{0}$ n, (b) 1.5 m, (c) $45\overline{0}$ n m, (d) 4.5×10^2 m n. 5

6. A force which does not act perpendicularly on a lever arm produces a torque equal to the product of (a) the force and the distance of its point of application from the axis of rotation, (b) the force and the perpendicular distance from the axis of rotation to the line of action of the force, (c) the force and the radius described by the lever arm, (d) the force, the distance of its point of application from the axis of rotation, and the sine of the angle the line of action of the force makes with the lever arm. 6

7. A load of $50\overline{0}$ n is attached to a uniform bar weighing $5\overline{0}$ n midway between the two ends. The supporting forces at the ends of the bar (a) are equal, (b) have a sum equal to $55\overline{0}$ n, (c) are $55\overline{0}$ n each, (d) are 275 n each. 7

8. Two boys are carrying a load suspended from the center of a bar. For the stronger boy to carry the larger share (a) the load may be shifted to a position closer to the weaker boy, (b) the stronger boy may move closer to the load, (c) the load may be shifted closer to the stronger boy, (d) the stronger boy may use both hands to hold the rod while the weaker boy uses only one hand. 8

9. As a truck moves across a bridge, the downward force due to its weight (a) acts against both supporting piers, (b) is parallel to the upward forces exerted by the piers, (c) remains equally divided between the two piers, (d) is parallel to the weight of the bridge itself. 9

10. The center of gravity of a rod weighing 80.0 n is $25\overline{0}$ cm from end A. If a force of $30\overline{0}$ n acts downward at a position $10\overline{0}$ cm from end A and $30\overline{0}$ cm from end B, the support at end A must exert an upward force of (a) 125 n, (b) 250 n, (c) 255 n, (d) 275 n. 10

DIRECTIONS: In the blank space at the right of each statement, write the word or expression which BEST completes the meaning.

11. If an upward force equal to its weight is applied at the center of gravity of a bar, the bar is in ..(11).. equilibrium. 11

12. If an upward force equal to its weight is applied at other than the center of gravity of a bar, the bar is in ..(12).. equilibrium. 12

13. Parallel forces act on an object at (the same place, different places) ..(13)... 13

14. Parallel forces act in the ..(14).. directions. 14

15. The point from which the lengths of all the torque arms are measured is called the ..(15)... 15

16. A common unit of torque in the MKS system is the ..(16)... 16

17. Two forces of equal magnitude which act in opposite directions in the same plane but not along the same line constitute a(n) ..(17)... 17

18. A couple can be balanced only by another couple which has a torque equal in ..(18a).. but opposite in ..(18b)... ....18a

....18b

DIRECTIONS: (A) Construct the diagrams for the following problems to a suitable scale using a well-sharpened pencil and straight edge. Label fully. (B) Place the answers to the following problems in the spaces provided at the right.

19. In the block above, diagram the following arrangement: A uniform bar AB, weighing 6.00 n, is 1.00 m long and rotates about a fixed point C, which is $\overline{4}0$ cm from B. A weight of 3.00 n is attached at A and an unknown weight is attached at B to produce equilibrium.

20. Using C as the axis of rotation, what is the counterclockwise torque? 20

21. What is the clockwise torque? 21

22. If the bar is in equilibrium, how must these two torques compare? 22

23. What is the magnitude of the weight attached at B? 23

24. In the block above, diagram the following arrangement: A bar AB, weighing $10\bar{0}$ n, is 5.00 m long. Its center of gravity is 2.00 m from A. A weight of $80\bar{0}$ n is attached at A and a weight of $60\bar{0}$ n is attached at B.

25. What is the magnitude of the upward equilibrant force? 25

26. How far from A must the equilibrant force act? 26

NAME PERIOD DATE INSTRUCTOR'S COMMENTS

5A EXERCISE Circular Motion

SECS. 5.1 - 5.4

DIRECTIONS: Write the answers to the following in the spaces provided. Where appropriate, make complete statements.

1. Describe the motion of an object having uniform velocity. 1

2. Describe the motion of an object having uniform acceleration. 2

3. How is motion along a curved path produced? 3

4. What are the two independent motions of a bullet fired from a gun? 4

5. How is the force of gravity counteracted in firing a gun at a target? 5

6. Neglecting air resistance, what is the shape of the trajectory of a projectile fired at an upward angle? 6

7. Why may the force of gravity be neglected in horizontal circular motion? 7

8. When a ball is moving in a vertical circle, how do the centripetal force and the weight of the ball affect the force the string exerts on the ball? 8

9. What is meant by the *critical velocity* of a ball moving in a vertical circle? 9

10. How does the mass of the ball affect the critical velocity? ....

....10

11. To an observer in a rotating system who considers himself or herself to be stationary, what is *centrifugal force*? ....

....

....11

12. To a stationary observer of a rotating system, why is centrifugal force a fiction? ....

....

....12

13. Define *frame of reference.* ....

....13

14. Distinguish between an inertial frame of reference and a noninertial frame of reference. ....

....

....

....14

15. Relate *frame of reference* to *centrifugal force.* ....

....

....15

DIRECTIONS: In the blank space at the right of each statement, write the word or expression which BEST completes the meaning.

16. Over a short distance the path of a high-velocity projectile is nearly a(n) ..(16a).. line, but over a greater distance its path is a(n) ..(16b).. line. ....16a 16b

17. An object moving at constant speed along a curved path of constant radius is undergoing ..(17).. motion. 17

18. The velocity of the object of No. 17 does not change in ..(18a).., but it does change uniformly in ..(18)... ....18a 18b

19. This uniform change in velocity constitutes a uniform ..(19)... 19

20. The angle between the instantaneous velocity vector and the instantaneous acceleration vector is ..(20)... 20

21. The acceleration involved in uniform circular motion is ..(21).. acceleration. 21

22. The force producing this acceleration is called ..(22).. force. 22

23. This force varies ..(23a).. with the mass of the body in circular motion and ..(23b).. with the radius of its path. ....23a 23b

24. This force also varies ..(24a).. with the ..(24b).. of the speed of the body along the circular path. ....24a 24b

25. Centripetal force is needed to keep a body in uniform circular motion because the body has ..(25)...

.... 25

26. If centripetal force ceases to act on a body in uniform circular motion, the body will move at ..(26a).. speed along a path which is ..(26b)...

.... 26a

.... 26b

27. At the critical velocity, the expression for the centripetal force on the ball is ..(27a).. and the expression for the weight of the ball is ..(27b)...

.... 27a

.... 27b

28. At the critical velocity, No. 27a ..(28).., No. 27b.

.... 28

29. Because the term ..(29a).. appears in the numerator of each expression, the critical velocity does not depend on the ..(29b).. of the object describing the motion.

.... 29a

.... 29b

30. The earth is usually considered as a(n) ..(30a).. frame of reference, while a car rounding a curve is a(n) ..(30b).. frame of reference.

.... 30a

.... 30b

NAME PERIOD DATE INSTRUCTOR'S COMMENTS

5B EXERCISE Rotary and Simple Harmonic Motions

SECS. 5.5 - 5.12

DIRECTIONS: Continue the following definitions in the spaces provided, forming accurate and complete sentences.

1. *Circular motion* is 1
2. *Rotary motion* is 2
3. *Uniform rotary motion* is 3
4. *Variable rotary motion* is 4
5. *Angular velocity* is 5
6. An *angle of one radian* is that angle which, ... 6
7. *Linear acceleration* is 7
8. *Angular acceleration* is 8
9. *Rotational inertia* is 9
10. *Precession* is 10

DIRECTIONS: In the space at the right, place the letter or letters indicating EACH choice which forms a correct statement.

11. Rotary motion is shown by (a) a turning electric fan, (b) a counter-balancing weight on a rotating automobile wheel, (c) a bicycle pedal as the bicycle is pumped, (d) an operating electric drill. 11
12. A unit of angular velocity may be (a) m/sec, (b) rad/sec, (c) cm/sec, (d) sec^{-1}. 12
13. One radian equals (a) approximately 57.3°, (b) 360°/2π, (c) 180°/3, (d) circumference/radius. 13
14. For an angular velocity vector (a) the length of the vector indicates the rate of change of angular velocity, (b) the direction of the vector is toward the center of the axis of rotation, (c) the length of the vector varies directly with the rate of rotation, (d) the direction of the vector parallels the axis of rotation. 14
15. Angular acceleration may (a) change the rate of rotation, (b) change the direction of the axis of rotation, (c) change both the rate of rotation and the direction of the axis of rotation, (d) not change the rate of rotation without a compensating change in the direction of the axis of rotation. 15

16. The rotational inertia of a body takes into account (a) the shape of a rotating body only, (b) the mass of a rotating body only, (c) both the shape and mass of a rotating body, (d) neither the shape nor mass of a rotating body. 16

17. A wheel, spinning rapidly and supported at one end of its horizontal axle by a cord (a) will react to the force of gravity and the axle will move into a vertical position, (b) will slowly rotate about the axis marked by the cord, (c) helps us understand the precession of the earth's axis, (d) is an example of circular motion. 17

18. In simple harmonic motion (a) the body goes above the equilibrium position as far as it goes down, (b) the force exerted on the body is inversely proportional to its displacement from its equilibrium position, (c) the force is directed from the equilibrium position in the direction of motion of the particle, (d) the acceleration of the particle is zero at the equilibrium position. 18

19. The period of a pendulum is (a) directly proportional to the mass of the bob, (b) inversely proportional to the amplitude of the swing, (c) directly proportional to the square root of its length, (d) inversely proportional to the square root of the acceleration of free fall. 19

20. A baseball bat (a) may vibrate like a pendulum, (b) does not sting the hands if a ball meets it at the center of percussion and the hands are turning the bat at the center of oscillation, (c) has a period of vibration equal to that of a simple pendulum with a length equal to the distance between the centers of oscillation and percussion, (d) has an interchangeable point of suspension and center of oscillation. 20

DIRECTIONS: In the parentheses at the right of each word or expression in the second column, write the letter of the expression in the first column which is MOST CLOSELY related.

a. The maximum displacement
b. The minimum displacement
c. Motion in which a body continually moves back and forth over a definite path in irregular time intervals
d. A special type of periodic motion
e. Any real object that can vibrate like a pendulum
f. Corresponds to one-half revolution of the reference circle
g. Instantaneous distance from the midpoint of vibration
h. Motion in which a body continually moves back and forth over a definite path in equal time intervals
i. Time required for a complete vibration
j. Number of vibrations per second
k. Corresponds to a complete revolution on the reference circle
l. A small, dense mass suspended by a cord of negligible mass

periodic motion ( ) 21
simple harmonic motion ( ) 22
complete vibration ( ) 23
displacement ( ) 24
amplitude ( ) 25
period ( ) 26
frequency ( ) 27
equilibrium position ( ) 28
simple pendulum ( ) 29
physical pendulum ( ) 30

NAME PERIOD DATE INSTRUCTOR'S COMMENTS

6 EXERCISE Conservation of Energy and Momentum

SECS. 6.1 - 6.16

DIRECTIONS: In the space at the right, place the letter or letters indicating EACH choice which forms a correct statement.

1. In physics, *work* is done **(a)** in lifting an object from the floor to the table, **(b)** in supporting an object on your shoulder, **(c)** in preparing school lessons, **(d)** in pushing an object across the floor. 1

2. The two factors which determine the *amount of work* done are **(a)** the magnitude of the force exerted and the weight of the object moved, **(b)** the distance the object is moved and the time required, **(c)** the displacement of the object and the magnitude of the force in the direction of the displacement, **(d)** the magnitude of the force in the direction of the displacement and the time required. 2

3. The *work done by a varying force* **(a)** equals Fs, **(b)** equals the area under a curved line on a graph of magnitude of force acting in the direction of motion versus distance through which the force acts, **(c)** may be approximated by the use of suitable rectangles, **(d)** may be determined by using calculus. 3

4. In physics, *power* is **(a)** the capacity for doing work, **(b)** the time rate of doing work, **(c)** the product of a displacement and the force in the direction of the displacement, **(d)** Fs/t. 4

5. *Energy* is **(a)** the capacity for doing work, **(b)** acquired by an object raised to an elevated position, **(c)** acquired by an object which is set in motion, **(d)** measured in work units. 5

6. *Kinetic energy* is **(a)** energy due to the position of a mass, **(b)** energy due to the motion of a mass, **(c)** energy due to the orientation of particles in a mass, **(d)** proportional to the mass of the object concerned. 6

7. *Potential energy* **(a)** is energy due to the position of a mass, **(b)** is energy due to the motion of a mass, **(c)** is energy due to the orientation of particles in a mass, **(d)** of position is proportional to the mass of the object concerned. 7

8. The *law of conservation of mechanical energy* **(a)** is illustrated by a swinging pendulum, **(b)** states that the sum of the potential and kinetic energies of an ideal energy system remains constant, **(c)** involves frictional forces, **(d)** involves systems where the work done is dependent on the path length. 8

9. *Machines* may be used **(a)** to multiply force, **(b)** to multiply speed, **(c)** to multiply force and speed simultaneously, **(d)** to change the direction of a force. 9

10. The *efficiency* of a machine is **(a)** the ratio of total work output to total work input, **(b)** the ratio of the useful work output to total work input, **(c)** always less than 100%, **(d)** $F_w h/F_a s$. 10

DIRECTIONS: In the blank space at the right of each statement, write the word or expression which BEST completes the meaning.

11. The unit of work in the MKS system is the ..(11)... 11

12. The force required to stretch a spring is directly proportional to the ..(12)... 12

13. Work in rotary motion is the product of ..(13a).. and ..(13b)... ....13a

..........13b

14. The unit of power in MKS units which equals one joule/second is the ..(14)... 14

15. Power in rotary motion is the product of ..(15a).. and ..(15b)... ....15a

..........15b

16. In raising the hammer of a pile driver, ..(16).. must be exerted over a given distance. 16

17. While the hammer is being raised to its elevated position, ..(17).. is being done on the hammer. 17

18. In its elevated position, the ..(18).. of the hammer is greater. 18

19. As the hammer is released and falls, its ..(19a).. is converted to ..(19b)... ....19a

..........19b

20. The ..(20).. of the hammer enables it to do work in driving a pile into the ground. 20

21. The rate at which the pile driver works is its ..(21)... 21

22. For linear motion, E_k = ..(22a).. in which *m* is the ..(22b).. of the moving body and *v* is its ..(22c)... ....22a

..........22b

..........22c

23. For rotary motion, E_k = ..(23a).. in which *I* is the ..(23b).. of the rotating body and ω is its ..(23c)... ....23a

..........23b

..........23c

24. For a stretched spring, E_p = ..(24a).. in which *k* is the ..(24b).. of the spring and *s* is the ..(24c).. the spring is stretched. 24a

..........24b

..........24c

25. Forces involved in the conservation of mechanical energy are called ..(25).. forces. 25

26. The force of friction is an example of a(n) ..(26).. force. 26

27. An inclined plane multiplies force at the expense of ..(27)... 27

28. The two basic types of machines are the ..(28a).. and the ..(28b)... ....28a

..........28b

29. Work output and work input would be exactly equal in a(n) ..(29)... machine. 29

30. A perpetual motion machine would need an efficiency of ..(30)... 30

DIRECTIONS: **Place the answers to the following problems in the spaces provided at the right.**

31. A pupil weighing $7\overline{5}0$ n ran up the stairs from the first floor to the third floor in the school building, a vertical distance of 8.00 m. How much work was done? 31

32. The physics instructor timed the pupil with a stop watch and found that 14.0 sec running time was required. What was the power? 32

33. What potential energy is acquired by a hammer, mass $75\overline{0}$ g, when raised 0.35 m? 33

34. If a ball has a mass of 0.50 kg and a velocity of 8.5 m/sec, what is its kinetic energy? 34

35. What power is required to raise $950\overline{0}$ n of coal from a mine 60.0 m deep in 175 sec? 35

DIRECTIONS: Write the answers to the following in the spaces provided. Where appropriate, make complete statements.

36. What is *momentum?* ....

.... 36

37. What is *impulse?* ....

.... 37

38. What is the relationship between impulse and change in momentum? ....

.... 38

39. What is the *law of conservation of momentum?* ....

.... 39

40. What is the relationship between Newton's third law of motion and the law of conservation of momentum?

.... 40

41. Give an example of an inelastic collision in one dimension. ....

.... 41

42. Give an example of a nearly elastic collision in two dimensions. ....

....

.... 42

43. What is *angular impulse?* 43

44. What is *angular momentum?* ....

.... 44

45. A man stands with arms outstretched on a rotating stool. Explain what happens if he lowers his arms.

....

.... 45

DIRECTIONS: Place the answers to the following problems in the spaces provided at the right.

46. What is the momentum of an automobile, mass $150\overline{0}$ kg, moving at a velocity of 20.0 m/sec northward? 46

47. For how many seconds must a force of $75\overline{0}$ n northward act to impart the momentum to the automobile of No. 28? 47

48. A ball, mass 25.0 g, velocity 30.0 cm/sec southward, collides with another ball, mass 10.0 g, moving along the same line with a velocity of 15.0 cm/sec southward. After collision, the 25.0-g ball has a velocity of 22.0 cm/sec southward. What is the velocity of the 10.0-g ball? 48

NAME PERIOD DATE INSTRUCTOR'S COMMENTS

7A EXERCISE Molecules and Atoms - Solids

SECS. 7.1 - 7.11

DIRECTIONS: Write the answers to the following in the spaces provided. Where appropriate, make complete statements.

1. Give two conclusions which can be drawn from a sequence of experiments in which a rock salt crystal is crushed and the properties of the powder are examined; the powder is dissolved in water and the properties of the solution are examined; the water is subsequently evaporated and the residue is examined.

   a. 1a

   b. 1b

2. What kinds of information are included in the modern atomic theory? ....

   ....

   2

3. What is a molecule? ....

   3

4. What is an atom? ....

   4

5. What is the meaning of the term *phase* in describing matter? ....

   5

6. Name the three phases of matter and describe the particle spacing in each.

   a. 6a

   b. 6b

   c. 6c

7. What are two basic aspects of the kinetic theory?

   a. 7a

   b. 7b

8. Compare the magnitudes of the molecular forces in the three phases of matter. ....

   8

9. How does the force between like molecules vary with the distance between them? If the distance between ....

   ....

   ....

   ....

   ....

   9

10. Describe the motion of molecules of solids, liquids, and gases.

    a. 10a

    b. 10b

    c. 10c

DIRECTIONS: In the space at the right, place the letter or letters indicating EACH choice which forms a correct statement.

11. The particles of a solid (a) do not move, (b) are held in relatively fixed positions, (c) have a smaller amplitude of vibration at lower temperatures, (d) lose kinetic energy as the temperature is raised. 11

12. *Diffusion* (a) occurs rapidly in solids, (b) is evidence of the vibratory motion of the particles of solids, (c) proceeds slowly in solids because the particles are closely packed, (d) does not occur in solids because of the orderly particle arrangement. 12

13. *Cohesion* is the force of attraction between molecules (a) of the same material, (b) which can be shown with polished metal blocks, (c) of unlike materials, (d) which acts as a binding force. 13

14. *Adhesion* is the force of attraction between molecules (a) of the same material, (b) of unlike materials, (c) which causes one material to stick to another, (d) which acts at a distance. 14

15. *Tensile strength* (a) is a property of solids, (b) of copper is greater than that of steel, (c) depends on the cohesive forces between molecules, (d) is the force required to break a rod or wire having unit cross-sectional area. 15

16. *Ductility* (a) is a property of metals, (b) is dependent on the force of adhesion between molecules, (c) enables a metal to be drawn into a wire, (d) is determined by the cohesive force existing between the atoms of a metal. 16

17. *Malleability* (a) is dependent on cohesion, (b) is the property of metals enabling them to be pounded out into thin sheets, (c) of gold is not very pronounced, (d) depends on the ability of atoms to hold together while being shifted from one position in the crystal pattern to another. 17

18. *Elasticity* (a) of solids depends on molecular forces, (b) cannot restore a wire to its original length after an applied force has stretched it, (c) measurements may involve stress and shear, (d) is a property of materials which return to their original shape after removal of a distorting force. 18

19. *Hooke's law* (a) states that the ratio of stress to strain is a constant, (b) is the principle on which a spring balance operates, (c) deals with the inverse relationship between amount of distortion and distorting force, (d) applies even though the elastic limit is exceeded. 19

20. *Young's modulus* (a) is the ratio of stress/strain, (b) is the ratio of distorting force per unit area to the relative amount of distortion thus produced, (c) is reasonably the same for a given solid material regardless of its size or shape, (d) may be expressed in n/m^2. 20

DIRECTIONS: **Place the answers to the following problems in the spaces provided at the right.**

21. The atomic mass of a silver atom is 106.9041 u.
   a. What is its mass in kilograms? 21a
   b. What is its mass number? 21b

22. What force is required to break a steel rod 0.50 centimeter in diameter? The tensile strength of steel is 2.90×10^8 n/m^2. 22

23. A spring balance is stretched 0.010 m by a force of 5.0 n. How far will it be stretched by a force of $10\bar{0}$ n if the elastic limit is not exceeded? 23

24. From a steel beam 5.00 m long, cross-sectional area 8.10 cm^2, hangs a load of 3750 n. How far does the beam stretch if the elastic modulus for steel is 20.0×10^{10} n/m^2? 24

NAME PERIOD DATE INSTRUCTOR'S COMMENTS

7B EXERCISE Liquids and Gases

SECS. 7.12 - 7.24

DIRECTIONS: In the parentheses at the right of each word or expression in the second column, write the letter of the expression in the first column which is MOST CLOSELY related.

a. Conversion of vapor molecules to liquid molecules
b. Conversion of solid molecules to liquid molecules
c. Indicates that liquid molecules move
d. Conversion of liquid molecules to solid molecules
e. Depends on cohesion and surface tension
f. Conversion of solid or liquid molecules to vapor molecules
g. Depends on adhesion and cohesion
h. Conversion of solid molecules to vapor molecules
i. Conversion of liquid molecules to vapor molecules
j. Indicates that solid molecules move
k. Depends on adhesion and surface tension
l. Conversion of solid molecules to liquid molecules by increased pressure and subsequent reconversion to solid molecules when pressure is decreased

Brownian movement	( )	1
capillarity	( )	2
condensation	( )	3
evaporation	( )	4
freezing	( )	5
melting	( )	6
meniscus	( )	7
regelation	( )	8
sublimation	( )	9
vaporization	( )	10

DIRECTIONS: In the space at the right, place the letter or letters indicating EACH choice which forms a correct statement.

11. In a *liquid*, (a) the particles are close together, (b) the kinetic energy of the particles completely overcomes the attractive forces between them, (c) the particles are in motion, (d) the motion of the particles is slow. 11

12. *Diffusion* in liquids (a) is faster than in solids, (b) occurs counter to the force of gravity, (c) is never complete, (d) is possible because liquid particles have both mobility and an open arrangement. 12

13. *Surface tension* (a) causes a liquid to behave as if it has a thin surface film, (b) is the result of attractive forces between the molecules of a liquid, (c) causes liquid films to be flexible, (d) causes a free liquid to assume a spherical shape. 13

14. *Capillarity* is (a) the depression of a liquid in a small-diameter tube which the liquid wets, (b) the elevation of a liquid in a small-diameter tube which the liquid does not wet, (c) directly proportional to the tube diameter, (d) reduced by an increase in temperature. 14

15. During *melting* (a) the potential energy of the particles increases, (b) the kinetic energy of the particles increases, (c) the energy supplied overcomes the forces holding the particles in fixed positions, (d) the temperature of a crystalline solid rises. 15

16. During *melting* (a) most substances contract, (b) most substances expand, (c) the separation of particles changes because of a change in their potential energy, (d) water molecules move closer together. 16

17. In a *gas*, (a) the particles are close together, (b) the kinetic energy of the particles completely overcomes the attractive forces between them, (c) the particles are in motion, (d) the motion of the particles is rapid. 17

18. During *vaporization* (a) the identity of the particles remains the same, (b) the energy of the particles remains the same, (c) the separation of the particles remains the same, (d) the temperature of the particles remains the same. 18

19. *Equilibrium vapor pressure* (a) is added pressure exerted by vapor molecules in equilibrium with liquid, (b) is independent of the identity of the molecules, (c) increases with a decrease in the temperature of the molecules, (d) is independent of the kinetic energy of the vapor molecules. 19

20. *Boiling* (a) occurs when the vapor pressure of a liquid equals the pressure on its surface, (b) is rapid vaporization which disturbs the liquid, (c) is a constant-temperature process, (d) is a constant-energy process. 20

DIRECTIONS: In the blank space at the right of each statement, write the word or expression which BEST completes the meaning.

21. Since water clings to a clean glass rod, the adhesion of water molecules to glass is (less than, equal to, greater than) ..(21).. the cohesion of water molecules. 21

22. Since mercury does not cling to a clean glass rod, the adhesion of mercury molecules to glass is (less than, equal to, greater than) ..(22).. the cohesion of mercury molecules. 22

23. The cleaning action of a detergent is partly due to its ability to lower the ..(23).. of water. 23

24. When viewed from above, the surface of water in a glass container is slightly ..(24)... 24

25. When viewed from above, the surface of mercury in a glass container is slightly ..(25)... 25

26. Water in a glass capillary tube creeps up the walls because of ..(26a).. and produces a curved liquid surface; this surface tends to be flattened by ..(26b)... These two forces raise the water above its surrounding level until they are counterbalanced by the ..(26c).. of the elevated liquid. ....26a ....26b 26c

27. The melting point and freezing point of pure crystalline solids are (the same, different) ..(27).. temperature(s) for a given pressure. 27

28. For most substances, an increase in pressure (lowers, raises) ..(28a).. the freezing point because it (aids, hinders) ..(28b).. the movement of the particles into a more compact regular pattern. ....28a 28b

29. The freezing point of a liquid is (lowered, raised) ..(29a).. by the addition of a dissolved substance because its presence (aids, hinders) ..(29b).. crystal formation. ....29a 29b

30. A gas completely fills its container because it has the property of ..(30)... 30

31. When a rubber balloon is inflated, the walls stretch because of increased ..(31).. exerted by the gas molecules. 31

32. At a given temperature, the rates of diffusion of gases and their densities are (directly, inversely) ..(32a).. related because heavier molecules move more ..(32b)... ....32a 32b

33. In a closed bottle containing a liquid, evaporation and condensation (continue, cease) ..(33).. to occur. 33

34. At one atmosphere pressure, boiling occurs at the ..(34a).. boiling point; if the pressure is increased, the boiling temperature is ..(34b).., while if the pressure is lowered, the boiling temperature is ..(34c)...

....34a

....34b

....34c

35. In general, solids dissolved in liquids ..(35a).. the boiling temperature, while gases dissolved in liquids ..(35b).. the boiling temperature.

....35a

....35b

NAME PERIOD DATE INSTRUCTOR'S COMMENTS

8A EXERCISE Temperature and Heat Measurement

SECS. 8.1 - 8.7

DIRECTIONS: In the blank space at the right of each definition, write the term which is defined.

1. The total potential and kinetic energy associated with the random motion and arrangement of the particles of a material. 1
2. Thermal energy which is transferred from one body at higher temperature to another at lower temperature. 2
3. The physical quantity which is proportional to the average kinetic energy of translation of the particles in matter. 3
4. The temperature at which water, ice, and water vapor can continuously coexist. 4
5. 4.18605 joules. 5
6. The quantity of heat needed to raise the temperature of a body 1°. 6
7. The heat capacity per unit mass of a material. 7
8. $Q_{lost} = Q_{gained}$ 8
9. A technique for measuring a quantity of heat in transit from one substance to another. 9
10. A thermally insulated metal cup. 10

DIRECTIONS: In the blank space at the right of each statement, write the word or expression which BEST completes the meaning.

11. The property of a body which indicates its ability to give up heat to, or absorb heat from another body is its ..(11)... 11
12. The unit of temperature difference is the ..(12)... 12
13. The temperature at which a constant-temperature process occurs can be used as a(n) ..(13).. in establishing a temperature scale. 13
14. The Kelvin temperature of the triple point of water is ..(14)... 14
15. On the Celsius scale the temperature at which alcohol freezes is -115 ..(15a).., the temperature at which alcohol boils is 78 ..(15b).., and the interval between these temperatures is ..(15c)... ....15a ....15b 15c
16. The difference between the melting point and boiling point of mercury is 395.5C°. This temperature interval is ..(16).. K°. 16
17. At 0°K the kinetic energy of the particles of matter is ..(17)... 17
18. In assigning numbers on the Kelvin temperature scale, both a fixed point, the ..(18a).. of water, and a measurable physical property of a substance which is ..(18b).. to the Kelvin temperature are required. ....18a 18b
19. The most consistent results in temperature measurement are obtained with a constant- ..(19).. thermometer. 19

20. Mercury-in-glass thermometers are in practical use because the expansions of mercury and glass are fairly ..(20).. over the useful temperature range of such instruments. 20

21. A quantity of heat is measured by the ..(21).. it produces. 21

22. A mass of aluminum cools more (slowly, rapidly) ..(22a).. than an equal mass of zinc because the heat capacity of aluminum is (lower, higher) ..(22b).. than that of zinc. 22a

..........22b

23. In the equation $Q = mc\Delta T$, Q will be given in calories if m is in ..(23a).., c is in ..(23b).., and ΔT is in ..(23c)... ....23a

..........23b

..........23c

24. The specific heat of most substances is (less than, the same as, more than) ..(24).. the specific heat of water. 24

25. An important source of error in high-school-laboratory heat experiments is the ..(25).. to the surroundings. 25

DIRECTIONS: **Place the answers to the following problems in the spaces provided at the right.**

26. Convert to Kelvin scale temperatures:
   a. 58°C 26a
   b. -142°C 26b

27. What is the specific heat of silver if 112 cal is required to raise the temperature of 80.0 g of silver from 25.0°C to 50.0°C? 27

28. What is the final temperature of the mixture of 60.0 g of water at 70.0°C and 30.0 g of water at 0.0°C? 28

29. Calculate the mass of water at 20.0°C to which $1\overline{0}0$ g of zinc (specific heat = 0.092 cal/g C°) at a temperature of 150.0°C can be added in order to produce a final temperature of 40.0°C. 29

30. A calorimeter (specific heat = 0.090 cal/g C°) has a mass of 145 g. It contains $14\overline{0}$ g of water at 20.0°C. The final temperature resulting when a $15\overline{0}$-g mass of tungsten at 330.0°C is added to the water in the calorimeter is 30.0°C. Calculate the specific heat of tungsten. 30

NAME PERIOD DATE INSTRUCTOR'S COMMENTS

8B EXERCISE Thermal Expansion

SECS. 8.8 - 8.14

DIRECTIONS: In the parentheses at the right of each word or expression in the second column, write the letter of the expression in the first column which is MOST CLOSELY related.

a. 0°C
b. 273°K, 1 atm
c. 760 mm mercury
d. $\Delta l/l\Delta T$
e. $V' = Vp/p'$
f. $V' = Vp'/p$
g. $\Delta V/V\Delta T$
h. $V' = VT_K/T'_K$
i. $V' = VT'_K/T_K$
j. $V' = V(p/p')(T'_K/T_K)$
k. $V' = V(p'/p)(T_K/T'_K)$

coefficient of linear expansion ( ) 1
coefficient of volume expansion ( ) 2
Charles' law ( ) 3
standard temperature ( ) 4
Boyle's law ( ) 5
standard pressure ( ) 6
STP ( ) 7
general gas law ( ) 8

DIRECTIONS: In the space at the right, place the letter or letters indicating EACH choice which forms a correct statement.

9. When the temperature of a typical solid is raised, **(a)** it will increase in length, **(b)** its thickness will be unchanged, **(c)** the amplitude of vibration of its atoms and molecules is increased, **(d)** the average distance between its atoms and molecules is unaffected. 9

10. The change in length of a typical solid with a change in temperature **(a)** is directly proportional to its length, **(b)** depends on the nature of the solid, **(c)** is inversely proportional to the coefficient of linear expansion, **(d)** is directly proportional to the temperature change. 10

11. When the temperature of a typical liquid is raised, **(a)** it will increase in volume, **(b)** the thermal energy of its molecules remains the same, **(c)** the amplitude of vibration of its molecules is unchanged, **(d)** the average distance between its molecules is decreased. 11

12. The change in volume of a typical liquid with a change in temperature **(a)** is directly proportional to its volume, **(b)** depends on the nature of the liquid, **(c)** is directly proportional to the coefficient of volume expansion, **(d)** is inversely proportional to the temperature change. 12

13. Water **(a)** expands when its temperature is raised from 0°C to 4°C, **(b)** expands when its temperature is raised from 4°C to 100°C, **(c)** has a point of maximum density at 4°C, **(d)** has a point of minimum density at 100°C. 13

14. As the temperature of water is raised from 0°C to 4°C, **(a)** the distance between the molecules increases, **(b)** the open crystal fragments begin to collapse, **(c)** the speed of the molecules increases, **(d)** the effect of the collapsing crystal structure is more significant than the change in speed of the molecules. 14

15. The change in volume of a typical gas with a change in temperature at constant pressure **(a)** is directly proportional to its volume, **(b)** depends on the nature of the gas, **(c)** is inversely proportional to the kinetic energy of its molecules, **(d)** is directly proportional to the Kelvin temperature change. 15

16. An ideal gas (a) consists of molecules which experience attractive forces, (b) consists of infinitely small molecules, (c) consists of perfectly elastic molecules, (d) is a hypothetical gas. 16

DIRECTIONS: In the blank space at the right of each statement, write the word or expression which BEST completes the meaning.

17. The coefficient of linear expansion of most solids (does not vary, varies slightly, varies greatly) ..(17).. with temperature. 17

18. The effect of No. 17 (may be, is never) ..(18).. neglected. 18

19. The magnitude of the coefficient of linear expansion (depends, does not depend) ..(19).. on whether length is measured in meters or centimeters. 19

20. In designing and building devices which will undergo temperature changes, allowances must be made not only for changes in ..(20a).. due to expansion and contraction but for different ..(20b).. of expansion and contraction of different materials. ....20a 20b

21. Mercury rises in the tube of a glass thermometer when the temperature rises because the coefficient of volume expansion of mercury is (lower than, the same as, higher than) ..(21).. the coefficient of volume expansion of glass. 21

22. When the temperature of a mass of water is raised above 4°C, the effect of collapsing crystal structure is (less than, equal to, more than)..(22a).. the effect of increasing molecular speed, and the volume of the water (decreases, remains the same, increases) ..(22b)... same, increases) ..(22b)... ....22a 22b

23. No ice forms on the surface of a fresh-water pond until the temperature of the entire pond is ..(23).. or below. 23

24. The coefficient of volume expansion for gases is ..(24).. the volume at 0°C. 24

25. If two quantities are in direct proportion, their ..(25a).. is a constant and their graph is a(n) ..(25b)... ....25a 25b

26. At constant volume, the pressure of a given mass of gas varies ..(26).. with the Kelvin temperature. 26

27. If two quantities are in inverse proportion, their ..(27a).. is a constant and their graph is a(n) ..(27b)... ....27a 27b

28. At constant temperature, the density of a given mass of gas varies ..(28).. with the pressure. 28

29. At constant pressure, the density of a given mass of gas varies ..(29).. with the Kelvin temperature. 29

30. Under usual temperature and pressure conditions, real gases approximate the behavior of the ideal gas because the molecules are so far apart that the effects of their ..(30a).. and the ..(30b).. between them are negligible. ....30a 30b

DIRECTIONS: Place the answers to the following problems in the spaces provided at the right.

31. A copper wire is $30\bar{0}$ m long at 20.0°C. If the temperature rises to 45.0°C, what is its increase in length? 31

32. What is the increase in volume of $25\bar{0}$ ml of mercury at 0.0°C when it is heated to 30.0°C? 32

33. A quantity of air occupies $80\overline{0}$ ml at 50.0°C. What will be its volume at 150.0°C if its pressure remains unchanged? 33

34. A mass of gas occupies 25.0 liters at a pressure of $75\overline{0}$ mm of mercury. With the temperature constant, calculate the volume the gas will occupy when subjected to a pressure of $80\overline{0}$ mm. 34

35. A $50\overline{0}$-ml sample of gas is collected at a temperature of 25.0°C and a pressure of $76\overline{0}$ mm. What will be the volume of the gas at -25.0°C and a pressure of $80\overline{0}$ mm? 35

NAME PERIOD DATE INSTRUCTOR'S COMMENTS

8C EXERCISE Change of Phase

SECS. 8.15 - 8.25

DIRECTIONS: Write the answers to the following in the spaces provided. Where appropriate, make complete statements.

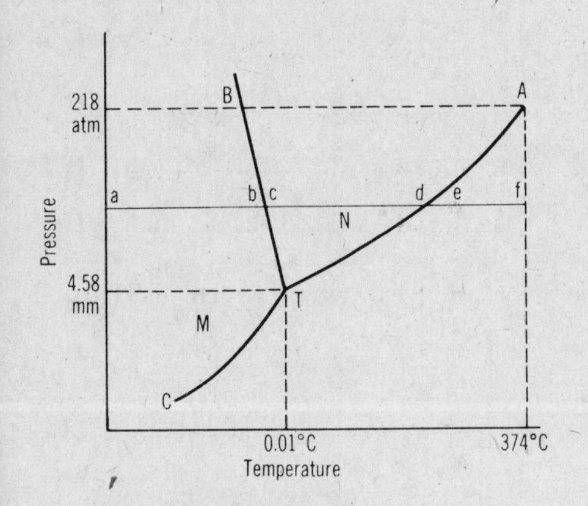

Fig. 8C-1 Temperature-pressure curves for pure water

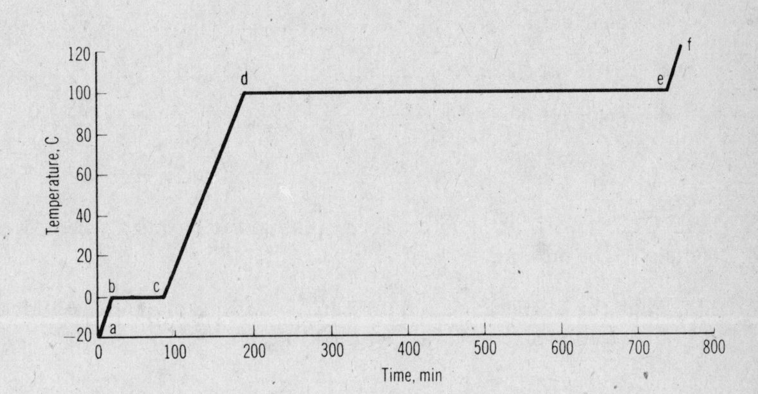

Fig. 8C-2 Temperature vs time of heating for a constant heat input of 100 cal/min, starting with 100 g of ice at −20.0°C

1. On Fig. 8C-1, label appropriate areas: solid, liquid, vapor; label appropriate points: triple point, critical point.
2. Where on Fig. 8C-1 do the following conditions prevail?
   a. Water vapor and water are in equilibrium. 2a
   b. Water vapor and ice are in equilibrium. 2b
   c. Ice and water are in equilibrium. 2c
   d. Ice, water, and water vapor are in equilibrium. 2d
3. In Fig. 8C-1, which curve is the equilibrium vapor pressure curve for water? 3
4. a. Can a solid at the temperature and pressure conditions represented by M, Fig. 8C-1, sublime? ....

   ....

   .... 4a

   b. Can a liquid at the temperature and pressure conditions represented by N, Fig. 8C-1, vaporize? ....

   ....

   .... 4b
5. On Fig. 8C-1, points b and c are the same point, as are points d and e. On Fig. 8C-2, points a, b, c, d, e, f have the same characteristics as the similarly designated points on Fig. 8C-1. Label appropriate portions of the graph in Fig. 8C-2: solid (ice), liquid (water), vapor (steam).
6. On Fig. 8C-2, label appropriate portions of the graph: melting of ice, boiling of water.
7. In Fig. 8C-2, what kind of potential and/or kinetic energy change is represented by segments ab, cd, and ef?

   ....

   7

8. In Fig. 8C-2, what kind of potential and/or kinetic energy change is represented by segments bc and de? ....

....

.... 8

9. What constant-temperature processes are shown in Fig. 8C-2? ....

.... 9

10. What quantity of heat is required for each segment of the graph in Fig. 8C-2? Show the smooth form computation.

    a. Segment ab 10a

    b. Segment bc 10b

    c. Segment cd 10c

    d. Segment de 10d

    e. Segment ef 10e

DIRECTIONS: In the blank space at the right of each statement, write the word or expression which BEST completes the meaning.

11. When the temperature and pressure conditions of a pure substance are established, its ..(11a).. and ..(11b).. are determined. 11a

    ....11b

12. The discontinuity between phases in equilibrium is known as a(n) ..(12)... 12

13. In the triple-point diagram for a typical crystalline substance, the solid-liquid curve has a(n) ..(13).. slope. 13

14. At ordinary temperatures the vapor pressure of solids is (less than, equal to, greater than) ..(14a).. that of liquids. Consequently, solids evaporate (slower than, at the same rate as, faster than) ..(14b).. liquids. 14a

    ....14b

15. A mixture of two liquids having different boiling points usually has a boiling temperature (the same as, different from) ..(15).. that of the lower boiling liquid. 15

16. When water boils below 100°C under reduced pressure, the heat of vaporization is (less than, equal to, more than) ..(16).. 539 cal/g. 16

17. When a given mass of ice becomes water at 0°C, the volume ..(17a)..; when a given mass of water becomes steam at 100°C at constant pressure, the volume ..(17b)... ....17a

    ....17b

18. Above its critical temperature, a substance can exist only in the ..(18).. phase. 18

19. At the critical point, the density of the liquid phase is (less than, equal to, greater than) ..(19a).. the density of the vapor phase, and the heat of vaporization is ..(19b)... ....19a

    ....19b

20. At temperatures above the critical temperature, the vapor phase of a substance is usually called a(n) ..(20a)..; while at temperatures below the critical temperature, it is called a(n) ..(20b)... ....20a

    ....20b

21. The temperature of the human body is partially controlled by the heat-absorbing process of ..(21a).., but this process is not very effective under conditions of (low, high) ..(21b).. humidity. 21a

    ....21b

22. For a gas to be a useful refrigerant, it must ..(22).. readily by pressure alone at room temperature. 22

DIRECTIONS: In the parentheses at the right of each word or expression in the second column, write the letter of the expression in the first column which is MOST CLOSELY related.

a. −119°C for oxygen
b. Amount of heat needed to vaporize a unit mass of liquid at its boiling point
c. Separation of liquids having different boiling points
d. Lowering the temperature of a substance below the normal freezing point without solidification
e. Amount of heat needed to melt a unit mass of a substance at its melting point
f. Lower limit of the liquid-gas evaporation curve
g. Pressure at the critical point
h. Evaporation, then condensation in a separate vessel
i. Pressure at the triple point
j. Upper limit of the liquid-gas evaporation curve

critical point ( ) 23
critical pressure ( ) 24
critical temperature ( ) 25
distillation ( ) 26
fractional distillation ( ) 27
heat of fusion ( ) 28
heat of vaporization ( ) 29
supercooling ( ) 30

DIRECTIONS: Place the answers to the following problems in the spaces provided at the right.

31. Calculate the mass of ice at 0.0°C that must be added to $40\bar{0}$ g of water at 80.0°C so that the final water temperature is 15.0°C. 31

32. How many calories of heat are needed to change 25.0 g of ice at −10.0°C to water at 100.0°C? 32

33. How many grams of alcohol can be vaporized at 78.5°C by 9.18×10^3 calories? 33

34. Steam (10.0 g) at 100.0°C is passed into cool water at 20.0°C. The resulting final water temperature is 60.0°C. What was the mass of the cool water? 34

35. What final temperature results from mixing 30.0 g of ice at 0.0°C, 20.0 g of water at 0.0°C, and 10.0 g of steam at 100.0°C? 35

NAME PERIOD DATE INSTRUCTOR'S COMMENTS

9 EXERCISE Heat and Work

SECS. 9.1 - 9.7

DIRECTIONS: Write the answers to the following in the spaces provided. Where appropriate, make complete statements.

1. What is *thermodynamics?* 1

2. What causes heat transfer from one body to another? 2

3. State the *first law of thermodynamics.* 3

4. What is an *adiabatic process?* 4

5. In what two ways may *heat be converted to useful work?*

   a. 5a

   b. 5b

6. What is an *isothermal process?* 6

7. What is an *ideal heat engine?* 7

8. What change in operating conditions increases the efficiency of an ideal heat engine? 8

9. State the *second law of thermodynamics.* 9

10. State the *law of entropy.* 10

DIRECTIONS: In the blank space at the right of each statement, write the word or expression which BEST completes the meaning.

11. The mechanical equivalent of heat in the MKS system is ..(11)... 11

12. The first law of thermodynamics is a special case of the law of ..(12)... 12

13. When heat is converted to another form of energy, or vice versa, there is (great, some, no) ..(13).. loss of energy. 13

14. In the Joule experiment, no heat enters or leaves the insulating jar. The water-churning process is, therefore, a(n) ..(14).. one. 14

15. Work done by a gas is considered positive when the gas ..(15a).. and negative when the gas ..(15b)... ....15a

..........15b

16. When the volume of a gas increases isothermally, its pressure ..(16a).. and its temperature ..(16b)... ....16a

..........16b

17. During an isothermal expansion, the internal energy ..(17a).. and the potential energy ..(17b)... ....17a

..........17b

18. When the volume of a gas increases adiabatically, its pressure ..(18a).. and its temperature ..(18b)... ....18a

..........18b

19. During an adiabatic expansion, the internal energy ..(19a).. and the potential energy ..(19b)... ....19a

..........19b

20. The source of the heat equivalent of the work done by an ideal gas during isothermal expansion is its ..(20a).., while the source of the heat equivalent of the work done by an ideal gas during adiabatic expansion is its ..(20b)... ....20a

..........20b

21. The c_p of a gas is always ..(21).. the c_v. 21

22. The reason for No. 21 is that ..(22a).. must include work done on the movable part of the gas container, while ..(22b).. does not. 22a

..........22b

23. The second law of thermodynamics makes it impossible to attain a temperature of ..(23)... 23

24. Entropy is the amount of energy that cannot be converted into mechanical ..(24)... 24

25. Natural processes tend to increase ..(25).. in the universe. 25

DIRECTIONS: Place the answers to the following problems in the spaces provided at the right.

26. In the Joule experiment, a paddle wheel was rotated in a calorimeter cup containing water. The total water equivalent was $2\bar{0}0$ g. The rotation was produced by a falling 20.0-kg mass coupled by pulleys to the paddle wheel. If the mass dropped 5.00 m and the resulting temperature rise in the water was 1.15 C°, find the mechanical equivalent of heat. Assume that the mass has no kinetic energy at the bottom of its fall. 26

27. A gas expands from a volume of 2.00 m^3 to 6.00 m^3 and does 6912 joules of work against a constant outside pressure. Find the outside pressure. 27

28. A gas which has a volume of $24\bar{0}$ m^3 at 0.0°C is heated to 182°C at a constant pressure of 9.25 n/m^2. What is the external work done? 28

29. Find the efficiency of an ideal heat engine operating between temperatures of $-10\bar{0}^{\circ}$C and $50\bar{0}^{\circ}$C. 29

30. How much would the efficiency of an ideal heat engine be improved if its lower temperature were changed from 80.0°C to 40.0°C while the upper temperature remained at $30\bar{0}^{\circ}$C? 30

NAME PERIOD DATE INSTRUCTOR'S COMMENTS

10 EXERCISE The Nature of Waves SECS. 10.1 - 10.16

DIRECTIONS: In the space at the right, place the letter or letters indicating EACH choice which forms a correct statement.

1. Energy can be transferred (a) by the movement of materials, (b) by the movement of heated gas, (c) through matter by mechanical waves, (d) through space by electromagnetic waves. 1

2. A wave (a) is a disturbance that moves only through solids, liquids, and gases, (b) produces a transfer of energy without a transfer of matter, (c) involves a quantity which changes in magnitude with respect to time at a given location, (d) involves a disturbance which changes in magnitude from place to place at a given time. 2

3. To produce a mechanical wave we need (a) an energy source, (b) a medium which behaves like an array of spring-connected particles, (c) to produce a displacement of some sort in matter, (d) a form of matter in which the displacement of one particle has no effect on adjacent particles. 3

4. A pulse moving along a spiral spring (a) carries particles of matter along with it, (b) is a method of energy transfer, (c) can be a crest or trough in longitudinal wave motion, (d) can be a compression or a rarefaction in transverse wave motion. 4

5. A periodic wave (a) is related to the simple harmonic motion of the wave source, (b) that is transverse may be generated in a horizontal spring attached to a weight which vibrates vertically, (c) requires no continuing supply of energy, (d) that is longitudinal may be generated in a horizontal spring attached to a weight which vibrates horizontally. 5

6. When a transverse pulse is reflected at the fixed termination of a medium, (a) the shape of the pulse is changed, (b) action and reaction forces are involved, (c) a crest is reflected as a crest, (d) a crest is reflected as a trough. 6

7. The impedance (a) of a medium is the ratio of the applied wave-producing force to the resulting displacement velocity, (b) of a terminating medium is infinite in cases of total in-phase reflection, (c) of a terminating medium exactly matches that of the transmitting medium in cases of zero energy transfer between the media, (d) mismatch of two wave-propagation media may be reduced by an impedance transformer. 7

8. Superposition (a) involves two or more waves moving simultaneously through the same medium, (b) allows complex waves to be analyzed in terms of simple wave combinations, (c) produces constructive interference when the crest of one wave coincides with the trough of a second wave, (d) produces destructive interference when the compression of one wave coincides with the rarefaction of a second wave. 8

9. A standing wave is generated by two wave trains (a) of different wavelength but of the same frequency and amplitude traveling in the same direction, (b) of different frequency but of the same wavelength and amplitude traveling in the same direction, (c) of different amplitude but of the same wavelength and frequency traveling in the same direction, (d) of the same wavelength, frequency, and amplitude traveling in opposite directions. 9

10. In a standing wave pattern (a) the particles vibrate in simple harmonic motion, (b) the amplitude of motion for all vibrating points is the same, (c) energy is transferred along the vibrating string, (d) there are alternate loops and nodes. 10

DIRECTIONS: Write the answers to the following in the spaces provided. Where appropriate, make complete statements.

11. What type of particle vibration occurs in *transverse waves?* ....

....11

12. What type of particle vibration occurs in *longitudinal waves?* ....

....12

13. What is a *crest?* ....

....13

14. What is a *trough?* ....

....14

15. What is a *compression?* ....

....

....15

16. What is a *rarefaction?* ....

....

....16

17. What is the *speed* of a wave? ....

....17

18. What are the characteristics of particles of a vibrating medium which are *in phase?* ....

....

....18

19. What is the *frequency* of a wave? ....

....19

20. How is the *period* of a wave related to its frequency? ....

....20

21. What is the *wavelength* of a wave? ....

....

....21

22. How is the power transmitted by a wave system related to the amplitude and frequency of the waves?

....

....

....22

DIRECTIONS: In the parentheses at the right of each word or expression in the second column, write the letter of the expression in the first column which is MOST CLOSELY related.

a. Movement through a uniform medium in a straight line
b. hertz
c. $f\lambda$
d. Periodic transverse straight waves
e. Change in wave speed on passage from one medium to another
f. $1/f$
g. vT
h. Periodic transverse circular waves
i. Spreading of a wave disturbance beyond the edge of a barrier
j. Occurs at the boundary of the transmitting medium
k. Mutual effect of two waves in the same medium
l. Reduction in wave amplitude due to dissipation of wave energy

v ( ) 23
λ ( ) 24
f ( ) 25
T ( ) 26
damping ( ) 27
rectilinear propagation ( ) 28
reflection ( ) 29
refraction ( ) 30
diffraction ( ) 31
interference ( ) 32

NAME PERIOD DATE INSTRUCTOR'S COMMENTS

11 EXERCISE Sound Waves

SECS. 11.1 - 11.16

DIRECTIONS: In the space at the right, place the letter or letters indicating EACH choice which forms a correct statement.

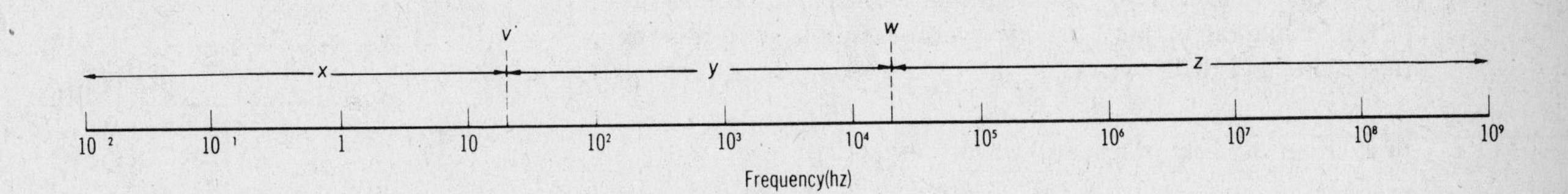

Fig. 11-1

1. On the sonic spectrum frequency scale shown in Fig. 11-1, **(a)** *y* represents the audio range, **(b)** *x* represents the region of infrasonic waves, **(c)** *z* represents the region of ultrasonic waves, **(d)** $x + y + z$ represents the region of sound. 1

2. Sounds are produced by **(a)** matter at rest, **(b)** matter in vibration, **(c)** solids only, **(d)** solids, liquids, or gases. 2

3. As a steel strip vibrates from left to right, it is **(a)** doing work on the gas molecules to the right, **(b)** removing energy from the gas molecules to the right, **(c)** producing a maximum compression in the molecules to the right at its equilibrium position, **(d)** producing no change in the molecules to the right at its point of minimum displacement. 3

4. Sound waves are **(a)** longitudinal waves in space, **(b)** longitudinal waves in matter, **(c)** transverse waves in space, **(d)** transverse waves in matter. 4

5. The most common transmitting medium for sound vibrations is **(a)** metal wires, **(b)** earth, **(c)** water, **(d)** air. 5

6. Sound waves are transmitted better through **(a)** rarefied gases than dense gases, **(b)** gases than through liquids, **(c)** gases than through solids, **(d)** solids and liquids than through gases. 6

7. Sound travels **(a)** with the same speed as light, **(b)** faster than light, **(c)** slower than light, **(d)** at a rate of approximately 331.5 m/sec in air at 0°C. 7

8. The speed of sound in air **(a)** decreases with a rise in temperature, **(b)** is not affected by temperature change, **(c)** increases approximately (0.6m/sec)/C°, **(d)** is approximately 346 m/sec at 25°C. 8

DIRECTIONS: In the blank space at the right of each statement, write the word or expression which BEST completes the meaning.

9. The time rate at which sound energy flows through a unit area normal to the direction of propagation is its ..(9)... 9

10. The physiological effect of the sound characteristic of No. 9 is ..(10)... 10

11. The intensity of a sound in a uniform medium at 1.5 km from a point source is ..(11).. as great as at 0.5 km. 11

12. The intensity level of sound during conversation is ..(12)... 12

13. On the sonic spectrum frequency scale, Fig. 11-1, the lower limit of audibility for most people is represented by point ..(13a).. with a value of ..(13b).. hertz, while the upper limit of audibility is represented by point ..(13c).. with a value of ..(13d).. hertz. ....13a ....13b ....13c 13d

14. The minimum intensity level for audible sounds is the threshold of ..(14a).., while the upper intensity level for audible sounds is the threshold of ..(14b)... ....14a 14b

15. The number of hertz of a sound wave is its ..(15)... 15

16. The physiological effect of the sound characteristic of No. 15 is ..(16)... 16

17. The Doppler effect is associated with the variation in ..(17).. heard when a source of sound and the ear are moving relative to each other. 17

18. As a locomotive approaches a crossing, the pitch of its horn heard by the engineer is ..(18).. than the pitch of the horn heard by the crossing guard. 18

19. As the locomotive passes the crossing, the pitch of the horn heard by the engineer (becomes higher, remains the same, becomes lower) ..(19a).., while the pitch of the horn heard by the crossing guard (becomes higher, remains the same, becomes lower) ..(19b)... ....19a 19b

20. The Doppler effect observation of light from many distant stars which suggests that the universe is expanding is called the ..(20)... 20

DIRECTIONS: Write the answers to the following in the spaces provided. Where appropriate, make complete statements.

21. When a guitar string vibrates as a whole,

a. what is characteristic of the frequency of the tone produced? 21a

b. what name is given to this tone? 21b

22. If a fundamental tone has a frequency of $44\overline{0}$ hz, its first harmonic has a frequency of 22

23. The fourth harmonic of the fundamental of No. 22 has a frequency of 23

24. What are *forced vibrations?* 24

25. What are *sympathetic vibrations?* 25

26. What is the approximate wavelength of the fundamental resonant frequency of a closed tube 35 cm long? 26

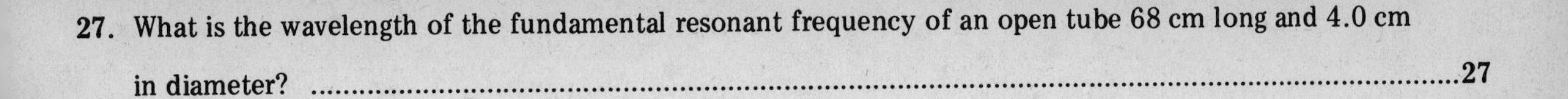

27. What is the wavelength of the fundamental resonant frequency of an open tube 68 cm long and 4.0 cm in diameter? 27

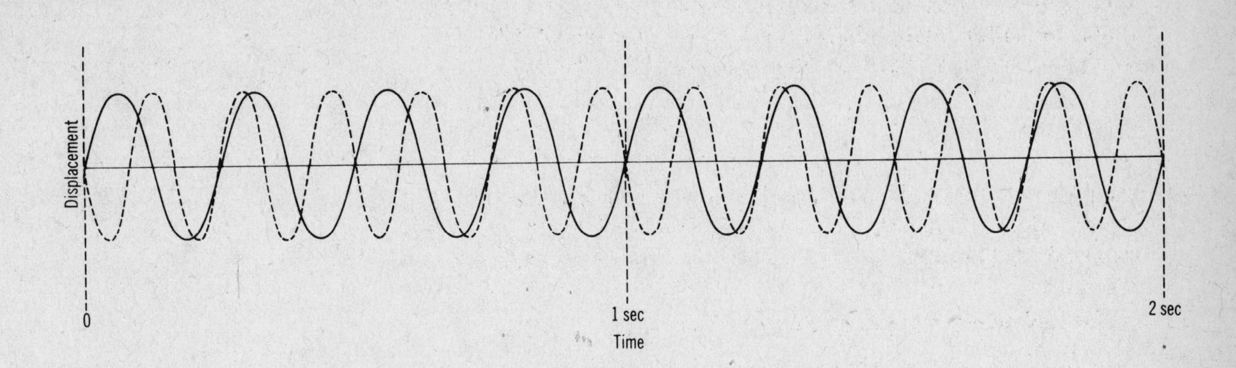

Fig. 11-2

28. Figure 11-2 shows the superposition of two waves, one having a frequency of 4 hz, the other of 6 hz. Draw the resultant wave on Fig. 11-2.

29. How many beats per second result from the wave superposition of No. 28? 29

30. What is the average frequency of the wave superposition of No. 28? 30

DIRECTIONS: In the space at the right, place the letter or letters indicating EACH choice which forms a correct statement.

31. The vibration of a string (a) is a longitudinal wave motion, (b) will decrease in amplitude unless energy is continually supplied to it, (c) sets up a transverse wave of the same frequency as that of the string in the surrounding air, (d) transfers energy to molecules of gases in the air. 31

32. A string which vibrates in four segments produces (a) its fundamental, (b) its first four harmonics, (c) its fundamental and fourth harmonic, (d) its fourth harmonic. 32

33. The *quality* of a sound depends on (a) the intensity of the fundamental, (b) the number and prominence of the harmonics present, (c) the frequency of the fundamental, (d) the characteristics of the instrument producing it. 33

34. The vibrating frequency of a string may be increased by (a) decreasing its length, (b) decreasing its tension, (c) plucking it more rapidly, (d) lowering its temperature. 34

35. The vibrating frequency of a string is inversely proportional to its (a) length, (b) diameter, (c) density, (d) tension. 35

NAME PERIOD DATE INSTRUCTOR'S COMMENTS

12A EXERCISE The Nature of Light

SECS. 12.1 - 12.13

DIRECTIONS: In the space at the right, place the letter or letters indicating EACH choice which forms a correct statement.

1. Rectilinear propagation, reflection, refraction, interference, and diffraction are all properties of (a) sound waves but not water waves, (b) water waves but not sound waves, (c) light waves only, (d) waves in general. 1

2. The corpuscular theory assumes that (a) light consists of streams of tiny particles coming from a luminous source, (b) particles of light travel at such high speeds that their paths are straight lines, (c) particles of light are perfectly elastic and rebound in regular fashion from a reflecting surface, (d) particles of light move faster through air than through water. 2

3. The corpuscular theory failed during the nineteenth century because (a) it could not explain rectilinear propagation satisfactorily, (b) the speed of light was shown to be slower in water than in air, (c) interference and diffraction cannot be explained very well by the behavior of particles, (d) particles of light were found to undergo inelastic collisions. 3

4. The wave theory assumes that (a) light is a train of waves having wave fronts in the same direction as the paths of the light rays, (b) each point on a wave front may be regarded as a new source of disturbance, (c) the speed of light in optically dense media is less than that in air, (d) a ray of light is the line of direction of waves sent out from a luminous source. 4

5. The electromagnetic theory (a) was developed by James Clerk Maxwell, (b) describes the manner in which radiated energy is propagated in free space, (c) predicts that heat radiations travel in space with the speed of light, (d) states that the energy of electromagnetic radiations is equally divided between an electric field and a magnetic field. 5

6. The electromagnetic spectrum consists of a range of radiation frequencies extending from (a) 10^{-1} to 10^{-25} hz, (b) 3×10^{7} to 3×10^{-17} hz, (c) 10^{1} to 10^{25} hz, (d) 3×10^{17} to 3×10^{-7} hz. 6

7. The photoelectric effect demonstrates that (a) the rate of emission of photoelectrons is directly proportional to the intensity of the light falling on the emitting surface, (b) the rate of emission of photoelectrons is inversely proportional to the frequency of the light falling on the emitting surface, (c) an increase in light intensity results in an increase in the velocities of emitted photoelectrons, (d) an increase in light intensity results in an increase in the number of photoelectrons emitted per second, these electrons having the same group of discrete velocities. 7

8. Following the discovery of the photoelectric effect, the wave theory became inadequate because it could not explain why (a) the magnitude of the photoelectric current is proportional to the incident light intensity, (b) all substances have characteristic cutoff frequencies above which emission does not occur no matter how intense the illumination, (c) there is a time lag between the illumination of a surface and the ejection of a photoelectron, (d) a very feeble light containing frequencies above a minimum value causes the ejection of photoelectrons. 8

9. The quantum theory assumes that a transfer of energy between light and matter occurs only in discrete quantities proportional to (a) the intensity of the light, (b) the frequency of the radiation, (c) the quantity of the matter, (d) the temperature of the matter. 9

10. The modern view of the nature of light assumes that (a) light behaves in some circumstances like waves, (b) light behaves in some circumstances like particles, (c) light energy is transported in photons, (d) photons are guided along their path by a wave field. 10

DIRECTIONS: In the blank space at the right of each statement, write the word or expression which BEST completes the meaning.

11. The transmission of light does not require the presence of ..(11)... 11

12. The main person supporting the corpuscular theory of light was ..(12a).., while the wave theory of light was upheld chiefly by ..(12b)... ....12a

....12b

13. Light is ..(13a).. which a human observer can ..(13b)... ....13a

....13b

14. The emission of electrons by a substance when illuminated by electromagnetic radiation is known as the ..(14)... 14

15. The fact that photoelectric emission consists of ejected electrons was first established by measuring the ..(15).. of the negative electricity. 15

16. Energy required to overcome the forces binding an electron within the surface of a material is known as the ..(16).. of that surface. 16

17. In any photoelectric situation the electrons ejected with maximum energy have their origin in the ..(17).. of atoms. 17

18. The negative potential on the collector of a photoelectric cell at which photoelectric current drops to zero is called the ..(18).. potential. 18

19. The cut-off potential for a given photoelectric system measures the kinetic energy of the ..(19).. photoelectrons ejected from the emitting surface. 19

20. The velocity of photoelectrons expelled from an emitting surface is independent of the ..(20).. of the light source. 20

21. The cut-off potential for a given photoelectric system is independent of the ..(21).. of the incident radiation. 21

22. The cut-off potential for a given photoelectric system depends only on the ..(22).. of the incident radiation. 22

23. For any emitting surface there is a characteristic frequency of illumination ..(23a).. which no photoelectrons are ejected; this is known as the ..(23b).. frequency of the emitter. 23a

....23b

24. To increase the maximum kinetic energy of photoelectrons, one must ..(24).. the frequency of the radiation illuminating the emitting surface. 24

25. The wave theory of light assumes that the light energy is distributed uniformly over the advancing ..(25)... 25

26. According to the wave theory, radiation of any frequency should produce photoelectric emission provided its ..(26).. is high enough. 26

27. The beginning of the quantum theory is closely associated with the work of ..(27a).. and ..(27b)... ....27a

....27b

28. Light quanta are known as ..(28)... 28

29. In the expression $E = hf$, if E is expressed in joules and f in hertz, h will have the dimensions ..(29)... 29

30. The energy of photons is directly proportional to the ..(30).. of the light. 30

31. According to the electromagnetic theory, electrons moving about the nucleus of an atom experience an acceleration and must ..(31).. energy. 31

32. The unique assumption of the Bohr model of the atom is that an electron cannot experience a(n) ..(32).. while occupying a discrete orbit. 32

33. The photon behaves as if it has a(n) ..(33).. equal to h/λ. 33

34. Any matter particle having a mass m and a velocity v may be considered to have a(n) ..(34).. equal to h/mv. 34

35. A practical application of the inverse photoelectric effect is the production of ..(35)... 35

36. A beam of light shines first on a totally reflecting surface and then on a totally absorbing surface. The pressure of the light on the first surface is ..(36).. that on the second. 36

NAME PERIOD DATE INSTRUCTOR'S COMMENTS

12B EXERCISE Illumination

SECS. 12.14 - 12.18

DIRECTIONS: In the parentheses at the right of each word or expression in the second column, write the letter of the expression in the first column which is MOST CLOSELY related.

a. Several rays of light coming from a point
b. Transmits light but diffuses it
c. Seen because of reflected light
d. A group of closely spaced rays
e. Emits light because of energy of its accelerated particles
f. Some rays of light are excluded
g. A single line of light
h. Several rays of light proceeding toward a point
i. Transmits light readily
j. Point at which light rays converge
k. Does not transmit light
l. All rays of light are excluded

luminous ( ) 1
illuminated ( ) 2
transparent substance ( ) 3
opaque substance ( ) 4
ray ( ) 5
beam ( ) 6
diverging pencil ( ) 7
converging pencil ( ) 8
umbra ( ) 9
penumbra ( ) 10

DIRECTIONS: In the space at the right, place the letter or letters indicating EACH choice which forms a complete statement.

11. In the practical study of light, all of the following are measured except (a) luminous intensity, (b) luminous flux, (c) luminous velocity, (d) illumination. 11

12. The luminous intensity of a 40-watt incandescent lamp is approximately (a) 35 candles, (b) 40 candles, (c) 80 candles, (d) 20 candles. 12

13. The efficiency of incandescent lamps in terms of the ratio of light output to wattage rating (a) increases with wattage increase, (b) decreases with wattage increase, (c) is constant for a given filament metal, (d) appears to be independent of the wattage. 13

14. The photometric quantity most closely related to power in mechanical systems is (a) luminous intensity, (b) luminous flux, (c) illumination, (d) candle power. 14

15. A 100-watt incandescent lamp emits luminous flux at the rate of approximately (a) 100 lm, (b) 130 lm, (c) 1600 lm, (d) 314 lm. 15

16. A 16.0-candle source at the center of a sphere of 1.00-meter radius provides illumination on the spherical surface of (a) 16.0 lm/m^2, (b) 4π lm/m^2, (c) 4.00 lm/m^2, (d) 201 lm/m^2. 16

17. If the spherical radius in No. 16 is increased to 4.00 meters, the illumination on the spherical surface becomes (a) 64.0 lm/m^2, (b) 4.00 lm/m^2, (c) 50.2 lm/m^2, (d) 1.00 lm/m^2. 17

18. The maximum illumination obtained from a 40.0-candle source 3.00 meters away is (a) 4.44 cd, (b) 13.3 lm/m^2, (c) 4.44 lm/m^2, (d) 120 cd. 18

19. If the luminous intensity of the source in No. 18 is tripled, the maximum illumination 3.00 meters away is (a) 13.3 cd, (b) 13.3 lm/m^2, (c) 40.0 lm/m^2, (d) 120 cd. 19

20. The Joly photometer (a) measures directly the intensity of a light source, (b) is more sensitive than the grease-spot photometer, (c) uses two blocks of paraffin separated by a metal plate, (d) compares the illumination from an unknown source to that from a standard source. 20

DIRECTIONS: In the blank space at the right of each statement, write the word or expression which BEST completes the meaning.

21. As the temperature of a hot body increases, the amount of visible radiation emitted ..(21)... 21

22. Using exponential notation and one significant figure, the speed of light in a vacuum in m/sec is approximately ..(22)... 22

23. The quantitative study of light is called ..(23)... 23

24. The unit of luminous intensity of a source of light is the ..(24)... 24

25. Approximately ..(25).. percent of the energy supplied to a 100-watt incandescent lamp is radiated in the visible region of the electromagnetic spectrum. 25

26. The portion of the total energy radiated per unit of time from a luminous source which can produce the sensation of sight is known as ..(26a).., the unit being the ..(26b)... ....26a 26b

27. A sphere with radius *r* has a surface area expressed by the formula ..(27)... 27

28. A luminous source having an intensity of 1 candle radiates luminous flux at the rate of ..(28)... 28

29. The illumination on a surface pertains to the ..(29).. of the luminous flux on the surface. 29

30. The MKS unit of illumination is the ..(30)... 30

31. The inverse square law applies to calculations of illumination from a point source on surfaces ..(31).. to the beam. 31

32. In the general case, the illumination on a surface varies ..(32a).. with the square of the distance from the luminous source and ..(32b).. with the ..(32c).. of the angle between the luminous flux and the normal to the surface. ....32a ....32b 32c

33. The inverse square law does not hold for calculations of illumination from a fluorescent tube because a fluorescent tube does not approximate a(n) ..(33)... 33

34. The instrument used to measure the candle power of a light source by comparing its intensity with that of a standard source is called a(n) ..(34)... 34

35. Laboratory comparison measurements of the intensities of light sources make use of the fact that when the comparing apparatus is correctly positioned, the intensities of the two light sources are ..(35a).. proportional to the ..(35b).. of their ..(35c).. from the screen. ....35a ....35b 35c

36. A(n) ..(36).. photometer is used commercially to measure the luminous intensity of electric lamps. 36

37. A "light meter" or "exposure meter" measures the amount of ..(37)... 37

DIRECTIONS: Place the answers to the following problems in the spaces provided at the right.

38. A table is located 1.50 m directly below a $13\overline{0}$-candle lamp. What is the illumination on the surface of the table? 38

39. A Bunsen photometer is equally illuminated when it is 35 cm from a standard lamp of 18 candles and 55 cm from a lamp of unknown intensity. What is the intensity of the unknown lamp? 39

40. A screen located 2 m from a luminous source has 9 times the illumination of a second screen illuminated by the same source. How far is the second screen from the luminous source? 40

NAME PERIOD DATE INSTRUCTOR'S COMMENTS

13 EXERCISE Reflection

SECS. 13.1 - 13.12

DIRECTIONS: (A) In the blank space at the right of each statement, write the word or expression which BEST completes the meaning. (B) Use a well-sharpened pencil and straight edge in all constructions.

1. The returning of light to the first medium from the boundary between two media is called ..(1)... 1

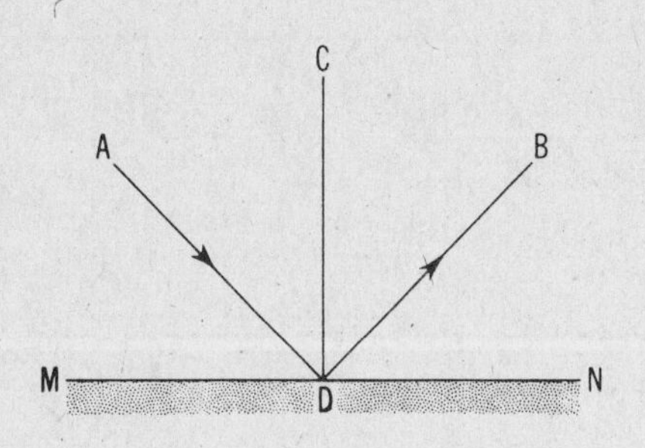

Fig. 13-1

2. In Fig. 13-1, the reflecting surface is represented by the line segment ..(2)... 2
3. The ray of light incident on the reflecting surface is shown as ..(3)... 3
4. The *normal* drawn from the point of incidence is ..(4)... 4
5. The *angle of incidence* is ..(5)... 5
6. The *angle of reflection* is ..(6)... 6
7. The angle of reflection is equal to the ..(7)... 7
8. In Fig. 13-2, label the *center of curvature* as C, the *vertex* as V, and the *principal focus* as F.
9. Construct the *principal axis* as a line segment PV and a secondary axis as SS′.

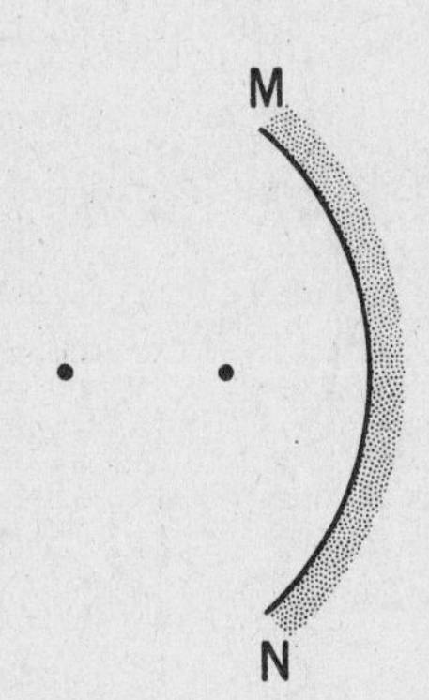

10. A *normal* to the surface of a concave mirror is a(n) ..(10a).. drawn from a point of ..(10b)... ....10a 10b
11. The *focal length* of the mirror is the distance from the ..(11a).. to the ..(11b)... ....11a 11b
12. In Fig. 13-3, construct the two line segments representing rays which will locate the image of point A. Label the image as point A′.
13. Construct the normal as a dashed line from the point of incidence of the ray already drawn in Fig. 13-3 which is not perpendicular at the point of incidence.

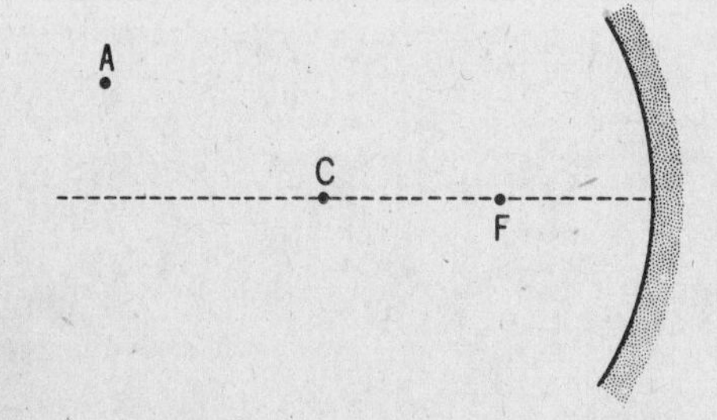

Fig. 13-3

14. A ray parallel to the principal axis of a spherical mirror will be reflected through the ..(14)... 14
15. A ray which follows the path of a ..(15).. will be reflected back upon itself. 15

DIRECTIONS: (A) Write the answers to the following in the spaces provided. Where appropriate, make complete statements. (B) Use a well-sharpened pencil and straight edge in all constructions.

16. Describe the image formed by a concave mirror of an object located between the center of curvature and the principal focus. ....

.... 16

17. How does the size of the image formed by a concave mirror compare with that of an object located at the center of curvature? ....

.... 17

18. What are two distinct differences between virtual images and real images, as formed by curved mirrors?

a. 18a

b. ....

.... 18b

19. In Fig. 13-4, complete the construction to locate the image A′B′ of the object AB.

20. Describe the image formed in Fig. 13-4. ....

.... 20

21. In Fig. 13-5, complete the construction to locate the image A′B′ of the object AB.

22. Describe the image formed in Fig. 13-5. ....

.... 22

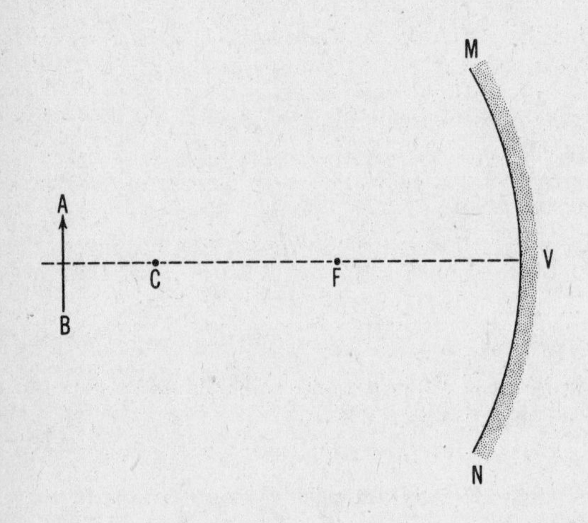

Fig. 13-4

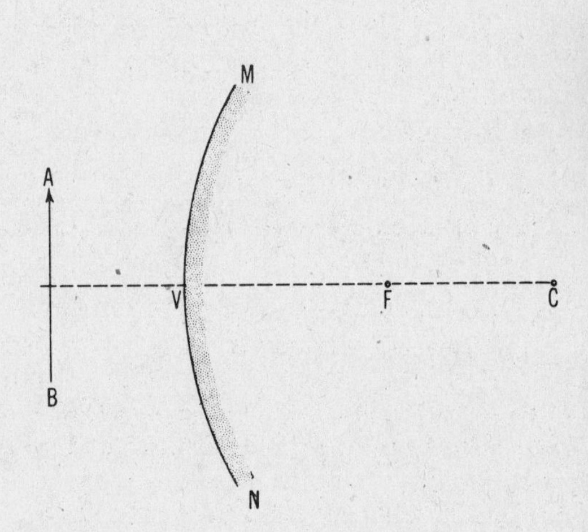

Fig. 13-5

23. An object 7.0 cm high is $3\overline{0}$ cm in front of a concave mirror which has a focal length of $1\overline{0}$ cm. How far from the mirror is the image formed and how high is it? ....

.... 23

24. A 4.0-cm high object is placed $1\overline{0}$ cm in front of a concave mirror. The mirror has a focal length of 25 cm. Where, with respect to the mirror, is the image formed; how far from the mirror is it; and how high is it?

.... 24

NAME PERIOD DATE INSTRUCTOR'S COMMENTS

14A EXERCISE Optical Refraction

SECS. 14.1 - 14.14

DIRECTIONS: In the blank space at the right of each statement, write the word or expression which BEST completes the meaning.

1. The speed of light in glass or water is (faster than, the same as, slower than) ..(1).. the speed of light in air. 1
2. A property of a transparent material which is an inverse measure of the speed of light through the material is called ..(2)... 2
3. The bending of light rays in passing obliquely from one medium into another of different optical density is called ..(3)... 3

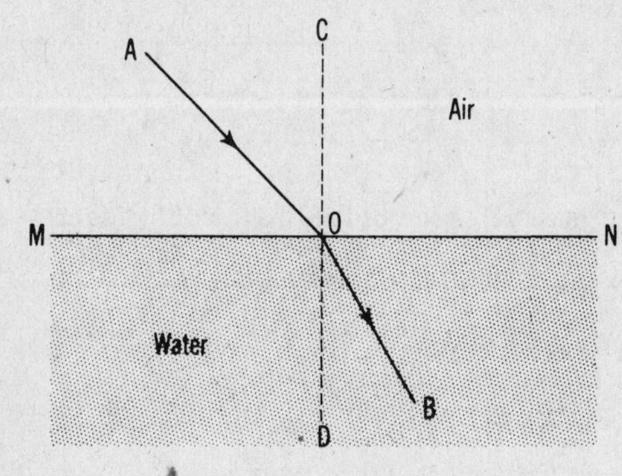

Figure 14A-1

4. In Fig. 14A-1, which illustrates the path of a light ray from air into water, AO is known as the ..(4).. ray. 4
5. Ray OB is known as the ..(5).. ray. 5
6. The line CO is the ..(6a).. drawn to the point of refraction, and OD is the ..(6b)... ....6a

   ....6b
7. Angle ..(7).. is the *angle of incidence*. 7
8. The *angle of refraction* is ..(8)... 8
9. A ray of light passing from one medium into another along the normal (is, is not) ..(9).. refracted. 9
10. The ratio of the speed of light in a vacuum to its speed in a substance is known as the ..(10).. of that substance. 10

DIRECTIONS: (A) Write the answers to the following in the spaces provided. Where appropriate, make complete statements. (B) Use a well-sharpened pencil and a straight edge in all constructions.

11. What are the three laws of refraction?

a. ....

....

....11a

b. ....

....

....11b

c. ....

....

....

....

....11c

12. How did Snell define the index of refraction? ....

....

....

....12

13. Why is the atmospheric refraction of light from the sun gradual rather than distinct? ....

....

....

....

....13

14. What is the size of the angle of refraction when the critical angle of incidence is reached? ....

....

....

....14

15. What occurs if the critical angle is exceeded? ....

....

....15

16. Given a right-angle crown-glass prism with two sides equal, what path will a ray of light follow if incident perpendicular to one of the equal surfaces? ....

....

....

....

....16

17. Complete the construction in Fig. 14A-2, locating the image of the object shown.

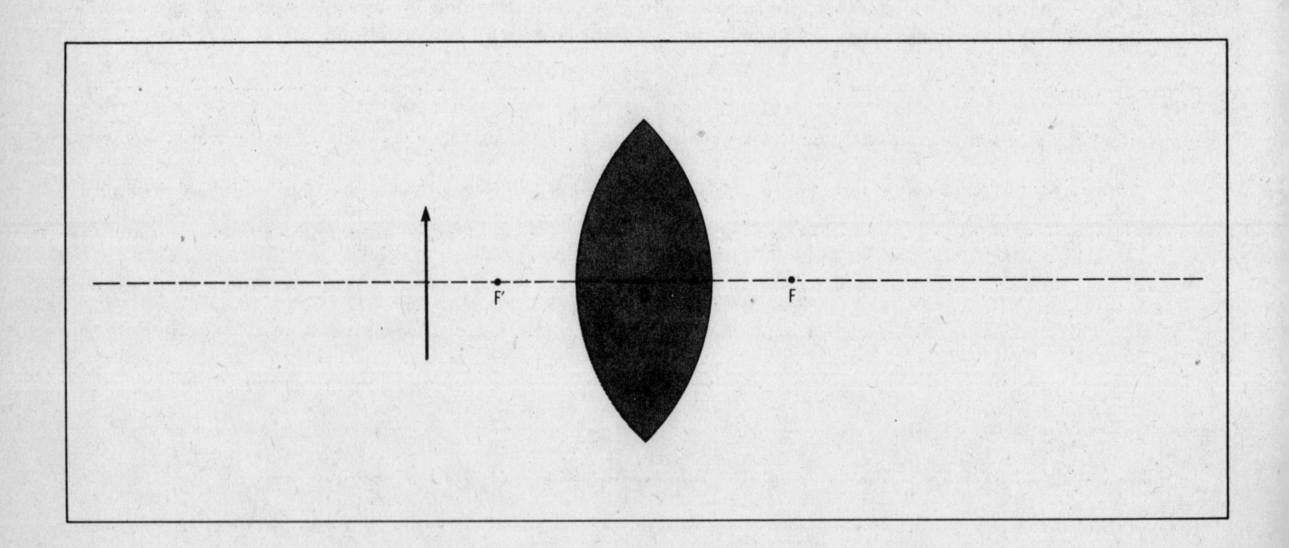

Figure 14A-2

18. Using a centimeter rule, measure the focal length of the lens and the object distance in Fig. 14A-2. Calculate the image distance and compare with your construction. 18

19. Measure the object height in Fig. 14A-2. Calculate the image height and compare with your construction.

....19

20. Complete the construction in Fig. 14A-3, locating the image of the object shown.

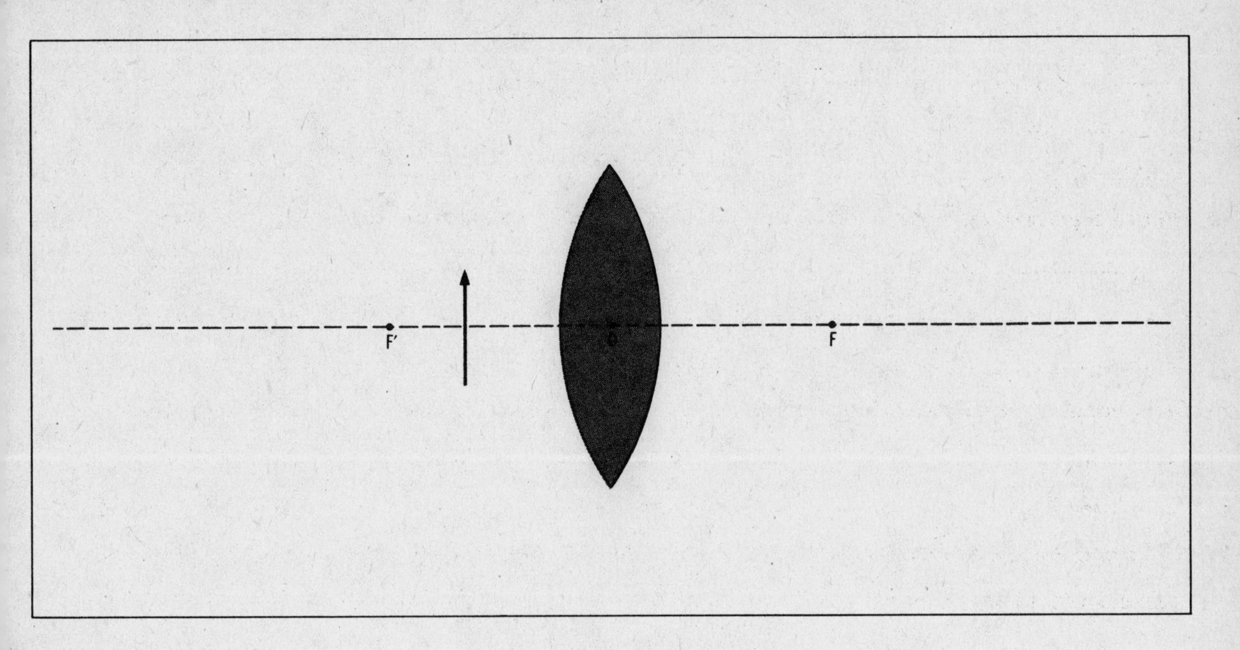

Figure 14A-3

21. What position relationship does the objective lens of a compound microscope have to the object and the image formed? ....

....

....

....21

22. List one similarity and one difference between prism binoculars and a refracting telescope. ....

....

....

....

....

....22

NAME PERIOD DATE INSTRUCTOR'S COMMENTS

14B EXERCISE Dispersion

SECS. 14.15 - 14.22

DIRECTIONS: In the parentheses at the right of each word or expression in the second column, write the letter of the expression in the first column which is MOST CLOSELY related.

a. Two colors which combine to form white
b. Combining primary colors
c. Continuous band of colors
d. Lower limit of visibility
e. Light composed of several colors
f. Nonfocusing of light of different colors
g. Red, green, blue
h. Light consisting of a single color
i. Upper limit of visibility
j. Cyan, magenta, yellow
k. Combining primary pigments
l. Property of light reaching the eyes

solar spectrum ( ) 1
monochromatic light ( ) 2
polychromatic light ( ) 3
color ( ) 4
complementary colors ( ) 5
primary colors ( ) 6
primary pigments ( ) 7
additive process ( ) 8
subtractive process ( ) 9
chromatic aberration ( ) 10

DIRECTIONS: In the blank space at the right of each statement, write the word or expression which BEST completes the meaning. Questions 11-15 refer to Fig. 14B-1 which shows the distribution of radiant energy from an ideal radiator.

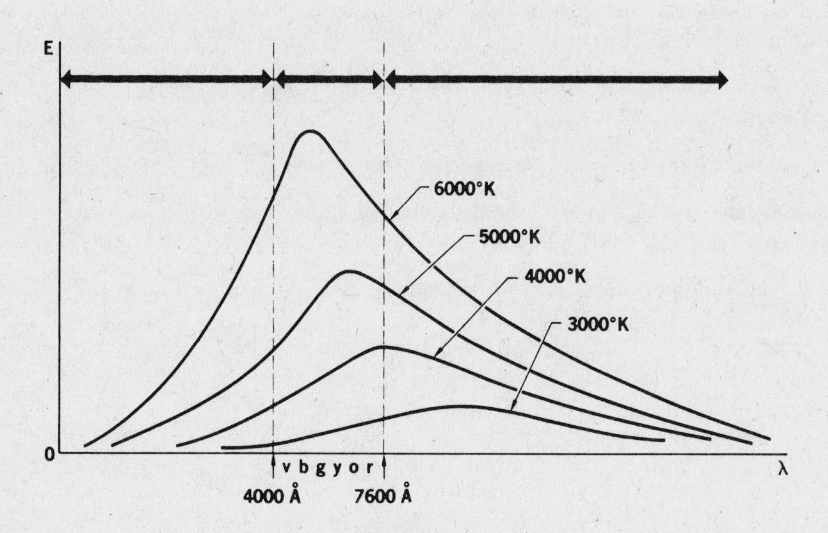

Figure 14B-1

11. Label appropriate regions: *visible, infrared, ultraviolet.*

12. The wavelength of light at the lower limit of visibility is ..(12a).., while that at the upper limit of visibility is ..(12b)... This represents a range of frequencies of about a(n) ..(12c)...

....12a

....12b

....12c

13. An increase in the temperature of the radiating body (increases, decreases, does not affect) ..(13).. the amount of visible radiation. 13

14. The temperature of the tungsten filament in a lamp is about 3000°K. At this temperature, most of its radiation is in the ..(14).. region. 14

15. Approximately how many times greater is the amount of energy detected as yellow light from a 6000°K source compared with that from a 3000°K source? 15

16. The dispersion of sunlight into several colors was described by ..(16)... 16

17. When white light is dispersed by a prism, ..(17a).. light is refracted least while ..(17b).. light is refracted most. 17a

..........17b

18. Color bears the same relationship to light as ..(18).. does to sound. 18

19. The color of an opaque object is not the property of the ..(19a).. we see, but is the property of the ..(19b).. which reach our eyes. 19a

..........19b

20. An opaque object described as *blue* reflects the ..(20)... portion of the solar spectrum. 20

21. If examined under a red light in a darkened room, the object of No. 20 will appear to be ..(21)... 21

22. An opaque object described as *white* ..(22).. all colors of the sunlight incident upon its surface. 22

23. An opaque object described as *black* ..(23).. all colors of the sunlight incident upon its surface. 23

24. A ..(24).. transparent object transmits all colors. 24

25. If the red light is subtracted from dispersed white light, the remaining colors will combine to form ..(25).. light. 25

26. If the blue light is subtracted from dispersed white light, the remaining colors will combine to form ..(26).. light. 26

27. If the green light is subtracted from dispersed white light, the remaining colors will combine to form ..(27).. light. 27

28. The primary pigments are the ..(28).. of the primary colors. 28

29. If white light is incident on a pigment, the light subtracted is the ..(29).. of the light reflected. 29

30. A pigment which subtracts blue light from white light appears ..(30a).. due to the reflection of both ..(30b)... ....30a

..........30b

31. A pigment which subtracts red light from white light appears ..(31a).. due to the reflection of both ..(31b)... ....31a

..........31b

32. If the two pigments of Questions 30 and 31 are mixed, the only color not absorbed is ..(32)... 32

33. The relationship between light wavelengths and color sensations (is, is not) ..(33).. fully understood by scientists. 33

34. According to the Young-Helmholtz color vision theory, the retina of the eye has three types of nerve receptors, each with a maximum sensitivity to a(n) ..(34).. color. 34

35. Other experiments with color vision indicate the eye may only need information about the relative balance of ..(35).. wavelengths reflected from what is seen. 35

NAME PERIOD DATE INSTRUCTOR'S COMMENTS

15 EXERCISE Interference, Diffraction, and Polarization

SECS. 15.1-15.12

DIRECTIONS: In the space at the right, place the letter or letters indicating EACH choice which forms a correct statement.

1. Constructive interference between two light rays results in (a) a loss of energy, (b) destruction of the wave front, (c) energy reinforcement, (d) bright areas of light. 1

2. Interference phenomena (a) provided Newton with his strongest arguments in favor of the corpuscular theory, (b) are observed in soap films and oil slicks, (c) require a wave model of light, (d) are difficult to observe without special apparatus. 2

3. Thin-film interference results from the (a) refraction of some of the incident light at both the upper and lower surfaces of the film, (b) reflection of some light at the upper surface and the refraction of some light at the lower surface, (c) reflection of some light at both the upper and lower surfaces, (d) transmission of light through the film. 3

4. A thin film observed by reflected monochromatic light appears (a) bright where the thickness is one-quarter wavelength, (b) dark where the thickness is one-quarter wavelength, (c) bright where the thickness is one-half wavelength, (d) dark where the thickness is one-half wavelength. 4

5. The spreading of light into the region behind an obstruction (a) is made evident by interference effects, (b) is most pronounced when the obstruction is large with respect to the wavelength of the light, (c) is known as diffraction, (d) was first explained by Fresnel. 5

6. In the case of a transmission grating situated in a light beam, (a) the ruled lines act as transmission slits, (b) new wavelets are generated at the narrow spaces between the ruled lines, (c) no diffraction effects are produced if the ruled lines are very close together, (d) spectra are produced that are generally less intense than those formed by prisms. 6

7. Light radiated from luminous bodies is unpolarized because (a) the radiation occurs in all directions, (b) the primary radiators oscillate independently, (c) the oscillation planes of the wave trains are randomly oriented, (d) the magnitudes of the vertical and horizontal components of the light vectors are equal to zero. 7

8. Light can become plane-polarized (a) through interactions with matter, (b) by reflection from a surface, (c) by refraction through some crystals, (d) by selective absorption of light in some crystals. 8

9. The two plane-polarized beams of light that emerge from an illuminated calcite crystal cannot interfere with each other because (a) they are transmitted through the crystal at different speeds, (b) they have a common source, (c) their planes of polarization are perpendicular to each other, (d) their separate intensities are below the interference threshold. 9

10. The scattering phenomenon is an example of (a) diffraction, (b) interference, (c) polarization, (d) none of the above. 10

DIRECTIONS: In the blank space at the right of each statement, write the word or expression which BEST completes the meaning.

11. The mutual effect of two beams of light which results in the loss of energy in certain areas and reinforcement of energy in others is known as ..(11)... 11

12. The phenomenon of No. 11 was first demonstrated with light by ..(12a).. in the year ..(12b)... ....12a

    ....12b

13. Light waves reflected from the upper surface of a thin soap film (do, do not) ..(13).. undergo a phase inversion. 13

14. Light waves which pass into a thin soap film and are reflected from its lower surface (do, do not) ..(14).. undergo a phase inversion. 14

15. A thin film appears bright by reflected monochromatic light where its thickness is a(n) ..(15).. number of quarter wavelengths. 15

16. A thin film appears dark by reflected monochromatic light where its thickness is a(n) ..(16).. number of quarter wavelengths. 16

17. A thin wedge of air between two glass plates yields a regular interference pattern. The glass plates are said to be ..(17)... 17

18. Light that spreads into a region behind an obstruction is said to be ..(18)... 18

19. The distance between adjacent ruled lines on a diffraction grating is called the ..(19)... 19

20. The light waves passing through a diffraction grating which constructively interfere and produce a first order image are out of phase by ..(20).. wavelength(s). 20

21. Interference and diffraction phenomena provide the best evidence of the ..(21).. characteristics of light. 21

22. When light is polarized, the oscillations are confined to a(n) ..(22).. perpendicular to the line of propagation. 22

23. Since light can be polarized, the wave-like character is considered to be ..(23)... 23

24. Sound is a(n) ..(24).. disturbance and thus cannot be polarized. 24

25. Tourmaline, which transmits light in one plane of polarization and absorbs light in other polarization planes, is said to have the property of ..(25)... 25

26. *Polaroid* film polarizes light passing through it by the method of ..(26)... 26

27. The polarization of reflected light is complete when the incident light strikes the reflecting surface at the particular angle of incidence known as the ..(27).. angle. 27

28. A calcite crystal placed on a line of print differs from a glass plate in that ..(28).. refracted image(s) of the print can be seen. 28

29. A doubly refracting crystal transmits ..(29a).. beam(s) of polarized light, ..(29b).. of which conform(s) to Snell's law. 29a

    ....29b

30. When a beam of light passes through a calcite crystal, it is separated into two beams incapable of ..(30).. with each other. 30

31. If two light waves cannot be made to interfere with each other, their oscillations must lie in ..(31)... 31

32. Materials that become doubly refracting when under stress are said to be ..(32)... 32

33. When scattering occurs, the light sent off to the side consists mainly of the ..(33).. wavelengths. 33

34. As a consequence of the phenomenon of No. 33, the setting sun has a(n) ..(34).. appearance. 34

35. Cane sugar may be described as a(n) ..(35).. substance because it rotates the plane of polarized light transmitted through it. 35

NAME PERIOD DATE INSTRUCTOR'S COMMENTS

16 EXERCISE Electrostatics

SECS. 16.1 - 16.17

DIRECTIONS: In the parentheses at the right of each word or expression in the second column, write the letter of the expression in the first column which is MOST CLOSELY related.

a. A material through which an electric charge is not readily transferred
b. Change in potential per unit of distance
c. Used to observe the presence of an electrostatic charge
d. Acquires a charge opposite in sign to that of the body producing it
e. Force per unit positive charge at a point in an electric field
f. Acquires a charge of the same sign as that of the body producing it
g. Process of producing electric charges on an object
h. Exists in a region of space if an electric charge placed in that region is subject to an electric force
i. Quantity of charge per unit area
j. Commonly produced by friction between two surfaces in close contact
k. Work done per unit charge
l. Has a number of easily moved free electrons

electrification	( )	1
static electricity	( )	2
conductor	( )	3
insulator	( )	4
charged by conduction	( )	5
charged by induction	( )	6
electric field	( )	7
electric field intensity	( )	8
potential difference	( )	9
potential gradient	( )	10

DIRECTIONS: Complete the diagrams below to show the sequence of events when a negatively charged rod is brought toward an uncharged pith ball. Add the pith ball and silk thread in appropriate positions to (B), (C), and (D), and use + and - signs to show the location and nature of the charges on the pith ball in all four diagrams.

11.

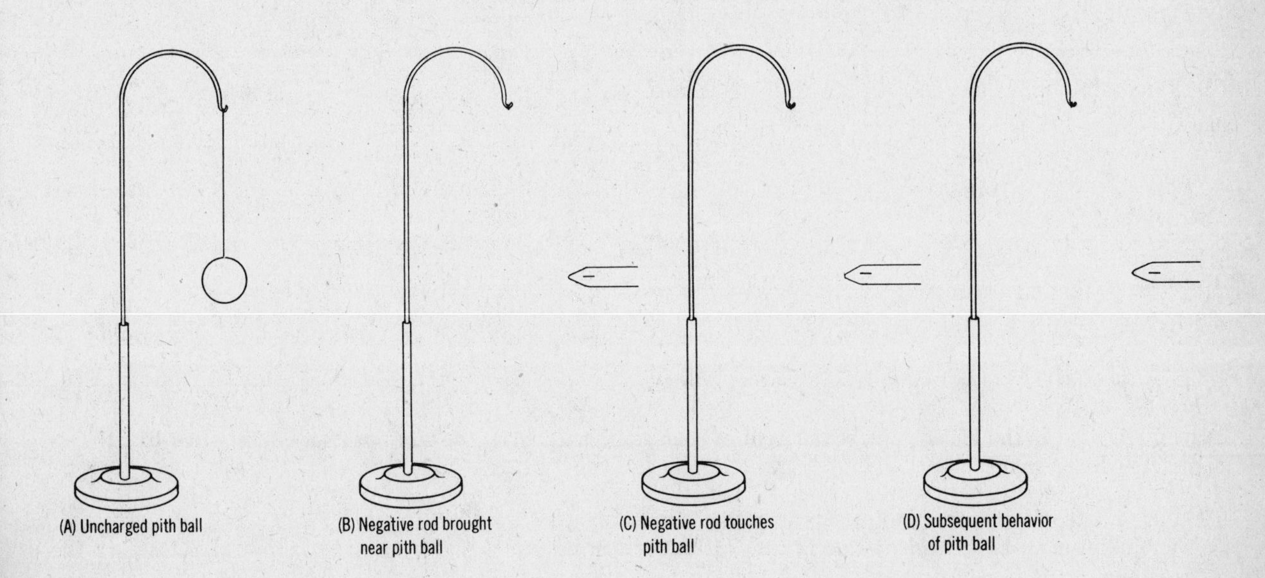

DIRECTIONS: Complete the diagrams below to show the steps in charging an electroscope negatively by induction. Add the two leaves to each electroscope and use + and – signs to show the location and nature of the charges.

12.

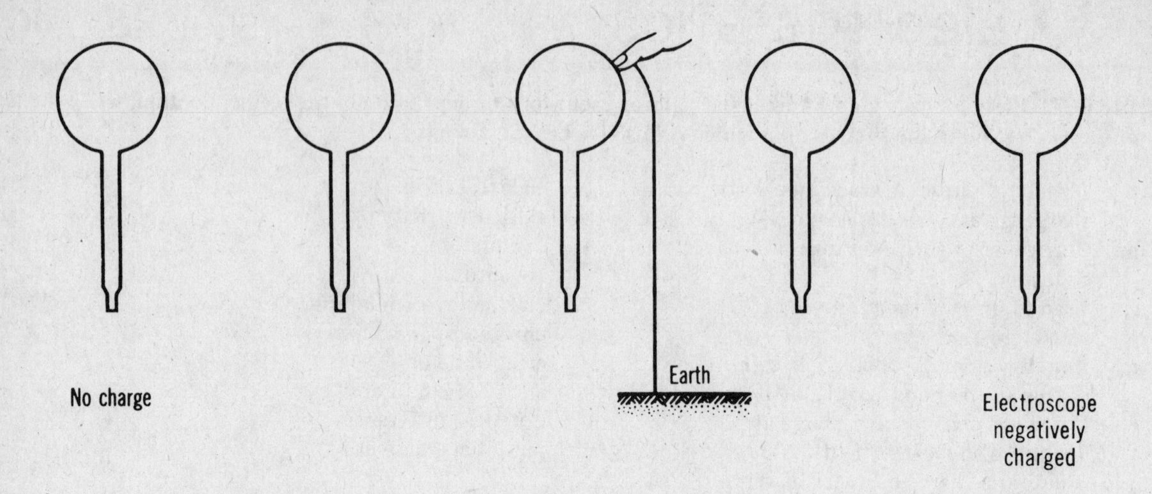

DIRECTIONS: Complete the following review table for metric prefixes.

13.

Letter prefix	k	m	M	p	μ
Name prefix					
Power of ten					

DIRECTIONS: Write statements or definitions of each of the following in the spaces provided. Make complete statements.

14. Basic law of electrostatics. ....

....14

15. Coulomb's law of electrostatics. ....

....

....15

16. Coulomb. 16

17. Volt. ....

....

....17

18. Farad. ....

....

....18

DIRECTIONS: In the blank space at the right of each statement, write the word or expression which BEST completes the meaning.

19. An electric line of force is a line so drawn that a(n) ..(19a).. to it at any point indicates the orientation of the ..(19b).. at that point. 19a

....19b

20. The electric field intensity is proportional to the number of ..(20a).. per ..(20b).. normal to the field.

....20a

....20b

21. The resultant force acting on a test charge placed at the midpoint between two equal charges is ..(21)...

....21

22. If a charge located in an electric field moves in response to the electric force, work is done by the ..(22a).. and ..(22b).. is removed from the system.

....22a

....22b

23. If a charge located in an electric field is moved against the electric force, work is done on it and ..(23).. is stored in the system.

....23

24. If work is done as a charge moves from one point to another in an electric field, these two points differ in ..(24)...

....24

25. If work is required to move a charge from one point to another in an electric field, these two points differ in ..(25)...

....25

26. The magnitude of this work is a measure of the ..(26)...

....26

27. The earth may be considered to be an inexhaustible ..(27a).. of electrons, or a limitless ..(27b).. for electrons; for practical purposes the potential of the earth is arbitrarily taken as ..(27c)...

....27a

....27b

....27c

28. The potential at any point in an electric field is the ..(28a).. between the point and the ..(28b).. taken as the zero reference potential.

....28a

....28b

29. All the electrostatic charge on an isolated conductor resides ..(29)...

....29

30. There can be no ..(30).. between two points on the surface of a charged isolated conductor.

....30

31. If points which have the same potential in an electric field near a charged object are joined, a(n) ..(31).. line or surface within the field becomes apparent.

....31

32. An electric field has (a, no) ..(32).. force component along an equipotential surface.

....32

33. A sharply pointed charged conductor may be discharged rapidly due to ..(33).. of the air.

....33

34. The slow leakage of charge from sharp projections of charged bodies is known as a brush or ..(34).. discharge.

....34

35. A combination of conducting plates separated by an insulator and used to store an electric charge is called a(n) ..(35)...

....35

36. The ..(36).. of a capacitor is the ratio of the charge on either plate to the potential difference between the plates.

....36

37. In practice, the dielectric constant of dry air at one atmosphere pressure is taken as ..(37),..

....37

38. If a substance having a dielectric constant of 2 is substituted for air separating the plates of a certain capacitor, the capacitance will be ..(38)...

....38

39. The capacitances of two capacitors connected in parallel are ..(39).. to give the total capacitance of the combination.

....39

40. For two capacitors connected in series, the total capacitance equals the ..(40a).. of the two capacitances divided by their ..(40b)...

....40a

....40b

NAME PERIOD DATE INSTRUCTOR'S COMMENTS

17A EXERCISE Direct-Current Circuits SECS. 17.1 - 17.5

DIRECTIONS: In the parentheses at the right of each word or expression in the second column, write the letter of the expression in the first column which is MOST CLOSELY related.

a. Flow of electrons continuously in one direction through a conductor
b. Conversion of mechanical energy into electric energy
c. One path for current
d. Measuring difference of potential
e. Two or more cells connected together
f. Measuring resistance
g. Average rate of motion of free electrons in the direction of the accelerating force
h. Two or more paths for current
i. Substance whose solution conducts electricity
j. Flow of electrons through a conductor
k. Measuring rate of flow of electricity
l. One cell

ammeter ( ) 1
voltmeter ( ) 2
electric current ( ) 3
direct current ( ) 4
drift velocity ( ) 5
electromagnetic induction ( ) 6
battery ( ) 7
electrolyte ( ) 8
series circuit ( ) 9
parallel circuit ( ) 10

DIRECTIONS: In the space at the right, indicate the meaning of each electric symbol shown below.

11. —(A)— 11

12. —/\/\/\/\— 12

13. —(V)— 13

14. —| |— 14

15. —(G)— 15

16. − + —| |— 16

17. 17

18. 18

19. − + ...19

20. − + ...20

DIRECTIONS: In the blank space at the right of each statement, write the word or expression which BEST completes the meaning.

21. During the time a capacitor is discharging, a(n) ..(21).. exists in the conducting circuit. 21

22. An electric current in a conductor is the ..(22).. of charge through a cross section of the conductor. 22

23. The MKS unit of current is the ..(23a).. which is defined as a current of ..(23b).. per second.

....23a

....23b

24. The direction in which electrons flow through an electric field in a metallic conductor is from ..(24a).. to ..(24b)...

....24a

....24b

25. Resistance is defined as the opposition to the ..(25)...

....25

26. The MKS unit of resistance is the ..(26)...

....26

27. In a closed-loop conducting path which includes a suitable source of electric current, ..(27a).. from the source is utilized in some device called the ..(27b)...

....27a

....27b

28. In the photoelectric cell, ..(28).. energy is transformed into electric energy.

....28

29. In the thermocouple, ..(29).. energy is transformed directly into electric energy.

....29

30. In the piezoelectric cell, ..(30).. energy is transformed into electric energy.

....30

31. Spontaneous ..(31a).. reactions are a source of continuous current in which ..(31b).. energy is transformed into ..(31c).. energy.

....31a

....31b

....31c

32. An electrochemical cell, such as a flashlight cell, which is replaced when its reactants are used up is a(n) ..(32).. cell.

....32

33. Electrochemical cells, such as those in an automobile battery, which can be repeatedly recharged, are known as ..(33).. cells.

....33

34. The cell in which the chemical energy of the reaction between hydrogen and oxygen to form water is converted directly into electric energy is a(n) ..(34).. cell.

....34

35. The electron-rich electrode of a dry cell is called the ..(35a).., while the electron-poor electrode is called the ..(35b)...

....35a

....35b

36. The open-circuit potential difference across a dry cell is known as the ..(36a).. of the cell and is defined as the ..(36b).. per unit charge supplied by the cell.

....36a

....36b

37. The *emf* of a battery composed of cells connected in ..(37).. is equal to the sum of the emfs of the individual cells.

....37

38. If three identical cells are connected in parallel, the current supplied by each cell is ..(38).. the total current in the circuit.

....38

39. A certain resistance load is designed to operate across a 4.5-volt d-c circuit and when in continuous operation draws 1.00 ampere of current. Draw the circuit diagram in the block below using an appropriate arrangement of No. 6 dry cells to supply 1.00 ampere of continuous current at 4.5 volts.

NAME	PERIOD	DATE	INSTRUCTOR'S COMMENTS

17B EXERCISE Series and Parallel Circuits

SECS. 17.6 - 17.13

DIRECTIONS: (A) Place the answers to the following problems in the spaces provided at the right. (B) Use a well-sharpened pencil and a straight edge in all diagrams.

1. What is the total resistance of a network of three resistors connected in *series* if each has a resistance of 6 ohms? 1

2. What is the equivalent resistance of a network of three resistors connected in *parallel* if each has a resistance of 6 ohms? 2

3. Given a *series* circuit with resistors R_1, R_2, and R_3 of 3 ohms, 4 ohms, and 5 ohms respectively, a. what is the combined resistance of R_1 and R_2? b. What is the total resistance of the circuit? ....3a 3b

4. If R_1, of $\overline{60}$ ohms, is connected in *parallel* with R_2, of $\overline{30}$ ohms, what equivalent resistance do they present to the circuit? 4

5. A *series* circuit contains two resistances, R_1 and R_2. Two cells connected in *series* serve as the source of emf. An ammeter is connected in the circuit to read the current through the external circuit, and a voltmeter is connected across the external circuit. In block A, draw the circuit diagram.

A

6. An electric circuit consists of two cells connected in *parallel* as a source of emf, resistors R_1 and R_2 in *parallel*, an ammeter to read total current, and a voltmeter to read the potential difference across the external circuit. In block B, draw the circuit diagram.

B

7. A circuit consists of a battery of four cells in *series*, a resistor R_1 connected in *series* with a *parallel* combination of R_2 and R_3. In block C, draw the circuit diagram.

C

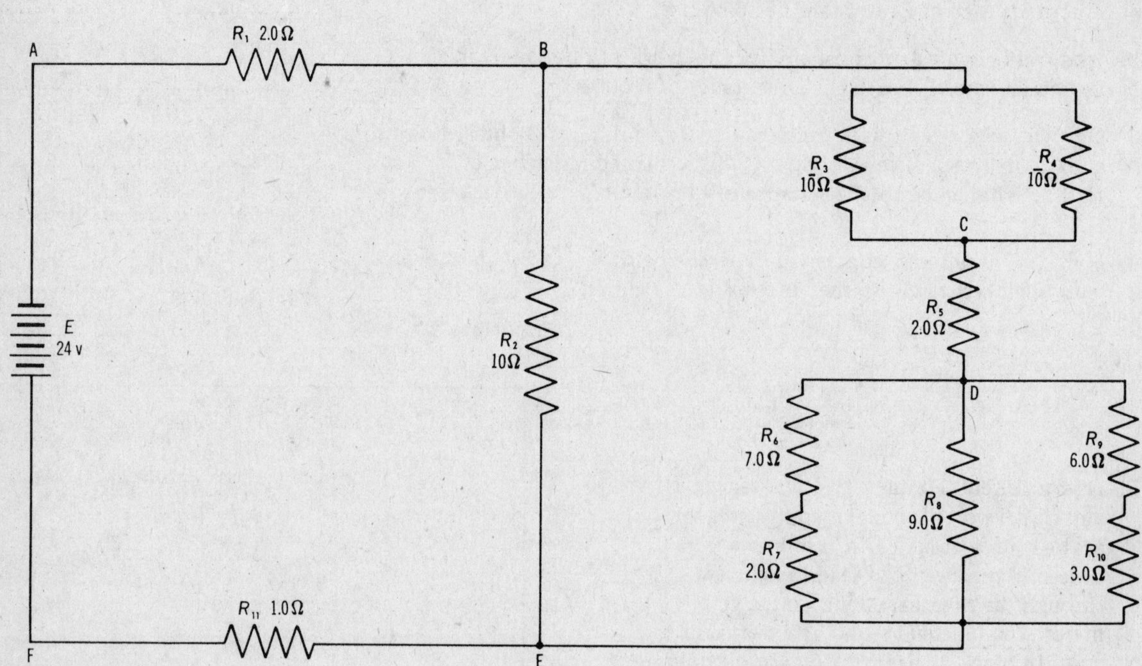

DIRECTIONS: Place the answers to the following problems in the spaces provided at the right. All refer to the circuit shown in the figure. Do all computations mentally if possible.

8. What is the equivalent resistance of R_3 and R_4? 8

9. What is the combined resistance of R_6 and R_7? 9

10. What is the combined resistance of R_9 and R_{10}? 10

11. What is the equivalent resistance from D to E? 11

12. What is the equivalent resistance from B to E? 12

13. What is the total resistance of the external circuit (from A to F)? 13

14. What is the total current in the external circuit? 14

15. What potential drop occurs across R_1? 15

16. What is the drop in potential from B to E? 16

17. How much current is in R_2? 17

18. How much current is in R_5? 18

19. What is the potential drop from B to C? 19

20. What is the current in R_3? 20

21. What is the voltage drop across R_5? 21

22. What is the current in each branch of the circuit from D to E? 22

23. What is the potential drop across R_6? 23

24. What potential drop occurs across R_7? 24

25. What is the potential drop across R_8? 25

26. What is the sum of the potential drops across R_9 and R_{10}? 26

27. Find the sum of the potential drops from A to B, B to E, and E to F. 27

28. What resistance is in parallel with R_2? 28

29. What is the sum of all currents between B and E? 29

30. Does Ohm's law apply to each part of the circuit as well as to the circuit as a whole? 30

NAME PERIOD DATE INSTRUCTOR'S COMMENTS

18A EXERCISE Heating Effects

SECS. 18.1 - 18.6

DIRECTIONS: Write the answers to the following in the spaces provided. Where appropriate, make complete statements.

1. What is the source of the electric energy expended in the external circuit as the electric charge moves through the circuit in response to a potential difference placed across it? ....

....1

2. In what form is the electric energy expended in an external circuit which consists of an ordinary resistor?

....2

3. State Joule's law. ....

....3

4. In the expression for Joule's law, $Q = \frac{I^2Rt}{J}$, show how Q may be expressed in kilocalories.

5. What is the relationship of the watt, joule, and second? ....

....5

6. What is the relationship between electric power expended by a current in a resistance and the current in the resistance? ....

....6

7. Is electric power dissipated within the source of emf in an electric circuit? Explain. ....

....7

8. What is the relationship between the total power consumed in an electric circuit, the power dissipated within the source, and the power expended in the load? ....

....

....

....8

9. Under what conditions is maximum power transferred to the load connected across a source of emf?

....

....

....9

10. In what unit is commercially sold electric energy measured? ....

....10

DIRECTIONS: In the blank space at the right of each statement, write the word or expression which BEST completes the meaning.

11. When one coulomb of electric charge is moved through a potential difference of one volt, one ..(11).. of work is done. 11

12. The equation for No. 11 is ..(12)... 12

13. One coulomb of electric charge transferred in one second constitutes a current of one ..(13)... 13

14. The equation for No. 13 is ..(14)... 14

15. Combining the equations of Nos. 12 and 14, the equation for the energy expended in the resistance of an external circuit in terms of the potential difference across the resistance, the current through the resistance, and the time the current flows is ..(15)... 15

16. The equation for the potential difference across the resistance of an external circuit in terms of the current in the circuit and the resistance of the circuit is ..(16)... 16

17. Substituting the equation of No. 16 in the equation of No. 15, the equation for the energy expended in the resistance of an external circuit in terms of the current in the resistance, the magnitude of the resistance, and the time the current flows is ..(17)... 17

18. In all cases when work is being done by an electric current, part of the work appears as ..(18).. because of the inherent resistance of the circuit. 18

19. Electric power is the ..(19).. at which electric energy is delivered to the circuit. 19

20. When the load resistance connected across a source of emf is equal to the internal resistance of the source, the source and the load are said to be ..(20)... 20

DIRECTIONS: Place the answers to the following problems in the spaces provided at the right.

21. How much heat is produced by a current of 6.0 amperes in a load resistance of $3\bar{0}$ ohms for 5.0 minutes? 21

22. The heating element of an electric stove draws 12.0 amperes when connected across a $12\bar{0}$-volt circuit. What is the resistance of the element? 22

23. How much heat will be produced by the heating element in Question 22 in 8.00 minutes? 23

24. How many kilograms of water can be heated from 20.0°C to 100.0°C by 165 kcal of heat? 24

25. What is the cost of the electricity used by the heating element in Question 22 at 3.0 cents per kilowatt hour? 25

NAME PERIOD DATE INSTRUCTOR'S COMMENTS

18B EXERCISE Electrolysis

SECS. 18.7 - 18.9

DIRECTIONS: In the parentheses at the right of each word or expression in the second column, write the letter of the expression in the first column which is MOST CLOSELY related.

a. Chemical energy to electric energy
b. Grams per coulomb
c. $Ag^{+} + e^{-} \rightarrow Ag^{0}$
d. Electric energy to chemical energy
e. Charged particle of matter
f. Negative terminal of an electrolytic cell
g. Coulombs
h. Positive terminal of an electrolytic cell
i. $Ag^{0} - e^{-} \rightarrow Ag^{+}$
j. Ratio of gram-atomic weight to ionic charge
k. Laws of electrolysis
l. Conducting solution of a cell

anode	( )	1
cathode	( )	2
chemical equivalent	( )	3
electrochemical cell	( )	4
electrochemical equivalent	( )	5
electrolyte	( )	6
electrolytic cell	( )	7
Faraday	( )	8
faraday	( )	9
ion	( )	10

DIRECTIONS: In the blank space at the right of each statement, write the word or expression which BEST completes the meaning.

11. Electron-transfer reactions that are spontaneous can be sources of ..(11)... 11

12. An arrangement in which a spontaneous electron-transfer reaction occurs is known as a(n) ..(12).. cell. 12

13. The products of a spontaneous electron-transfer reaction have (more, less) ..(13).. energy than the reactants. 13

14. Electron-transfer reactions that are not spontaneous can be forced to occur by supplying ..(14).. from an external source. 14

15. An arrangement in which a forced electron-transfer reaction occurs is known as a(n) ..(15).. cell. 15

16. The products of a forced electron-transfer reaction have (more, less) ..(16).. energy than the reactants. 16

17. When an electrolytic cell is in operation, the anode is given a(n) ..(17a).. charge and the cathode is given a(n) ..(17b).. charge. 17a

....17b

18. When the electrodes are charged, ..(18a).. ions migrate to the cathode, acquire ..(18b).. of high potential energy, and are discharged. 18a

....18b

19. Ions having a(n) ..(19a).. charge migrate to the anode, give up ..(19b).. of low potential energy, and are discharged. 19a

....19b

20. The conducting solution of an electrochemical or electrolytic cell contains a(n) ..(20a).. which furnishes positively and negatively charged ..(20b)... ....20a

....20b

21. The ..(21a).. of electrons by the cathode and the ..(21b).. of an equal number of electrons by the anode is, in effect, the ..(21c).. of electric charge through an electrolytic cell. 21a

.... 21b

.... 21c

22. The ..(22a).. of electric charge through a solution of a(n) ..(22b).., together with the resulting ..(22c).. changes is called ..(22d)... .... 22a

.... 22b

.... 22c

.... 22d

23. The products of an electrolysis depend on the kinds of ..(23a).., on the nature of the ..(23b).., and to some degree on the ..(23c).. of the source. 23a

.... 23b

.... 23c

24. Electrolytic cells are used in ..(24a).. compounds and in refining and plating ..(24b)... .... 24a

.... 24b

25. In an electroplating cell, the object to be plated is connected in the circuit as the ..(25)... 25

26. In an electroplating cell, the ..(26).. is made of the plating metal. 26

27. In an electroplating cell, the conducting solution contains a(n) ..(27).. of the metal to be plated. 27

28. The action at the cathode results in electrons being (acquired by, removed from) ..(28).. that electrode. 28

29. The action at the anode results in electrons being (acquired by, removed from) ..(29).. that electrode. 29

30. The metallic ions are removed from the plating solution at the ..(30a).., plating out as ..(30b).. of the metal. 30a

.... 30b

31. In effect, atoms of the metal are transferred from ..(31a).. to ..(31b).. during the electroplating process. 31a

.... 31b

32. In effect, electrons from the external circuit are transferred from ..(32a).. to ..(32b).. during the electroplating process. 32a

.... 32b

33. Very small amounts of impurities in copper cause a marked increase in its electric ..(33)... 33

34. In purifying copper by electrolysis, the equation for the anode reaction is ..(34a).., while the equation for the cathode reaction is ..(34b)... .... 34a

.... 34b

35. The important relationship between the quantity of electricity passing through an electrolytic cell and the quantity of a substance liberated by the chemical action was discovered by ..(35)... 35

36. The mass of an element deposited during electrolysis is proportional to the quantity of ..(36).. that passes through the cell. 36

37. The quantity of electric charge which will liberate 1 gram of hydrogen will also deposit ..(37a).. of silver and ..(37b).. of aluminum. 37a

....37b

38. From the atomic weights of these elements, it is apparent that each ion of silver gains ..(38a).. and each aluminum ion gains ..(38b).. when discharged by electrolytic action. 38a

....38b

39. The quantity of electricity required to deposit the chemical equivalent of an element in grams is approximately ..(39).. coulombs. 39

40. This quantity of electricity is called a(n) ..(40)... 40

NAME PERIOD DATE INSTRUCTOR'S COMMENTS

19A EXERCISE Magnetism

SECS. 19.1 - 19.10

DIRECTIONS: In the blank space at the right of each definition, write the term which is defined.

1. The property of materials that are strongly attracted by magnets. 1
2. The property of materials that are slightly attracted by very strong magnets. 2
3. The property of materials that are feebly repelled by very strong magnets. 3
4. An atom possessing the characteristics of a permanent magnet. 4
5. A group of atoms of a ferromagnetic material which forms a microscopic magnetic region. 5
6. The group of ferromagnetic substances on which a new magnet technology is based. 6
7. A theoretical N pole which repels an exactly similar pole placed 1 centimeter away with a force of 1 dyne. 7
8. A region in which a magnetic force can be detected. 8
9. A line in a magnetic field drawn so that a tangent to it at any point indicates the direction of the magnetic field. 9
10. The number of flux lines per unit area permeating a magnetic field. 10

DIRECTIONS: In the blank space at the right of each statement, write the word or expression which BEST completes the meaning.

11. Natural magnets consist of an iron ore called ..(11)... 11
12. Three common metallic elements that have ferromagnetic properties are ..(12)... 12
13. Magnetism is a property of an electric charge which is ..(13)... 13
14. An electron has two kinds of motion: it ..(14a).. about the nucleus of an atom and it ..(14b).. on its own axis. 14a

    ....14b
15. If only the planetary motions of electrons were involved in the magnetic character of substances, all would be ..(15)... 15
16. Electrons spinning in opposite directions may form ..(16a).. and neutralize their ..(16b).. character. 16a

    ....16b
17. The strong ferromagnetic properties of iron are explained by the four ..(17a).. electrons having ..(17b).. oriented spins in its atom. 17a

    ....17b
18. When a ferromagnetic material is subject to an external magnetic field, favorably oriented domains may ..(18a).. and other domains may become ..(18b)... ....18a

    ....18b
19. The temperature above which the domain regions of a ferromagnetic material disappear is called the ..(19)... 19

20. *Like* magnetic poles ..(20a)..; *unlike* poles ..(20b)... ....20a

....20b

21. The force between two magnetic poles is ..(21a).. proportional to the product of the strengths of the poles and ..(21b).. proportional to the square of their distance apart. 21a

....21b

22. The lines of flux perpendicular to a specified area in a magnetic field are referred to collectively as the ..(22)... 22

23. The MKS unit of magnetic flux is the ..(23)... 23

24. Flux density in the MKS system is expressed in ..(24)... 24

25. The force exerted by a magnetic field on a unit N pole situated in the field indicates the ..(25).. at that point. 25

DIRECTIONS: Write the answers to the following in the spaces provided. Where appropriate, make complete statements.

26. Arrange the following materials in order of increasing permeability: iron, air, aluminum, zinc. ....

....26

27. What physical change occurs when a bar of soft iron is placed in a magnetic field? ....

....27

28. Describe the polarity of a nail, the head of which is attracted to and in contact with the S pole of a strong bar magnet. ....

....28

29. What condition prevails when an increasing magnetizing force fails to produce greater magnetization of a ferromagnetic material? ....

....29

30. What is *hysteresis?* ....

....30

31. Hardened steel and soft iron are each taken through a complete cycle of magnetization. If the flux density at saturation is the same for both materials, which requires the greater energy dissipation? Why?

....

....

....31

32. What is the angle of declination at a point on an agonic line? ....

....32

33. What is the magnetic inclination at the North Magnetic Pole? ....

....33

34. What is the *magnetosphere?* ....

....

....34

35. What regions of the magnetosphere contain energetic protons and electrons trapped by the earth's magnetic field? 35

NAME PERIOD DATE INSTRUCTOR'S COMMENTS

19B EXERCISE Electromagnetism - d-c Meters

SECS. 19.11 - 19.18

DIRECTIONS: In the space at the right, place the letter or letters indicating EACH choice which forms a complete statement.

1. **An electron flow from north to south in a straight conductor supported *above* a compass (a) causes a deflection of the N pole of the compass toward the east, (b) produces no change in the compass-needle equilibrium, (c) causes a deflection of the S pole of the compass toward the east, (d) causes a deflection of the N pole of the compass toward the west.** 1

2. **An electron flow from south to north in a straight conductor supported *below* a compass (a) causes a deflection of the N pole of the compass toward the east, (b) produces no change in the compass-needle equilibrium, (c) causes a deflection of the S pole of the compass toward the east, (d) causes a deflection of the N pole of the compass toward the west.** 2

3. **The strength of a magnetic field around a conductor varies (a) directly with the distance from the conductor, (b) directly with the magnitude of the current, (c) inversely with the distance from the conductor, (d) inversely with the square of the distance from the conductor.** 3

4. **Ampère's rule for a straight conductor (a) requires that the direction of the current be known, (b) stipulates that the right thumb is to be extended in the direction of the electron flow, (c) enables one to predict the direction of the magnetic flux encircling a conductor, (d) can be used to show whether the forces acting on two parallel conductors are attractive or repulsive.** 4

5. **When a current is in a solenoid, (a) the two faces of the solenoid show magnetic polarity, (b) the core of each turn becomes a magnet, (c) the core of the solenoid becomes a magnetic tube through which nearly all of the flux passes, (d) the solenoid acts as a bar magnet.** 5

6. **Ampère's rule for a solenoid (a) requires that the direction of the current be known, (b) stipulates that the fingers of the right hand encircle the coil in the direction of the electron flow, (c) is not valid for a single turn, (d) indicates the end of the core which acts as the N pole.** 6

7. **When a soft iron rod is used as the core of a solenoid, (a) the flux density is not changed appreciably, (b) the solenoid becomes a strong electromagnet, (c) the permeability of the core is increased over that for air, (d) either the current must be reduced or some turns must be removed.** 7

8. **An electromagnet for lifting large quantities of scrap iron should have a core made of (a) air, (b) soft iron, (c) hardened steel, (d) wood.** 8

DIRECTIONS: In the blank space at the right of each statement, write the word or expression which BEST completes the meaning.

9. **The sensitivity of a galvanoscope can be increased by ..(9).. the number of turns of the conductor.** 9

10. A galvanometer can consist of a(n) ..(10a).. pivoted on jeweled bearings between the poles of a(n) ..(10b).. horseshoe magnet. 10a

..........10b

11. When a current is in the galvanometer coil, the coil becomes a(n) ..(11).. free to turn on its own axis. 11

12. When a current is in the coil of a galvanometer situated between the poles of a permanent magnet, a(n) ..(12a).. acts on the coil which rotates in an attempt to align its plane ..(12b).. to the direction of the field of the permanent magnet. 12a

..........12b

13. The final position of the galvanometer coil is reached when the torque acting on it is ..(13).. by the reaction of the control springs. 13

14. When the galvanometer coil reaches its equilibrium position, the opposing torques are ..(14a).. and the deflection angle of the coil is ..(14b).. to the current in it. 14a

..........14b

15. In order to convert a galvanometer movement for service as a d-c voltmeter, a high resistance must be added in ..(15).. with the moving coil. 15

16. To convert a galvanometer movement for service as a d-c ammeter, a very low resistance must be added in ..(16).. with the moving coil. 16

17. Because the total resistance of the ..(17a).. is very low, it is connected in ..(17b).. with the load; and because the total resistance of the ..(17c).. is very high, it is connected in ..(17d).. with the load. 17a

..........17b

..........17c

..........17d

18. The use of an ohmmeter in measuring resistance is actually a modified version of the ..(18).. method. 18

DIRECTIONS: Write the answers to the following in the spaces provided. Where appropriate, make complete statements.

19. For a voltmeter to have negligible effect on the current in a load across which it is connected, how should its resistance compare with that of the load? ....

..........19

20. What would be the effect on the total current of incorrectly connecting such a voltmeter in series with the load? ....

..........20

21. For an ammeter to have a negligible effect on the magnitude of current in a load when connected in series with it, how should its resistance compare with that of the load? ....

..........21

22. What would be the effect on the total current of incorrectly connecting such an ammeter in parallel with the load? ....

..........22

23. What would be the effect on the ammeter if such a mistake were made? ....

....

..........23

24. If a voltmeter designed to read from 0 to 150 volts has a sensitivity of 1000 ohms per volt, would it be a suitable meter to use across a load resistance of 8000 ohms? Explain. ....

....

....24

25. What would be the result of using the voltmeter of No. 24 to read the voltage across a load resistance of 100,000 ohms? ....

....25

NAME PERIOD DATE INSTRUCTOR'S COMMENTS

20A EXERCISE Induced Currents - Generators

SECS. 20.1-20.13

DIRECTIONS: In the space at the right, place the letter or letters indicating EACH choice which forms a correct statement.

1. Electromagnetic induction was discovered by (a) Faraday, (b) Henry, (c) Lenz, (d) Oersted. 1

2. A current is induced in a conducting loop (a) when the loop is poised in a strong magnetic field, (b) when the loop is moved in a magnetic field perpendicular to the flux, (c) when the loop is moved in a magnetic field parallel to the flux, (d) as a result of relative motion producing a change of flux linking the conductor. 2

3. The magnitude of current induced in a conducting coil in a magnetic field is increased by (a) increasing the number of turns comprising the coil, (b) increasing the relative motion between the coil and flux, (c) increasing the strength of the magnetic field, (d) decreasing the rate at which the flux linked by the conductor changes. 3

4. An emf is induced across a loop cutting through flux lines (a) only if the loop is a part of a closed conducting path, (b) even if the ends of the loop are open, (c) and is proportional to the relative motion between the loop and flux, (d) only if the ends of the loop are open. 4

5. A coil of 10^2 turns moving perpendicular to the flux of a magnetic field for 10^{-2} sec experiences a change in flux linkage of 5×10^{-5} weber. The emf induced across it is (a) 10^{-5} v, (b) -0.5 v, (c) 5×10^{-1} v, (d) 5×10^3 v. 5

6. A straight wire held in a north-south direction in a magnetic field in which the flux lines extend from west to east is pushed downward through the field. The north end of the wire (a) acquires a negative charge, (b) acquires a positive charge, (c) remains uncharged, (d) is more difficult to move than the south end. 6

7. A wire 5 cm long moving through a magnetic field of 10^{-2} weber/m^2 perpendicularly to the flux, with a velocity of 1 m/sec, has an emf induced across it of (a) 0.05 v, (b) 0.005 v, (c) 5×10^{-4} v, (d) 0.5 mv.. 7

8. Lenz's law (a) illustrates the conservation of energy principle, (b) is true of all induced emfs, (c) states that an induced current is in such direction as to produce a magnetic force that opposes the force causing the motion by which the current is induced, (d) indicates that work must be done to induce a current in a conducting circuit. 8

9. For electrons to acquire potential energy in a magnetic field, (a) they must move parallel to the flux, (b) they must move perpendicular to the flux, (c) the conductor must be moved in opposition to a magnetic force, (d) the region of the field into which the conductor moves must be weaker than the region behind it. 9

10. Work done by an induced current in an external load expends energy acquired by the electrons (a) in the internal circuit, (b) resulting from work done in moving the wire in the magnetic field, (c) in the external circuit, (d) as they fall through the potential difference across the load. 10

DIRECTIONS: In the blank space at the right of each statement, write the word or expression which BEST completes the meaning.

11. An electric generator converts ..(11a).. energy into ..(11b).. energy. 11a

....11b

12. The conducting loops across which an emf is induced form the ..(12).. of a generator. 12

13. As the armature of a simple a-c generator rotates through a complete cycle, there are ..(13).. reversals in direction of the induced current. 13

14. A current which has one direction during part of a generating cycle and the opposite direction during the remainder of the cycle is a(n) ..(14).. current. 14

15. The prefix used to designate direct-current properties is ..(15a).., while the prefix used to designate alternating-current properties is ..(15b)... ....15a

....15b

16. The magnitude of a varying voltage at any instant of time is called the ..(16).. voltage. 16

17. An armature loop rotating at a constant rate in a magnetic field of uniform flux density has a voltage induced across it the magnitude of which varies ..(17).. with respect to time. 17

18. In general, the instantaneous voltage across the loop is expressed as ..(18)... 18

19. Instantaneous current is expressed as ..(19)... 19

20. The output of a practical generator is increased by increasing the number of ..(20a).. on the armature, or increasing the ..(20b)... ....20a

....20b

21. The number of cycles of current or voltage per second is the ..(21)... 21

22. A commutator is a(n) ..(22a).. ring, each segment of which is connected to a(n) ..(22b).. of a corresponding armature ..(22c)... ....22a

....22b

....22c

23. If the armature turns of an a-c generator are connected to a commutator, the generator can supply ..(23).. current. 23

24. A d-c generator in which a portion of the induced power energizes the field magnets is said to be ..(24)... 24

25. The resistance of the armature turns of a generator is analogous to the ..(25).. of a battery. 25

DIRECTIONS: Write the answers to the following in the spaces provided. Where appropriate, make complete statements.

26. What are the essential components of an electric generator? ....

....

....26

27. What is the purpose of slip rings and brushes in an electric generator? ....

....

....27

28. For what is the generator rule commonly used, and what two considerations are taken into account in its application? ....

....

....

....28

29. To what does the *displacement angle* of a generator loop refer? ....

....

....29

30. Why is exciter voltage, rather than armature voltage, transferred by the slip rings and brushes in a large commercial generator? ....

....

....30

31. How does the nature of the current output of a d-c generator compare with that of an electrochemical source? ....

....

....31

32. How does an increase in the load affect the induced emf of series-wound and shunt-wound generators?

....

....

....32

NAME PERIOD DATE INSTRUCTOR'S COMMENTS

20B EXERCISE Motors - Inductance SECS. 20.14-20.22

DIRECTIONS: In the parentheses at the right of each word or expression in the second column, write the letter of the expression in the first column which is MOST CLOSELY related.

a. Squirrel-cage rotor
b. Connected to the load
c. A coil
d. Emf induced by generator action of a motor
e. Converts mechanical energy to electric energy
f. Connected to a current source
g. $L_1 + L_2 + L_3$
h. Converts electric energy to mechanical energy
i. Emf induced by motor action of a generator
j. Operates on either a-c or d-c power
k. $1/L_1 + 1/L_2 + 1/L_3$
l. Electric clock

electric motor ( ) 1
back emf ( ) 2
universal motor ( ) 3
induction motor ( ) 4
synchronous motor ( ) 5
primary coil ( ) 6
secondary coil ( ) 7
inductor ( ) 8
L_T ( ) 9
$1/L_T$ ( ) 10

DIRECTIONS: Complete the following statements forming accurate and complete sentences.

11. The *mutual inductance* of two circuits is 11

12. The mutual inductance of two circuits is *one henry* if 12

13. The self-inductance of a coil is the ratio of 13

14. The self-inductance of a coil is *one henry* if 14

15. The *turns ratio* of a transformer is the ratio of the number of turns in the 15

DIRECTIONS: In the blank space at the right of each statement, write the word or expression which BEST completes the meaning.

16. A current-carrying conductor poised in a magnetic field experiences a(n) ..(16).. force. 16

17. When the plane of a conducting loop is parallel to the magnetic flux, the magnitude of the torque on the loop is ..(17a)..; when the plane of the loop is perpendicular to the magnetic flux, the magnitude of the torque is ..(17b)... ....17a 17b

18. The motor rule: Extend the thumb, forefinger, and middle finger of the ..(18a).. hand at right angles to each other. Let the forefinger point in the direction of the ..(18b).. and the middle finger in the direction of the ..(18c)..; then the thumb points in the direction of the ..(18d)... ....18a ....18b ....18c 18d

19. As a motor gains speed, the back emf ..(19a).. and the circuit current ..(19b)...

....19a

....19b

20. In order to decrease the pulsation of the torque in practical d-c motors, the armature and field are composed of ..(20).. coils. 20

21. In an induction motor, the rotor must ..(21).. behind the field in order for a torque to be developed. 21

22. The synchronous motor is a constant ..(22).. motor. 22

23. The magnitude of the induced emf in mutual inductance is increased by ..(23a).. the relative motion between conductors and flux, by increasing the number of turns in the ..(23b).., and by placing a soft-iron core in the ..(23c)...

....23a

....23b

....23c

24. The magnitude of the induced emf in self-inductance is increased by ..(24).. the rate of change of current in the inductor. 24

25. The inductance of a coil depends on the number of ..(25a).., the ..(25b).. of the coil, the ..(25c).. of the coil, and the nature of the ..(25d)...

....25a

....25b

....25c

....25d

26. Because inductance in electricity is analogous to ..(26a).. in mechanics, inductance imparts a(n) ..(26b).. effect in a circuit having a varying current.

....26a

....26b

27. If the primary of a transformer has a larger number of turns than the secondary, the transformer is called a step- ..(27).. transformer. 27

28. The resistance of the wires in the primary and secondary coils of a transformer produces the energy loss called ..(28)... 28

29. The energy loss in reversing the magnetic polarity of the transformer core is the ..(29).. loss. 29

30. Closed loops of induced current circulating in a conducting mass in planes perpendicular to the magnetic flux are called ..(30)... 30

DIRECTIONS: Place the answers to the following problems in the spaces provided at the right.

31. The mutual inductance of two coils is 1.50 henrys. What is the average emf induced in the secondary if the primary current rises to 5.00 amperes in 0.0200 second? 31

32. The turns ratio of a transformer is 0.125. If the primary voltage is 115 volts, what is the secondary voltage? 32

NAME PERIOD DATE INSTRUCTOR'S COMMENTS

21A EXERCISE Alternating Current

SECS. 21.1 - 21.8

DIRECTIONS: In the blank space at the right of each statement, write the word or expression which BEST completes the meaning.

1. The alternating current in a resistance load is ..(1).. with the alternating voltage applied across it. 1
2. When current and voltage are in phase, the graph of instantaneous power as a function of time varies between ..(2a).. and some ..(2b).. maximum. .... 2a 2b
3. The effective value of current or voltage is the ..(3a).. of the mean of the instantaneous values ..(3b).. and is frequently called the ..(3c).. value. .... 3a .... 3b 3c
4. The common mechanism which is probably most versatile for making electric measurements is found in the ..(4)... 4
5. The watt-hour meter is essentially a small single-phase ..(5a).. that turns at a rate proportional to the ..(5b).. used. .... 5a 5b
6. Inductance produces a(n) ..(6).. phase angle in an a-c circuit. 6
7. The *power factor* in an a-c circuit is equal to the ..(7a).. of the phase angle between current and voltage, being ..(7b).. when the phase angle is zero and ..(7c).. when the phase angle is 90°. .... 7a .... 7b 7c
8. The nonresistive opposition to current in an a-c circuit is called ..(8a).. and is expressed in ..(8b)... .... 8a 8b
9. When the nonresistive opposition to current is due to inductance in the circuit, it is referred to as ..(9)... 9
10. The reactance in an a-c circuit due to inductance is ..(10a).. proportional to the product of the ..(10b).. and the inductance. .... 10a 10b
11. In a practical a-c circuit containing reactance and resistance, their joint effect is called ..(11)... 11
12. Impedance has a magnitude equal to ..(12a).. and a direction angle ϕ whose tangent is ..(12b)... .... 12a 12b
13. Capacitance produces a(n) ..(13).. phase angle in an a-c circuit. 13
14. Reactance due to capacitance in an a-c circuit is called ..(14)... 14
15. Reactance in an a-c circuit due to capacitance is ..(15a).. proportional to the product of the ..(15b).. and the capacitance. .... 15a 15b

DIRECTIONS: Write the answers to the following in the spaces provided. Where appropriate, make complete statements.

16. How is the *effective value* of an alternating current defined? ....

....

....

....16

17. What is the basic difference between the power considerations in d-c and a-c circuits? ....

....

....

....17

18. Distinguish between *apparent power* and *actual power* in an a-c circuit. ....

....

....

....18

19. What is the cause of a difference in phase between current and voltage in an a-c circuit? ....

....

....19

20. What are the common types of meter movements used in a-c measurements? ....

....

....20

21. Why may an iron-vane meter be used in either an a-c or a d-c circuit? ....

....

....

....21

22. In what respect is a hot-wire meter unique in its principle of operation? ....

....

....

....22

23. What is the form of the Ohm's law expression for an entire a-c circuit? 23

24. Briefly describe the construction of an impedance diagram. ....

....

....

....

....

....24

25. Explain the meaning of the expression $Z = 120\ \Omega \underline{|-27^\circ}$. ....

....

....

....25

NAME PERIOD DATE INSTRUCTOR'S COMMENTS

21B EXERCISE Series and Parallel a-c Circuits

SECS. 21.9 - 21.11

DIRECTIONS: **Place the answers to the following problems in the spaces provided at the right. In the spaces below Nos. 9-15, show the essential steps of your solution in computation form.**

1. A capacitor of 2.00 microfarads capacitance is placed in a circuit operating at a frequency of 1.00×10^3 hertz. What is the capacitive reactance in the circuit? 1

2. If the frequency in No. 1 is reduced to 10.0 hertz, what is the capacitive reactance of the circuit? 2

3. What is the inductive reactance of a coil of 2.00 henrys inductance when placed in a circuit operating at a frequency of 5.00×10^2 hertz? 3

4. If the frequency in No. 3 is reduced to 5.00 hertz, what is the inductive reactance in the circuit? 4

A circuit operating at 60.0 hertz has a resistance of 2.0×10^3 ohms, a capacitance of 1.00 microfarad, and an inductance of 10.0 henrys all connected in series.

5. In block A, draw the circuit diagram for the above circuit.

6. Determine the magnitude of the impedance. 6

7. Compute the phase angle. 7

8. In block B, construct the impedance diagram using a scale of 0.3 cm to 100 ohms.

A

B

An a-c generator develops $11\overline{0}$ volts at 60.0 hertz across a load consisting of 8.00 ohms resistance, 53.1 millihenrys inductance, and 189.7 microfarads capacitance connected in series.

9. What is the magnitude of the circuit impedance? 9

10. Compute the phase angle. 10

11. What is the magnitude of the circuit current? 11

12. What is the magnitude of the potential difference across the capacitor? 12

13. What is the magnitude of the potential difference across the inductor? 13

14. What is the magnitude of the potential difference across the resistance? 14

15. What is the magnitude of the potential difference across the entire load? 15

16. What is the phase relationship between the voltage across the load and the circuit current?

....

....

....16

17. In block C, construct the impedance diagram showing X_C, X_L, X, R, Z and Φ. Use a scale of 0.3 cm to 1 ohm.

18. In block D, construct the voltage diagram for the series circuit using a scale of 0.3 cm to 11 volts.

C

D

NAME PERIOD DATE INSTRUCTOR'S COMMENTS

21c EXERCISE Resonance

SECS. 21.12 - 21.16

DIRECTIONS: In the space at the right, place the letter or letters indicating EACH choice which forms a correct statement.

1. The inductive reactance of an inductor in an a-c circuit (a) increases with the time rate of change of current, (b) decreases with frequency, (c) varies inversely with frequency, (d) increases as the frequency is raised. 1

2. A graph of the inductive reactance of a coil as a function of frequency yields (a) a hyperbolic curve, (b) a linear curve, (c) a curve which starts at the origin of the coordinates regardless of the inductance of the coil, (d) a curve whose slope is constant. 2

3. The capacitive reactance of a capacitor in an a-c circuit (a) increases with frequency, (b) varies inversely with frequency, (c) is infinite at a frequency equal to zero, (d) is a negligible value at extremely high frequencies. 3

4. A graph of the capacitive reactance of a capacitor as a function of frequency yields (a) a hyperbolic curve, (b) a curve of constant slope, (c) a curve which starts at the origin of the coordinates regardless of the capacitance of the capacitor, (d) a curve that approaches zero at extremely high frequencies. 4

5. When an inductor and a capacitor are connected in series in an a-c circuit (a) the voltage across one cancels the voltage across the other, (b) the voltages across the two are of opposite polarity, (c) the same current must be in each, (d) at low frequencies the current in the inductor will be larger than that in the capacitor. 5

6. When an a-c circuit contains inductive reactance and capacitive reactance in series, (a) the frequency of the applied signal determines the reactance of the circuit, (b) the reactance of the circuit will be equal to zero at some particular frequency, (c) the impedance of the circuit is constant regardless of the frequency, (d) the circuit current is constant at all frequencies. 6

7. In a series circuit containing inductance, capacitance, and resistance, (a) the impedance equals zero at the resonant frequency, (b) the circuit is inductive at frequencies above the resonant frequency, (c) the circuit is capacitive at frequencies below the resonant frequency, (d) the reactance equals zero at the resonant frequency. 7

8. At the resonant frequency of a *L-R-C* series circuit (a) the phase angle is zero, (b) the voltage across the inductance equals zero, (c) the voltage across the capacitance is equal to the voltage across the inductance but is opposite in polarity, (d) the circuit current is maximum. 8

9. The property of a resonant circuit which discriminates between signal voltages of different frequencies is known as its (a) sensitivity, (b) power factor, (c) selectivity, (d) resonant frequency. 9

10. The characteristics of a series-resonant circuit depend primarily on (a) the ratio of the inductive reactance to the circuit resistance, (b) the ratio of the capacitive reactance to the circuit resistance, (c) the ratio of the resistance to the circuit impedance, (d) the magnitude of the signal voltage applied. 10

DIRECTIONS: In the blank space at the right of each statement, write the word or expression which BEST completes the meaning.

11. In *series resonance*, the impedance of the *L-R-C* series circuit is equal to the ..(11a).. of the circuit, and the voltage and circuit current are ..(11b)... ....11a 11b

12. When a signal voltage of constant magnitude and varying frequency is applied to a *L-R-C* series circuit, a distinct ..(12).. in circuit current occurs as the resonance frequency is approached. 12

13. At *resonance*, the phase angle is zero and the power factor is ..(13)... 13

14. The lower the resistance of a *L-R-C* series circuit, the ..(14).. is the current at resonance. 14

15. A resonant circuit responds to impressed voltages of different frequencies in a(n) ..(15).. manner. 15

16. The ratio X_L/R of a series-resonant circuit is commonly called the ..(16).. of the circuit. 16

17. *Resonance* is generally avoided in power circuits because resonant current surges would produce unusually high ..(17).. across the reactance components. 17

18. When a signal voltage is applied across an inductor and capacitor connected in parallel, the total current is the ..(18).. sum of the currents in the two branches. 18

19. A condition of *parallel resonance* is reached when the equivalent impedance of an *L-R-C* parallel circuit is ..(19)... 19

20. At the resonance frequency of a parallel-resonant circuit there is a very large current ..(20a).. between the inductance and capacitance and a line current just large enough to supply the ..(20b).. losses in the circuit. ....20a 20b

NAME PERIOD DATE INSTRUCTOR'S COMMENTS

22 EXERCISE Electronic Devices

SECS. 22.1 - 22.23

DIRECTIONS: In the blank space at the right of each definition, write the term which is defined.

1. The electrode of a vacuum tube which serves as the source of electrons. 1
2. The electrode of a vacuum tube which attracts electrons. 2
3. An electrode of a vacuum tube used to control the flow of electrons. 3
4. A three-electrode vacuum tube. 4
5. The escaping of electrons from a hot surface. 5
6. The negative charge in the interelectrode space of a vacuum tube. 6
7. The smallest negative grid voltage, for a given plate voltage, which causes the tube to cease to conduct. 7
8. The ratio of a small change in plate voltage to a small change in grid voltage of the opposite sense, which maintains a constant plate current. 8
9. The internal resistance to electron flow from cathode to plate of a vacuum tube. 9
10. The ratio of the small change in plate current to the small change in grid voltage producing it, when the plate voltage is held constant. 10

DIRECTIONS: In the blank space at the right of each statement, write the word or expression which BEST completes the meaning.

11. The number of substances suitable for use as emitters to produce satisfactory thermionic emission is limited by the ..(11).. required. 11
12. A triode contains a cathode, an anode, and a third electrode called a(n) ..(12a).. which is normally maintained at some ..(12b).. potential with respect to the cathode. ....12a 12b
13. The direction of the plate current inside a vacuum tube is from ..(13a).. to ..(13b)... ....13a 13b
14. The direction of the plate current outside the tube is from ..(14a).. to ..(14b)... ....14a 14b
15. When a signal voltage on the grid of a tube is at its positive maximum instantaneous value, the voltage on the plate is at its ..(15).. instantaneous value. 15
16. *Triodes* are not used as amplifiers at radio frequencies because of coupling effects between input and output circuits due to ..(16)... 16
17. *Electron emission* resulting from the bombardment of the plate of a vacuum tube by high-velocity electrons is called ..(17)... 17
18. By the insertion of a second grid, known as the ..(18a).., the frequency range of the tube is increased, but electrons are accelerated causing ..(18b).. at the plate. ....18a 18b
19. In the *pentode*, the effects of secondary emission are eliminated by the insertion of a third grid known as the ..(19)... 19

20. A television-picture tube is a special kind of vacuum tube known as a(n) ..(20)...

....20

21. A faint light incident on the photoemission cathode of a six-stage photomultiplier tube produces three photoelectrons per millisecond. Each electron impact produces three secondary electrons. The number of electrons arriving at the collector plate per millisecond has an order of magnitude of ..(21)...

....21

22. Semiconductors have a unique property which permits a relatively ..(22a).. flow of electrons in the forward direction with a small applied voltage, and a very ..(22b).. flow of electrons in the reverse direction with a much larger applied voltage.

....22a

....22b

23. The amplifying property of a transistor is a result of changes in ..(23a).. which occur in the different regions of the structure when proper ..(23b).. are applied.

....23a

....23b

24. The ratio of the output power of a P-N-P common-base amplifier circuit to the input power is called the ..(24).. of the circuit.

....24

25. For amplifying an input signal — where high current gain is desired, the common-..(25a).. circuit is preferred; where large power gain is desired, the common-..(25b).. circuit is preferable.

....25a

....25b

DIRECTIONS: Write the answers to the following in the spaces provided. Where appropriate, make complete statements.

26. What is the nature of the germanium crystal which is significant in its property of semiconduction? ....

....

....

....

....26

27. Why is antimony referred to as a *donor* element when minute traces are added to germanium? ....

....

....

....

....27

28. Why is aluminum referred to as an *acceptor* element when minute traces are added to germanium? ....

....

....

....

....28

29. What is the difference between N- and P-types of semiconductors? ....

....

....

....

....

....

....29

30. What is the P-N junction? ....

....

....

....30

NAME PERIOD DATE INSTRUCTOR'S COMMENTS

23 EXERCISE Atomic Structure SECS. 23.1 - 23.12

DIRECTIONS: In the space at the right, place the letter or letters indicating EACH choice which forms a correct statement.

1. Experimental evidence shows that a hydrogen atom is composed of (a) at least two parts, (b) parts with opposite electric charges, (c) parts having equal masses, (d) positive and negative particles, the negative particles having larger masses. 1

2. The Thomson experiment (a) involved a special cathode-ray tube, (b) established a value for the charge on an electron, (c) established a value for the mass of an electron, (d) established a value for the *e/m* ratio for an electron. 2

3. The Millikan experiment (a) involved cathode rays, (b) established a value for the charge on an electron, (c) made possible the computation of the mass of an electron, (d) established a value for the diameter of an electron. 3

4. The *electron* (a) is a negatively charged particle, (b) has about the same mass as a hydrogen atom, (c) was discovered by Millikan, (d) has a rest mass of 5.486×10^{-4} u. 4

5. The *proton* (a) is a neutral particle, (b) has about the same mass as a hydrogen atom, (c) was discovered by Thomson, (d) has an atomic mass of 1.00727663 u. 5

6. The *neutron* (a) is a neutral particle, (b) has about the same mass as a hydrogen atom, (c) was discovered by Chadwick, (d) has a mass of 9.10905×10^{-31} kg. 6

7. The *atomic nucleus* (a) was discovered by Rutherford, (b) is negatively charged, (c) is the dense central part of an atom, (d) contains an equal number of protons and neutrons. 7

8. The electron *shells* of an atom (a) are definite electron paths about a nucleus, (b) compose the electron cloud, (c) may be designated by the letters K to Q, (d) consist of groups of orbitals. 8

9. *Isotopes* (a) contain the same number of protons but differ in the number of neutrons, (b) contain the same number of neutrons but differ in the number of protons, (c) are atoms of the same element, (d) are atoms of different elements. 9

10. *Nuclides* (a) are varieties of atoms distinguished by their nuclear composition, (b) may be atoms of the same element, (c) may be atoms of different elements, (d) which are isotopes have the same atomic number. 10

DIRECTIONS: In the blank space at the right of each statement, write the word or expression which BEST completes the meaning.

11. An atom is composed of smaller particles which are arranged in (simple, complex) ..(11).. ways. 11

12. Different kinds of atoms are formed when their subatomic particles are ..(12).. in different ways. 12

13. The first subatomic particle to be identified and studied was the ..(13)... 13

14. The rays which cause the glass walls of an evacuated tube to glow with a greenish fluorescence are called ..(14).. rays. 14

15. By the use of a magnet, Perrin concluded that cathode rays were streams of ..(15).. coming from the negative electrode in an evacuated tube. 15

16. Because the nature of cathode-ray particles did not change when Thomson changed the composition of the cathode or the gas in his cathode-ray tubes, he concluded that cathode-ray particles are present in (no, some, all) ..(16).. forms of matter. 16

17. By studying how an electron beam is influenced by an electric field and a magnetic field, Thomson was able to determine the ..(17).. ratio of the electron. 17

18. An electron with a positive charge is called a(n) ..(18)... 18

19. When X rays of known energy produce a pair of electrons, the rest mass of the electrons can be calculated by using the equation, ..(19)... 19

20. Electrons (do, do not) ..(20).. account for a large share of the mass of substances. 20

21. Millikan determined the charge on a(n) ..(21a).. by means of the ..(21b).. experiment. 21a

.... 21b

22. An electron diffraction pattern indicates that electrons exhibit ..(22).. duality. 22

23. Rutherford discovered the atomic nucleus by bombarding a thin ..(23a).. with ..(23b)... .... 23a

.... 23b

24. The nucleus of an atom contains most of the mass of the atom yet is less than ..(24).. the diameter of the atom itself. 24

25. The number of protons in the nucleus of an atom is the ..(25).. of the atom. 25

26. An atom containing 13 protons, 14 neutrons, and 13 electrons has ..(26).. nucleons. 26

27. The average atomic mass of the atoms of an element based on a large number of samples found in nature is the ..(27).. of the element. 27

28. A mass spectrograph is used to measure the ..(28).. of ionized atoms with great precision. 28

29. The nuclear binding force is a very short-range force, being effective to a distance about ..(29).. times the radius of a proton. 29

30. The energy required to move an electron between two points which have a potential difference of one volt is one ..(30)... 30

DIRECTIONS: Place the answers to the following problems in the spaces provided at the right.

31. A platinum atom consists of 78 protons, 78 electrons, and 116 neutrons. What is its atomic number? 31

32. What is the mass number of the atom of No. 31? 32

33. What is the combined mass in atomic mass units of the particles composing the nucleus of the atom of No. 31? 33

34. If the actual mass of the nucleus in No. 31 is 193.9200 u, what is its nuclear mass defect in atomic mass units? 34

35. What is the nuclear binding energy of 1.6527 u in Mev? 35

NAME PERIOD DATE INSTRUCTOR'S COMMENTS

24 EXERCISE Nuclear Reactions

SECS. 24.1-24.12

DIRECTIONS: In the parentheses at the right of each word or expression in the second column, write the letter of the expression in the first column which is MOST CLOSELY related.

a. Emission of radiation by a nucleus forming a lighter nucleus
b. Formation of a more complex nucleus from simple nuclei
c. ${}^{4}_{2}He$
d. Ionized gas molecule
e. Unit of radioactivity
f. Addition of one particle to a nucleus with subsequent emission of a second particle
g. ${}^{0}_{-1}e$
h. Unit of exposure to gamma rays
i. Break-up of a heavy nucleus into nuclei of intermediate masses
j. Half-life
k. High energy photon
l. Change in the identity of a nucleus

alpha particle ( ) 1
beta particle ( ) 2
gamma ray ( ) 3
radioactive decay ( ) 4
nuclear bombardment ( ) 5
fission ( ) 6
fusion ( ) 7
nuclear transformation ( ) 8
curie ( ) 9
roentgen ( ) 10

DIRECTIONS: In the blank space at the right of each statement, write the word or expression which BEST completes the meaning.

11. Radioactivity is the spontaneous, uncontrollable ..(11a).. of an atomic nucleus with the ..(11b).. of particles and rays. 11a

....11b

12. The naturally occurring elements with atomic numbers greater than ..(12).. are all radioactive. 12

13. The property of radioactive elements which led to their discovery is the ability of their radiations to affect the ..(13a).. on a ..(13b)... ....13a

....13b

14. The length of time during which half a given number of atoms of a radioactive nuclide will decay is called its ..(14)... 14

15. The fraction of the original number of atoms of a radioactive nuclide remaining after the elapse of four half-lives is ..(15)... 15

16. List α particles, β particles, and γ rays in order of increasing speed. 16

17. List α particles, β particles, and γ rays in order of increasing penetrating power. 17

18. In a(n) ..(18a).. change, the composition of the substance is not changed; in a(n) ..(18b).. change, new substances with new properties are produced but the nuclei of the interacting atoms are unchanged. 18a

....18b

19. All forms of alpha, beta, and gamma emission are examples of ..(19).. change. 19

20. In symbols like ^1_1H which represents a(n) ..(20a).. and ^1_0n which represents a(n) ..(20b).., the subscript designates the ..(20c).. of the particle while the superscript indicates the number of ..(20d).. in the particle. ....20a ....20b ....20c 20d

21. In addition to nuclear binding energy, the neutron-proton ..(21).. affects nuclear stability. 21

22. The ratio of the number of nuclei decaying per second and the total number of original nuclei is called the ..(22)... 22

23. The first transformation by nuclear bombardment was produced in 1919 by ..(23)... 23

24. Cockcroft and Walton verified the equation ..(24).. by bombarding lithium with high-speed protons. 24

25. Neutrons can penetrate a nucleus more easily than other particles because they have no ..(25)... 25

26. A material such as graphite which slows down fast neutrons acts as a(n) ..(26)... 26

27. Elements such as plutonium (at. no. 94), curium (at. no. 96), and fermium (at. no. 100) are known as ..(27).. elements. 27

28. A thermonuclear reaction is another name for a(n) ..(28).. reaction. 28

29. The energy of the sun is believed to be released during the fusion of protons to form ..(29a).. and ..(29b)... ....29a 29b

30. Fusion reactions are more desirable than fission reactions for energy production because ..(30).. is required. 30

31. A reaction in which the material or energy starting the reaction is also one of the products and can cause similar reactions is a(n) ..(31).. reaction. 31

32. The amount of radioactive material required to sustain a chain reaction in a reactor is called the ..(32)... 32

33. A nuclear reactor is a device in which controlled fission produces new ..(33a).. substances and ..(33b)... ....33a 33b

34. The rate of the fission reaction in a nuclear reactor is regulated by the position of control rods which absorb ..(34)... 34

35. A reactor in which one fissionable material is produced at a greater rate than another fissionable material is consumed is known as a(n) ..(35).. reactor. 35

36. When the production of neutrons in a reactor equals the sum of the neutrons which are absorbed and those which escape from the reactor, the reactor is said to be ..(36)... 36

37. In order to have 100 neutrons produce new fissions in a nuclear reactor, a total of ..(37).. neutrons must be provided. 37

38. Radioactive isotopes are called ..(38)... 38

39. Radioactive isotopes are very useful as ..(39).. elements in determining the course of chemical reactions. 39

40. The age of prehistoric wood samples may be determined by measuring the amount of the radioactive isotope ..(40).. they presently contain. 40

DIRECTIONS: Complete the following nuclear equations.

41. ${}^{226}_{88}Ra \rightarrow {}^{222}_{86}Rn +$ ....

42. ${}^{234}_{91}Pa \rightarrow {}^{234}_{92}U +$ ....

43. ${}^{9}_{4}Be + {}^{4}_{2}He \rightarrow {}^{12}_{6}C +$ .... $+$ energy

44. ${}^{235}_{92}U + {}^{1}_{0}n \rightarrow {}^{138}_{56}Ba + {}^{95}_{36}Kr +$ .... $+$ energy

45. $4\,{}^{1}_{1}H \rightarrow$ .... $+ 2\,{}^{0}_{+1}e +$ energy

NAME PERIOD DATE INSTRUCTOR'S COMMENTS

25A EXERCISE Detection Instruments - Particle Accelerators SECS. 25.1-25.8

DIRECTIONS: In the parentheses at the right of each word or expression in the second column, write the letter of the expression in the first column which is MOST CLOSELY related.

a. Makes a permanent record
b. Contains a hollow copper cylinder
c. Series of charged metal plates
d. Particularly suited for study of high-energy particles
e. Contains photomultiplier tube
f. Invented by E.O. Lawrence
g. Movable piece of metal foil
h. Contains liquid helium
i. Zinc sulfide screen
j. Invented by C.T.R. Wilson
k. Contains liquid hydrogen
l. None of the other choices

bubble chamber ( )1
cloud chamber ( )2
electroscope ( )3
Geiger counter ( )4
ionization chamber ( )5
photographic emulsion ( )6
scintillation counter ( )7
solid-state detector ( )8
spark chamber ( )9
spinthariscope ( )10

DIRECTIONS: Write the answers to the following in the spaces provided. Where appropriate, make complete statements.

11. List three main characteristics of the particles and rays emitted in nuclear reactions.

a. 11a

b. 11b

c. 11c

12. For each of the characteristics of No. 11, identify the detection instruments which are applications by using their item numbers from Nos. 1 - 10.

a. 12a

b. 12b

c. 12c

13. What determines the rate of discharge of an electroscope? ....

.... 13

14. What enables current to flow between the cathode and anode of a Geiger tube? ....

....

.... 14

15. How are the tracks of alpha particles made visible in a diffusion cloud chamber? ....

....

.... 15

16. How are the tracks of protons made visible in a bubble chamber? ....

.... 16

17. In what way is the bubble chamber superior to the cloud chamber? ....

....

.... 17

18. What is the nature of the charged-particle track in a solid-state detector? ....

....18

19. Contrast the spinthariscope with the photomultiplier tube. ....

....

....19

20. How is the passage of a particle recorded on a photographic film? ....

....

....20

DIRECTIONS: In the blank space at the right of each statement, write the word or expression which BEST completes the meaning.

21. A moving belt is used in the ..(21)... 21

22. Particles are accelerated in a doughnut shaped tube in the ..(22)... 22

23. Dees are used in the ..(23)... 23

24. The oscillating voltage and magnetic field are varied in the ..(24)... 24

25. Drift tubes are characteristic of ..(25)... 25

26. For each of the following accelerators, give the maximum energy imparted to the accelerated particles.

    a. Van de Graaff generator 26a

    b. Betatron 26b

    c. Cyclotron 26c

    d. Synchrocyclotron 26d

    e. Synchrotron 26e

    f. Linear accelerator 26f

27. Most of the ionization of the atmosphere is produced by ..(27)... 27

28. The source of the high-speed particles which come into the atmosphere from outer space is probably (within, beyond) ..(28).. the solar system. 28

29. Because cosmic ray intensity varies with latitude, the particles are electrically ..(29a).. and are affected by the earth's ..(29b)... ....29a

....29b

30. Cosmic rays are thought to consist mostly of ..(30)... 30

NAME PERIOD DATE INSTRUCTOR'S COMMENTS

25B EXERCISE Fundamental Particles - Atomic Models

SECS. 25.9 - 25.15

DIRECTIONS: Write the answers to the following in the spaces provided. Where appropriate, make complete statements.

1. List the four types of interaction forces between particles in the order of decreasing strength. 1

2. Name the three conservation laws that are observed in nuclear reactions. 2

3. What happens when a particle and its antiparticle collide? 3

4. State the *law of parity.* 4

5. State the *uncertainty principle.* 5

6. What is *quantum mechanics?* 6

7. State the *exclusion principle.* 7

8. List three kinds of bonding between atoms. 8

9. Define *ionization energy.* 9

10. What is *electron affinity?* 10

DIRECTIONS: Complete the following table.

	Category	Name	Symbol	Rest Mass	Baryon Number	Lepton Number	Strangeness Number
11.		electron					
12.			n				
13.				2333	+1		−1
14.	boson						
15.	baryon			3280			
16.		neutrino_μ					
17.	meson			1074			
18.			Σ^+				
19.			$\overline{p}$				
20.			Ω^+				

DIRECTIONS: In the blank space at the right of each statement, write the word or expression which BEST completes the meaning.

21. The nuclear force is not effective beyond a distance of ..(21)... 21

22. The only force that can both attract and repel is the ..(22).. force. 22

23. The ratio of the gravitation force to the nuclear force is ..(23)... 23

24. Strangeness numbers are not conserved in ..(24).. interactions. 24

25. According to some scientists, all "fundamental" particles may, in turn, be composed of still more elementary bits of matter-energy called ..(25)... 25

26. The uncertainty principle destroyed the ..(26).. model of the atom. 26

27. The principle quantum number describes the ..(27).. of the orbit of the electron. 27

28. The Zeeman effect justifies the concept of the ..(28).. quantum number. 28

29. The two possible values of m_s are ..(29a).. and ..(29b)... ....29a

....29b

30. DeBroglie proposed the existence of ..(30).. waves. 30

DIRECTIONS: In the table below, place the values of n, l, m_l, and m_s for each electron in the neutral sulfur atom.

n	l	m_l	m_s

NAME PERIOD DATE INSTRUCTOR'S COMMENTS

1 EXPERIMENT Measuring Length

PURPOSE: (1) To learn how to make measurements of length with a meter stick, a vernier caliper, and a micrometer caliper. (2) To understand the relationship between the construction of a measuring instrument and the precision of the measurements made with it.

APPARATUS: Meter stick; 15 cm or 30 cm ruler; vernier caliper; micrometer caliper; rectangular wooden block; numbered metal cylinder of brass or aluminum; pieces of wire, gauge numbers 18, 22, and 30, for example; numbered rectangular metal blocks, small enough to fit into micrometer caliper.

INTRODUCTION:

1. Meter stick

In making measurements with a meter stick, keep the following suggestions in mind:

a. Since the end of the meter stick may be worn, start the measurement at some intermediate mark. (See Fig. 1-1.) Of course, the reading at the mark with which you start the measurement must be subtracted from the final reading.

b. Place the meter stick on edge, as in Fig. 1-1. This will avoid errors that are easily made when viewing the stick in the flat position shown in Fig. 1-2.

c. Estimate the reading to the nearest 0.5 mm. In other words, the last digit of your measurement, which is an estimate, should be 0 or 5.

2. Vernier caliper

A vernier caliper consists of two metric and two English scales, as shown in Fig. 1-3. The fixed scales have a jaw at one end, the metric fixed scale being subdivided into centimeters and millimeters. The other jaw is attached to the sliding or vernier scales. The metric vernier scale is ruled so that *nine* millimeters is divided into *tenths*. From the enlarged scale of Fig. 1-4, it is evident that one division of the metric vernier scale equals 0.9 mm. When the jaws of the caliper are closed, the zero of the fixed scale exactly coincides with the zero of the vernier scale.

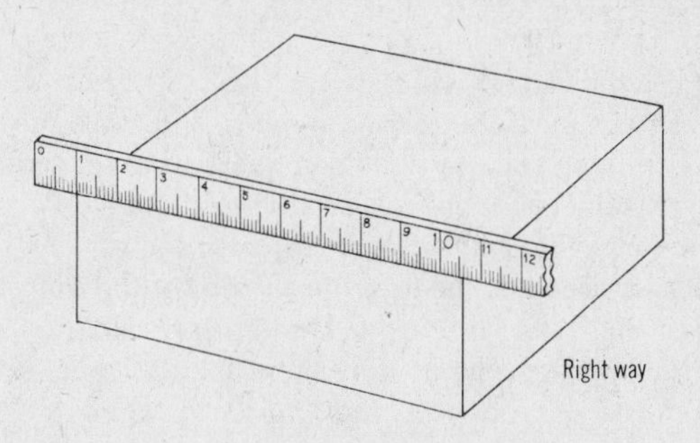

Figure 1-1

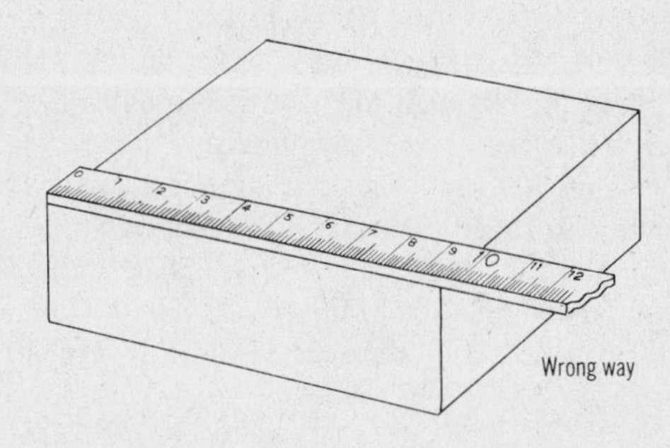

Figure 1-2

When the vernier scale B is moved to the right until its *first* division coincides with the one-millimeter mark on the fixed scale A, it has been moved exactly 0.1 mm, or 0.01 cm, and the jaws are separated by that amount. When the vernier scale is moved to the right until its second division coincides with the two-millimeter mark on the fixed scale A, it has been moved a total distance of 0.2 mm, or 0.02 cm. Moving the scale to the right 0.03 cm brings its third division opposite the three-millimeter mark on A, and so on. That particular division on the vernier scale which coincides with a line or division on the fixed scale reads *hundredths of a centimeter.*

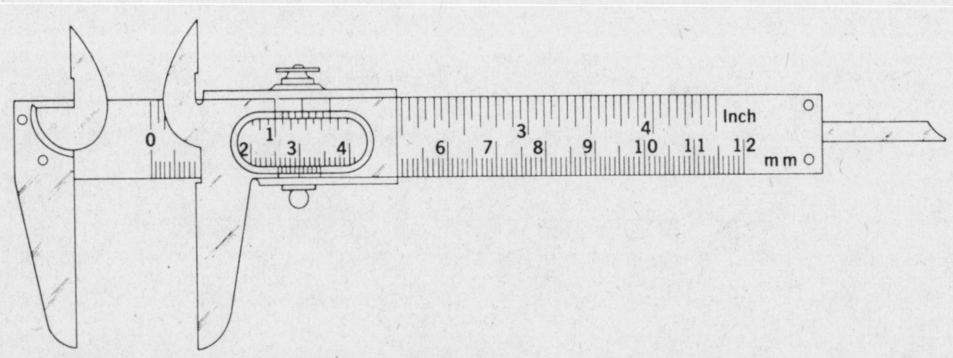

Figure 1-3

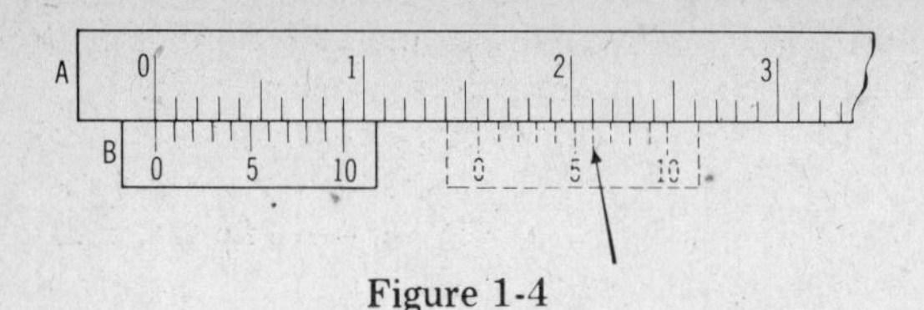

Figure 1-4

To use the caliper, separate the jaws, place the object to be measured between them, and close the jaws firmly on the object. Then tighten the set screw, if there is one, enough to keep the vernier scale from moving while it is being read. Suppose the vernier scale now occupies the position shown by the dotted lines of Fig. 1-4. The zero of the vernier scale shows that the jaws are separated by more than 1.5 cm and less than 1.6 cm. *Centimeters and tenths of centimeters are read on the fixed scale. Hundredths of centimeters are read from the vernier scale by locating that particular division on the vernier scale which coincides with a division on the fixed scale.* As indicated by the arrow, it is No. 6. Therefore, the correct reading is 1.560 cm. If no division on the vernier scale coincides exactly with a division on the fixed scale, then the last digit of the measurement should be a 5. Thus, the correct reading for Fig. 1-5 is 1.565 cm. To measure the inside diameter of a tube or hollow cylinder, the upper parts of the jaws are used. Keep in mind the fact that centimeters and tenths of centimeters are read from the fixed scale, while hundredths of a centimeter are read from the vernier scale.

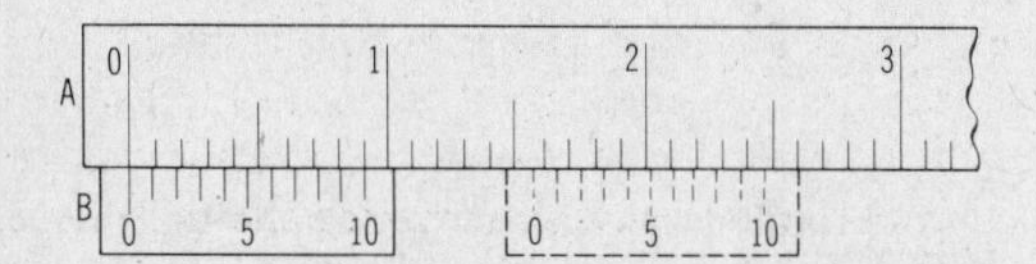

Figure 1-5

3. Micrometer caliper

A micrometer, Fig. 1-6, consists of several parts. The flat end of a set screw forms one surface against which the object to be measured rests. It is called the anvil. The end of the spindle forms the other surface. Within the caliper, the spindle is an accurately threaded screw which can be moved back and forth through the frame of the caliper as the thimble is turned.

If the micrometer is graduated in metric units, the threads on the spindle may have a pitch of exactly one millimeter, provided the edge of the thimble is divided into 100 equally-spaced divisions. This means that the opening at C is opened or closed *exactly one millimeter* when the thimble is given one complete turn in a clockwise or counter-clockwise direction. When the spindle rests firmly, but not too tightly against the anvil, the zero on the thimble should exactly coincide with the zero mark on the barrel of the frame. Turning the thimble through one-hundredth of a complete revolution changes the opening *one-hundredth of a millimeter.*

If the threads on the spindle have a pitch or interval of half a millimeter, then the thimble must be given two complete turns to open or close the opening at C by one millimeter. In such a case the thimble is divided into 50 equal divisions. Hence, twisting the thimble through one division opens or closes the space at C by one-hundredth of a millimeter. Millimeters are read from the scale on the barrel; hundredths of a millimeter are read from the scale on the thimble. For example, the caliper of Fig. 1-6 shows by the scale on the barrel that the opening at C is more than 6 mm and less than 7 mm. Furthermore, the thimble scale is closer to 6 mm than it is to 7 mm. Division number 38 on the thimble coincides with the horizontal line on the barrel. Therefore, the reading of the caliper is 6.380 mm. If the thimble scale had been closer to 7 mm than to 6 mm, the thimble must have been turned 50 scale divisions past 6 mm and then an additional 38 scale divisions. The reading would then be 6.880 mm. If the horizontal line on the barrel does not coincide with a division number on the thimble, then the measurement should be estimated to the nearest 0.001 mm.

As the caliper is closed, it takes only a slight amount of force to make the zero line of the thimble coincide with the horizontal line of the scale on the barrel. In making any measurement, this same force must be used in closing the caliper on the object being measured. It is easy to injure so delicate and sensitive an instrument. A mechanic tightens his micrometer so the object will slip "stickily," as a nail pulled along a magnet. Some calipers have a ratchet thimble which has enough friction to close the cali-

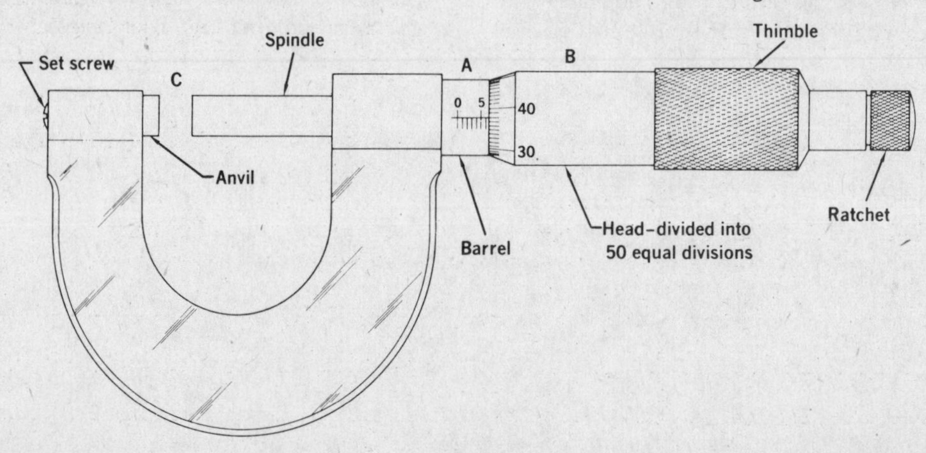

Figure 1-6

per, but turns when the caliper is closed. This protects the instrument.

Remember that if the edge of the thimble has 50 divisions, a reading of 30 on the thimble may be either 0.300 mm or 0.800 mm, depending on whether the thimble was turned through only part of a revolution after passing a particular millimeter division on the barrel, or whether it was turned one complete revolution plus part of another. In a similar manner, do not confuse 0.150 mm with 0.650 mm, 0.220 mm with 0.750 mm, or any other similar examples.

PROCEDURE:

1. Meter stick

Measure the width of the laboratory table and record these measurements to the nearest 0.5 mm. Repeat the measurement, but this time start the measurement at a different mark on the meter stick. Make a third measurement, this time from the opposite end of the table. The three readings should agree to within 1 mm.

Measure the length, width, and thickness of the rectangular block furnished by the instructor. Make each measurement three times as before. Average the readings and calculate the volume of the block. Observe the rules of significant figures in your calculations. Record all your data in the table below. Use centimeters as the unit of measurement. Be sure to write the appropriate unit after each measurement and calculation.

DATA

	Trial 1	*Trial 2*	*Trial 3*	*Average*
Width of table				
Length of block				
Width of block				
Thickness of block				
Volume of block				

2. Vernier caliper

Make three measurements of the length and diameter of the numbered cylinder furnished by the instructor. Record your measurements in the table. Calculate the volume of the cylinder, using significant figures correctly.

DATA

	CYLINDER NO.		
TRIAL	*Length* (cm)	*Diameter* (cm)	*Volume* (cm^3)
1			
2			
Average			

3. Micrometer caliper

Begin by closing the caliper gently to observe the zero reading. If the reading is considerably off zero, ask your instructor to adjust the caliper. Move it to zero and back a few times to get the "feel" of the instrument. Open the caliper and try to read it at some position set at random. Move the thimble again and take a second reading. Next place a piece of wire, gauge No. 18, for example, between the anvil and spindle of the caliper. Close the caliper on the wire by turning the thimble. Read the diameter of the wire to the nearest thousandth of a mm, convert to centimeters, and record in the data table. In a similar manner, measure and record the diameters of the other wires of different gauge numbers. Consult Table 21 in Appendix B for the actual values of the diameters. Calculate your absolute and relative errors for each trial and record these in the table.

DATA

MATERIAL	GAUGE NO.	*Diameter (experimental)* (cm)	*Diameter (from table)* (cm)	*Absolute error* (cm)	*Relative error* (%)

4. Comparative precision

Take the rectangular metal block and measure its dimensions successively with the metric ruler, the vernier caliper, and the micrometer caliper, using each instrument to the precision described earlier. Record your measurements in the table at the right and calculate the volume of the block from each set of measurements. Be sure to follow the rules of significant figures. Compare the results.

DATA

INSTRUMENT	BLOCK NO.			
	Length (cm)	*Width* (cm)	*Height* (cm)	*Volume* (cm^3)

QUESTIONS: Your answers should be complete statements.

1. Why should a meter stick be placed on edge as shown in Fig. 1-1 when making measurements?

....

.... 1

2. Why should measurements be started at a mark other than the end of the meter stick?

....

.... 2

3. What is the smallest marked metric unit on a meter stick? micrometer caliper?

....

.... 3

4. How many significant figures do the three dimensions of the rectangular block have? How does this affect the significant figures in the volume of the block?

....

....

....

....

.... 4

5. How could you use a meter stick to determine the thickness of a single sheet in a book? How would the number of significant figures in such a measurement compare with those in the measurement of the block?

....

....

....

.... 5

6. How can a vernier caliper be used to measure the inside diameter of a hollow cylinder?

....

.... 6

7. John measured the dimensions of a block as 1.5 cm × 1.3 cm × 1.0 cm and recorded the volume as 1.950 cm^3. Henry measured the dimensions as 1.485 cm × 1.314 cm × 0.986 cm, and recorded the volume as 1.92 cm^3. Which answer is more reliable in terms of the data from which it was calculated? Why?

....

....

.... 7

8. How do the precisions of the volumes in Step 4 compare with each other?

....

....

....

.... 8

NAME PERIOD DATE INSTRUCTOR'S COMMENTS

2 EXPERIMENT Measuring Mass

PURPOSE: (1) To learn how to use a platform balance. (2) To learn how to use an inertia balance.

APPARATUS: Platform balance and set of masses, or triple beam balance; inertia balance; several identical C-clamps of 100 to 150 g each; various solid objects with masses below 1 kg; supply of small nails; stopwatch with 10 sec sweep, or watch with sweep second hand.

INTRODUCTION:

1. Platform balance

The platform balance is a device used to determine the mass of an object by comparison with known masses. Figure 2-1 shows one type of platform balance commonly used in physics laboratories. Another type of platform balance, Fig. 2-2, has a triple beam. Various sliding masses are suspended from this beam, making the use of separate masses almost unnecessary. The center of the beam and the pans themselves rest upon knife-edged bearings which have little friction. There is a pointer attached to the beam to show when the pans are balanced.

In using a balance, first see that the sliding mass is at the extreme left, or at zero. Then touch one of the pans lightly to start the beam swinging. The pans should move far enough so the pointer will swing at least three spaces to either side of the center mark. If it swings farther to one side than it does to the other, the zero adjustment screw may be turned until the pointer moves as many spaces to the left as to the right. The balance is now in equilibrium and is ready for use.

The object whose mass is to be determined is always placed on the *left* pan of the balance. Select a mass which you judge to be greater than that of the object being measured. Place it on the *right* pan. If it is greater, remove it and add the *next-smaller* mass. If this mass is less than that of the object, continue to add other masses to the right pan. Try masses *successively* from the larger to the smaller, until the last mass added is the ten-gram mass. Then the sliding mass, which is used to indicate mass to the nearest tenth of a gram, should be moved far enough to the right to make the pointer swing as many spaces to the right as to the left. You do not need to wait for the pointer to come to rest. The mass of the object equals the sum of all the masses on the pan, plus the mass represented by the sliding mass.

To ensure obtaining the maximum precision of which the balance is capable, it must be level, and the pans must be free from dirt or foreign matter. Beakers and similar objects must be dry on the outside before they are placed on the pan of the balance.

If a mass is missing from the box, or if the balance is not working properly, notify your instructor before you begin.

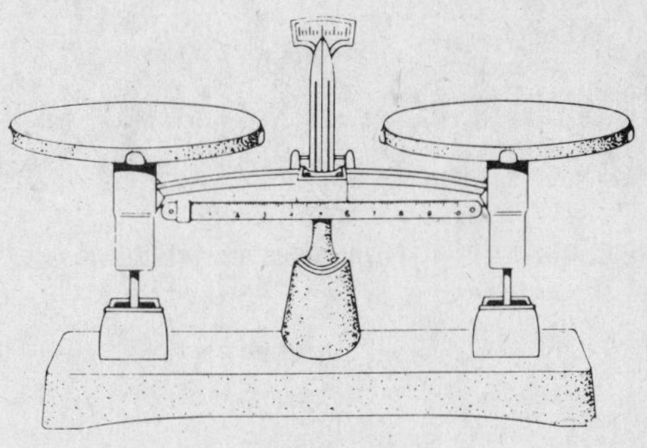

Figure 2-1

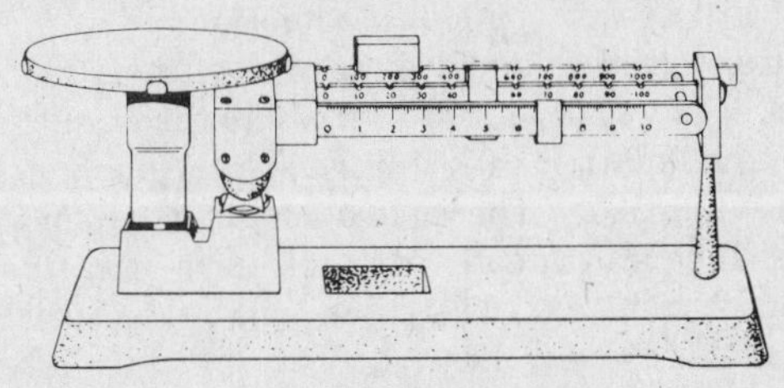

Figure 2-2

Be careful not to jar the balance when adding or removing the larger masses. Such jarring dulls the knife-edges and makes the balance less sensitive. Jarring may be avoided by supporting the pan with one hand while objects are being added to or removed from the pans. The object whose mass is being determined, and also the large masses, should be placed near the center of the balance pan.

When you are finished, see that all the masses are returned to their proper places in the box.

2. Inertia balance

The inertia balance is used to compare masses by measuring the frequency of vibration when a mass is placed on the pan of the balance and the pan is set in horizontal vibration. Figure 2-3 shows an inertia balance attached to a laboratory table. On such a

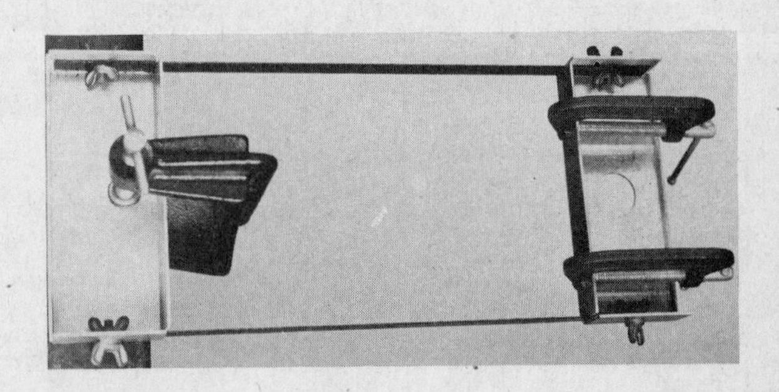

Figure 2-3

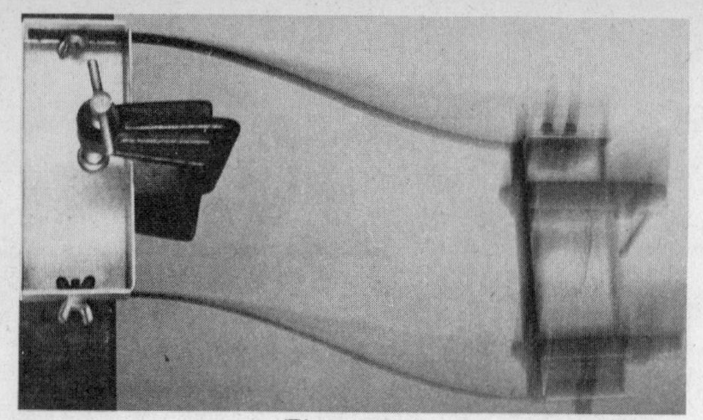

Figure 2-4

balance, masses and frequencies are related according to the equation

$$\frac{m_1}{m_2} = \frac{T_1^{\,2}}{T_2^{\,2}}$$

where m_1 and m_2 are different masses and T_1 and T_2 are their respective periods of vibration.

Since m_1 and m_2 include the mass of the balance pan, it is more convenient to measure the frequencies of several known masses and then plot a graph with the resulting data. The value of an unknown mass can then be read directly from the graph after its frequency is measured. This is the method that will be used in this experiment.

The inertia balance does not depend upon the force of gravitation. It is therefore a true mass-measuring device. Platform balances compare masses by comparing their weights. They do not work in the absence of gravity. They are faster and more convenient to use than inertia balances, however, and are used for all mass determinations in beginning physics.

PROCEDURE:

1. Platform balance

Determine the masses of several solid objects, reading the balance to its greatest possible precision. Ask your instructor to help you determine this.

Find the mass of a group of 20 small, identical nails. Repeat with 40, 60, 80, and 100 nails. In each case, calculate the mass of a single nail. Record all data in the table below. Plot a graph of the data on the nails and use the graph to predict the mass of a random number of nails, say 29 or 84. Check your prediction with an actual measurement.

DATA

Number of nails	*Mass* (g)	*Mass of single nail* (g)
Average		

2. Inertia balance

Fasten one of the C-clamps to the pan of the inertia balance. Set the balance in motion, using a small (about 2 cm) vibration amplitude. Get yourself synchronized with the motion of the balance before starting the stopwatch. Start the watch on the count of "zero" and read the time of 20 vibrations to the nearest 0.01 second. Repeat at least twice more and average. Compute the time of a single vibration. This is called the *period* of the balance. In the same way, find the period when 2, 3, and 4 of the clamps are fastened to the pan. Record all data below.

DATA

TRIAL	*No. of C-clamps*	*Vibrations*	*Period* (sec)
1			
2			
3			
4			
5			

Find the mass of a single C-clamp on the platform balance. Plot a graph with your data, using the masses as abscissas and the periods of vibration squared as ordinates. Connect the points with a smooth line.

Place an unknown mass on the pan of the inertia balance and measure its period of vibration. With the help of your graph, estimate the mass of the object. Check this reading against the mass as measured with the platform balance. Using the reading of the platform balance as the accepted value, calculate your relative error. Repeat this procedure with two more unknown masses.

If the inertia balance is equipped with a cylindrical mass and a hole in the pan through which the mass fits, measure the period of the balance while the mass is suspended in the hole from a string. Then place the mass in the pan and measure the period again. Compare the results.

DATA

TRIAL	*Period* (sec)	*Mass*		*Relative error*
		Graph (kg)	*Platform balance* (kg)	
1				
2				
3				
4				

QUESTIONS: Your answers should be complete statements.

1. Why is the unknown mass placed on the left-hand pan of a two-pan platform balance?

....

....

....

.... 1

2. Why are large masses used first in using a platform balance?

....

....

.... 2

3. Would an inertia balance yield the same results if it were used in a vertical position? Verify your answer by experimentation.

....

.... 3

4. Does the angle of vibration affect the results of an inertia balance? Support your answer with experimental data.

....

.... 4

5. Would an inertia balance work in an orbiting space vehicle?

....

....

.... 5

6. List some possible sources of error in the method of finding the mass of a single nail.

....

.... 6

7. How can the method of finding the mass of a single nail be used to find the mass of a single sheet of a ream of paper?

....

.... 7

NAME PERIOD DATE INSTRUCTOR'S COMMENTS

3 EXPERIMENT Measuring Time

PURPOSE: To learn how to measure short time intervals with a recording timer.

APPARATUS: Recording timer; timing tape; stopwatch with 10 sec sweep, or watch with sweep second hand; dry cell or other DC power supply; connecting wire; C-clamp; switch; string; set of weights.

INTRODUCTION: A recording timer is a device for measuring the time it takes an object to move a short distance. It consists of an electromagnet and a clapper very much like those in an electric doorbell (Fig. 3-1). A paper tape is inserted under the clapper, and a disc of carbon paper is placed face down between the clapper and the paper tape. Thus, each time the clapper strikes, it makes a dot on the paper tape. (In some timers, waxed paper tape is used, thus eliminating the necessity of carbon paper discs.)

When the paper tape is pulled through the timer, the distance between two dots is the distance the tape moved during one back and forth vibration of the clapper. The time required for a single vibration is called the *period* of the timer. Once the period is measured, it can be used to determine the time interval for any motion of the paper tape.

In this experiment, you will measure the period of a recording timer. This is called calibrating the timer. You will also use the timer to measure the speed of a falling object. In later experiments, you will use the timer to study other types of motion.

PROCEDURE:

1. Calibrating the timer

Fasten the recording timer to the table with a C-clamp and connect it to the dry cell. Choose a location that will enable you to pull a long section of paper tape through the timer in a straight line without hitting any obstacles. Insert the end of the tape between the clapper and the carbon paper disc as shown in Fig. 3-1, making sure that the tape can move freely. One student should hold the roll of tape by passing a pencil through the center as an axis. The other student should grasp the end of the tape in one hand and a watch in the other.

Turn on the timer by closing the switch. Then pull the tape steadily through the timer for 3 or 4 seconds, using the stopwatch to measure the time accurately. This is done by walking away from the timer at a steady pace. Turn off the timer and mark the first and last dots on the tape. Count the number of dots, ignoring any variation in the intervals between dots. The spaces between dots represent equal time intervals regardless of variations in the spaces. Repeat this procedure several more times. Record all data and compute the period of the timer. Make sure you have at least three trials that are in good agreement with each other. Find the average of your results and record in data table.

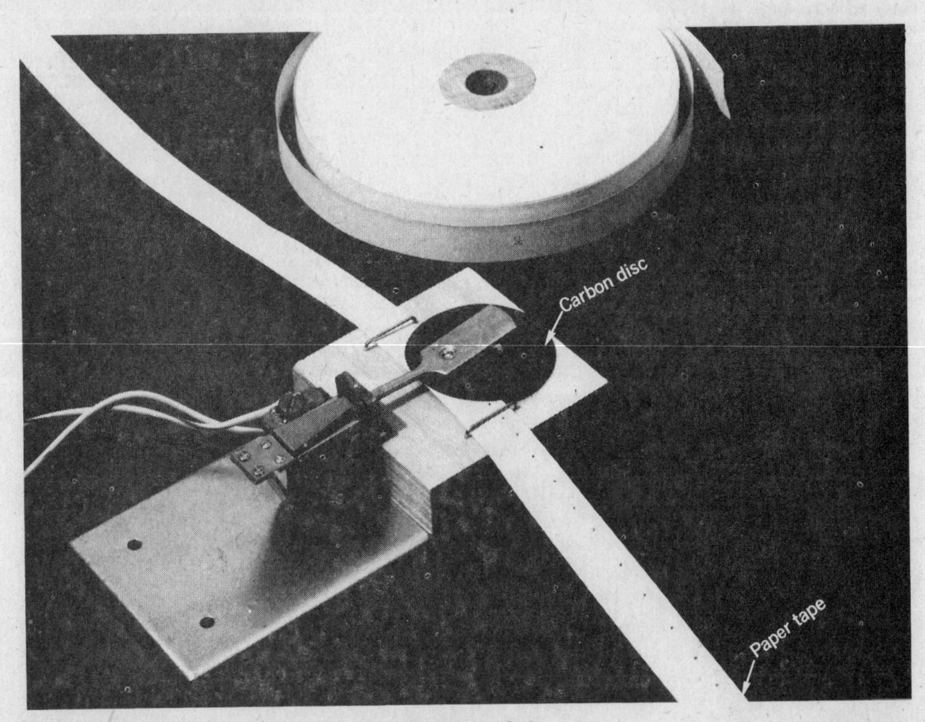

Figure 3-1

DATA

TRIAL	*Time* (sec)	*No. of dots*	*Frequency* (sec^{-1})	*Period* (sec)
1				
2				
3				
4				
Average				

2. Speed of a falling object

Set up the timer as shown in Fig. 3-2. Cut a piece of paper tape that is at least 20 cm longer than the distance between the table top and the floor. Pass the end of the tape through the timer and fasten a 20-g mass to the end of the tape by means of adhesive tape or by hooking the mass through the tape. Hold the mass at a convenient level near the top of the table. Start the timer and let the mass fall to the floor. Stop the timer when the mass hits the floor.

Count the number of dots on the tape and multiply it by the period of the timer to get the time of fall. Measure the distance between the level from which the mass was dropped and the floor. Record these data and use them to find the average speed of the mass. Repeat this procedure several times, using the same mass. Then run several trials with smaller and larger masses. Compare the results. Use the stopwatch to check the time calculated for several of your trials.

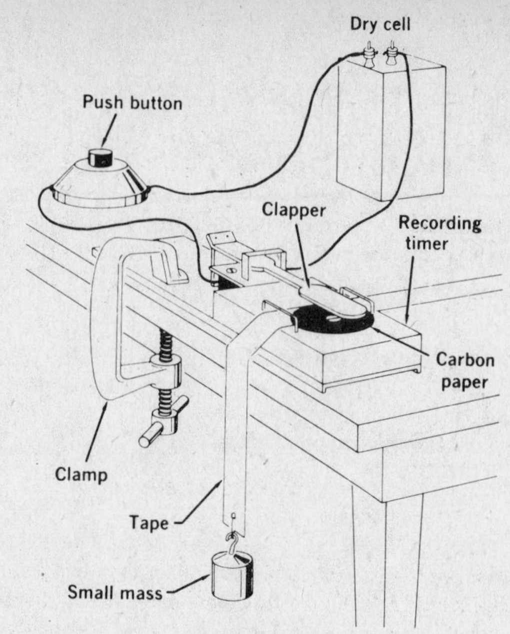

Figure 3-2

DATA

TRIAL	*Mass* (g)	*Time* (sec)	*Distance* (m)	*Average speed* (m/sec)
1				
2				
3				
4				

QUESTIONS: Your answers should be complete statements.

1. What is the order of magnitude of the period of the timer?

....

.... 1

2. How do the speeds of various falling masses compare with each other?

....

....

....

.... 2

3. What is a possible source of error in the measurement of the period of the timer?

....

....

....

.... 3

4. What does the spacing of the dots in Part 2 indicate?

....

.... 4

NAME PERIOD DATE INSTRUCTOR'S COMMENTS

4 EXPERIMENT Mass Density

PURPOSE: (1) To determine the mass density of a regular solid. (2) To find the volume of an irregular solid and determine its mass density. (3) To determine the mass density of water and alcohol.

APPARATUS: Platform balance, or balance with 0.01 g sensitivity; rectangular solid and cylinder; graduated cylinder, 100-ml or 250-ml capacity; irregular solids such as lumps of coal, rock, or metal; thread; thermometer; 2 beakers, 100-ml; buret, 50-ml; buret clamp; ring stand; alcohol.

INTRODUCTION: The mass of a unit volume of a substance is called its mass density. To find the mass density of any material experimentally, first find the mass of the material and then find its volume. By dividing the mass by the volume, the mass of a unit volume, which is the mass density, may be calculated. In this experiment, the volume of a rectangular solid will be calculated from its linear measurements. The volume of an irregular solid will be determined by the displacement method. The volume of two liquids will be measured by means of a buret. (The method of making buret readings is described in the Introduction.)

PROCEDURE:

1. *Regular solids*

If you use the rectangular solid and cylinder of Experiment 2, the volumes calculated in that experiment may be used as data here. Otherwise, measure the length, width, and thickness of the block and cylinder, as required, with a vernier caliper, record your data, and calculate the volumes.

Using the balance, find the mass of the block and cylinder to the precision warranted by the balance. Calculate the mass density of each, using significant figures. If the identity of the material is known, find the accepted value for its mass density from a handbook. Calculate the absolute error and the relative error.

DATA *Regular solids*

Volume (cm^3)	*Mass* (g)	*Mass density (experimental)* (g/cm^3)	*Mass density (accepted)* (g/cm^3)	*Absolute error* (g/cm^3)	*Relative error* (%)

2. *Irregular solids*

Find the mass of one of the irregular solids using the platform balance. Tie a piece of thread about 30 cm long to the solid. Fill a graduated cylinder about two-fifths full of water, and read the height of the water level as exactly as possible. Now lower the irregular solid into the water in the cylinder. Because matter has the property of impenetrability, the solid displaces a volume of water equal to its own volume. See Fig. 4-1. Take a second reading of the water level. What is the relationship between the difference in water level readings and the volume of the irregular solid? Calculate the mass density. Repeat the experiment with as many different solids as your instructor may direct. Compare the experimental values with accepted values.

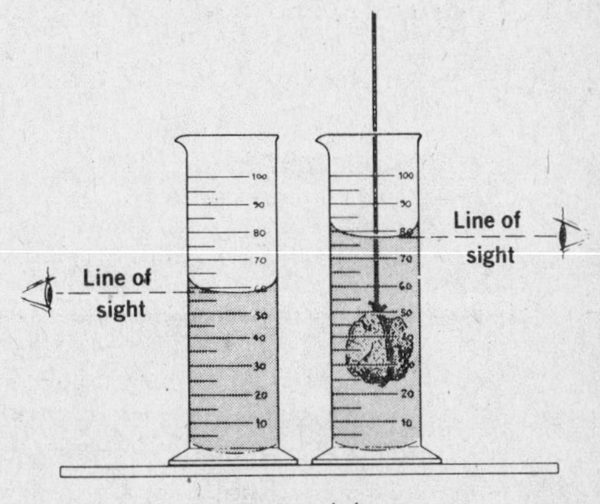

Figure 4-1

3. *Density of water*

Determine the mass of a clean, dry beaker.

Clamp a buret, Fig. 4-2, in a vertical position, and fill it with water to a height of about two centimeters above the zero mark. Then open the stopcock or pinch clamp a little, and draw enough water into the second beaker to bring the water level below the zero mark onto the graduated portion of the buret. This

DATA *Irregular solids*

MATERIAL	*Water level reading (initial)* (ml)	*Water level reading (final)* (ml)	*Volume of solid* (cm^3)	*Mass* (g)	*Mass density (experimental)* (g/cm^3)	*Mass density (accepted)* (g/cm^3)	*Absolute error* (g/cm^3)	*Relative error* (%)

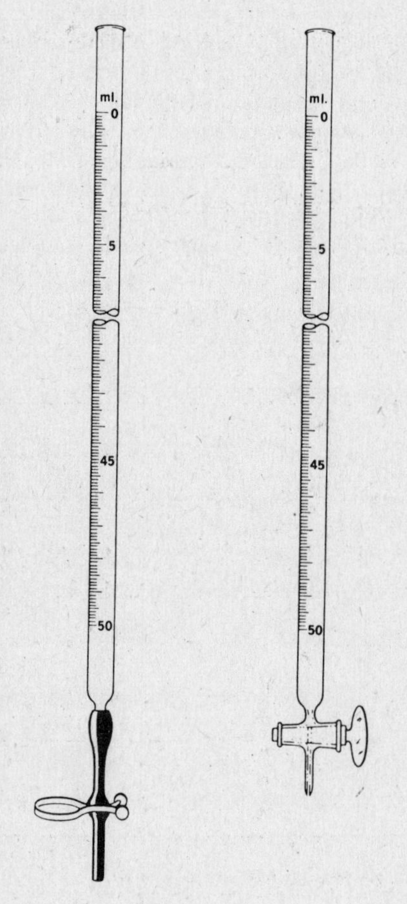

Figure 4-2

method of filling and adjusting the liquid level in the buret should expel all the air from the tip of the buret. (If all the air has not been expelled from the buret tip, it will be necessary to let more water run through the tip to expel the air. Then water should be added to the buret, so that it contains at least 30 ml.)

Record the initial reading of the buret to the nearest tenth of a milliliter if the masses are determined with a platform balance and to the nearest hundredth of a milliliter if a balance with a sensitivity of 0.01 g is being used. Slowly draw off about 40 ml of water into the beaker of known mass. Record the final buret reading to the same precision as you did the initial reading. Find the mass of the beaker and water. By subtraction, find the mass of the volume of water used.

Refill the buret, and repeat the experiment using 45 ml of water for the second trial. For each trial, calculate the mass in grams of one cubic centimeter of water. Take the temperature of the water, and look up Table 9 in Appendix B to find the accepted mass of one cubic centimeter of water at the observed temperature. Calculate your absolute error and relative error and record in the following table.

4. Density of alcohol

Drain the buret of water, and rinse it several times with small portions of alcohol. Then fill the buret with alcohol. Repeat the experiment with two alcohol samples of appropriate volume, and determine its mass density. Consult a handbook to find the accepted value of the mass density of alcohol at the observed temperature. Compute the absolute error and the relative error.

DATA *Density of water*

TRIAL	*Liquid used*	*Buret reading (initial)* (ml)	*Buret reading (final)* (ml)	*Volume of liquid* (cm^3)	*Mass of beaker* (g)	*Mass of beaker and liquid* (g)	*Mass of liquid* (g)
1							
2							
3							
4							

DATA *Density of alcohol*

Mass density of liquid (experimental) (g/cm^3)	*Temperature of liquid* (°C)	*Mass density of liquid (accepted)* (g/cm^3)	*Absolute error* (g/cm^3)	*Relative error* (%)

QUESTIONS: Your answers should be complete statements.

1. As used in this experiment, how does the accuracy of the displacement method of measuring volume compare with the direct calculation method?

....

.... 1

2. On what general property of matter does the displacement method of measuring volume depend?

....

.... 2

3. How could the volume of a cork be determined by water displacement?

....

....

....

....

....

.... 3

4. Why is a buret graduated from the top down?

....

....

.... 4

5. How are the densities of water and alcohol affected by changes in temperature?

....

....

.... 5

6. Why is it necessary to read the buret to 0.01 ml if a balance sensitivity of 0.01 g is used?

....

....

.... 6

NAME PERIOD DATE INSTRUCTOR'S COMMENTS

5 EXPERIMENT Acceleration-I

PURPOSE: (1) To study the change in acceleration of a moving mass when the applied force changes. (2) To study the change in acceleration produced by a given force when the mass changes.

APPARATUS: Recording timer; timing tape; connecting wire; dry cell or other DC power source; set of masses; pulley with table clamp; string; platform balance; metal cart.

INTRODUCTION: Newton's law of acceleration states that an unbalanced force applied to a mass produces acceleration according to the equation

$$F = ma$$

Because of friction, this law seems to contradict common experience. In driving a car, for example, a constant force is required to keep the car moving with a constant velocity. If the force is removed, the force of friction brings the car to a stop. In the absence of friction, however, the car would continue to move with constant velocity after the force is removed. The continued application of force would then result in acceleration.

In this experiment, you will study the law of acceleration, first by measuring the change in acceleration when a constant mass is subjected to a varying force, and then by noting the change in acceleration when the force remains constant and the mass varies. Friction will be minimized by using a rolling cart and by a small counter-weight.

PROCEDURE:

1. Constant mass, but varying force

Set up the apparatus as shown in Fig. 5-1. (The operation of the recording timer is explained in Experiment 3.) The timing tape is fastened to one end of the cart and the string is fastened to the other end.

Measure the mass of the cart on the platform balance. Then load it with the following masses: 50 g, 100 g, and 200 g. Add these masses to the mass of the cart and record the sum in your data table. Pass the string over the pulley and fasten a small mass to the end to offset the frictional force on the cart. The mass is correct when the cart moves uniformly forward when you give it a push. This counterweight should stay on the string throughout this part of the experiment, but it is not recorded in the data.

For the first trial, remove the 50-g mass from the cart and fasten it to the end of the string. Hold the cart while placing the tape in the timer. Start the timer and release the cart. Stop the cart when the 50-g mass hits the floor and then stop the timer. Remove the tape and label it "50-g trial." Compute the weight of the 50-g mass in newtons and record in your table.

Repeat the procedure with the 100-g mass (replacing the 50-g mass in the cart). Remove the tape and label it "100-g trial." Repeat with the other masses until you have 3 valid trials. Keep these tapes separate from those in the next step.

2. Constant force, but varying mass

For the three trials in this part of the experiment, keep a 100-g mass on the string (in addition to the counterweight). Without adding any masses to the cart, run the experiment and record the total mass and accelerating force under Trial 4 in the data table. For Trial 5, add a 500-g mass to the cart and for Trial 6 add an additional 1000 g. Be sure to include the 50-g mass on the string when recording the total mass for each of these trials. Label each tape.

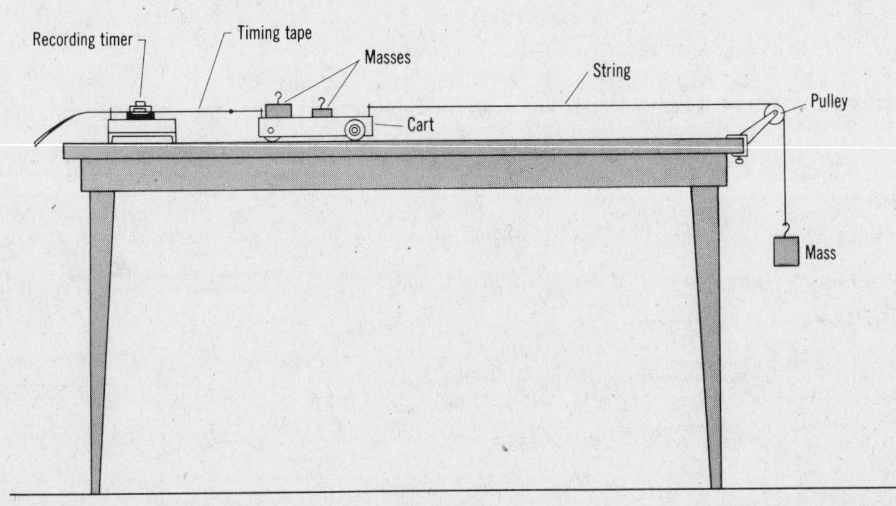

Figure 5-1

CALCULATIONS: Count off the dots on each tape and circle every 10th dot. Measure the distance between each of the circled dots and the succeeding dot. Compute the velocity in each case and record it in the data table. Then calculate the acceleration between successive velocities. Be careful that you do not measure the dot intervals that were recorded after the accelerating mass has hit the floor.

DATA *Period of timer* ________ *sec.*

TRIAL	Total mass (kg)	Accelerating force (n)	*Velocity* (m/sec)	*Acceleration* (m/sec^2)
1				
2				
3				
4				
5				
6				

GRAPH: Plot a graph of the relationship between acceleration and the accelerating force as shown by the data for the three trials of Part 1. Use the accelerating force as abscissas and the resulting acceleration as ordinates. Plot a second graph, using the data in the three trials of Part 2. In this graph, use the total mass as abscissas and the corresponding accelerations as ordinates.

QUESTIONS: Your answers should be complete statements.

1. Why does the mass in Step 1 remain constant even though various masses are removed from the cart during the trials?

....

....

.... 1

2. What do the graphs for this experiment tell you about acceleration?

....

....

.... 2

3. In what way is the motion of a freely falling object different from the motion of the carts in this experiment? How could the experiment be modified to study free fall?

....

....

....

....

....

....

.... 3

NAME PERIOD DATE INSTRUCTOR'S COMMENTS

6 EXPERIMENT Acceleration - II

PURPOSE: To study the motion of an object which is under the influence of the accelerating force of gravity.

APPARATUS: Packard's acceleration board; graph paper; carbon paper; level; Scotch tape; 2 wood blocks (to increase tilt of board).

Courtesy of Sargent-Welch Scientific Company

Figure 6-1

INTRODUCTION: Galileo studied acceleration by rolling a ball down an inclined plane. Such a ball is under the influence of the same accelerating force as is that of a freely falling object, namely the force of gravity. The inclined plane slows down the vertical motion of the ball, thus making it easier to measure.

In this experiment, the path of a ball rolling down an inclined plane is recorded on graph paper through the use of carbon paper. The ball is also given a horizontal velocity. Assuming that this horizontal velocity remains relatively unchanged for each trial, equal horizontal components of the recorded path can be used to measure equal periods of time. The downward component of the path for each time period can then be measured.

PROCEDURE: Set up the acceleration board as shown in Fig. 6-1. If special instructions are furnished with the apparatus, study them carefully. The following factors are important to the proper operation of the acceleration board:

(1) The board should be tilted at such an angle that the ball will stay within the width of the graph paper as it rolls down the board. Make several practice runs in order to obtain the necessary tilt angle.

(2) The ball guide and graph paper must both be square with the board. Use the level to check this requirement.

(3) The graph paper must be positioned so that the top line is on a horizontal line with the center of the ball as it leaves the ball guide. The Y-axis of the graph paper must also coincide with the point where the center of the ball leaves the guide. (This is *not* the point where the ball first strikes the paper. The ball will strike the paper slightly to the right of the Y-axis.)

After the board and graph paper are properly positioned, fasten the graph paper with Scotch tape in addition to using the clips on the board. Place a piece of carbon paper loosely over the graph paper. Let the left edge of the carbon paper extend beyond the left edge of the graph paper. Release the ball and check its path on the graph paper. If it is a smooth curve and is easily legible, run a second trial with a slightly increased tilt of the board. Then run a third trial on the same graph paper with a still greater tilt angle. If necessary, use the wood blocks to obtain a sufficient tilt. The lower part of the three paths should be separated from each other by about 3 centimeters in order to make meaningful measurements.

Use the first piece of graph paper for practice runs. After you have learned to position this paper properly and have noted the necessary tilt for each trial, tape a second piece of graph paper over the first for your final runs, using the practice paper as a guide.

CALCULATIONS: After three successful trials have been recorded, remove the graph paper from the board. Divide the horizontal spaces on the paper into seven equal segments and label them from 1 to 7. These spaces represent equal intervals of time and the labels correspond to the numbers in the t-column of the data table. s is the distance the ball travels along the Y-axis during increasing time intervals: for $t=1$, record the number of Y-spaces for the first time interval; for $t=2$, the number of Y-spaces during the first two time intervals; for $t=3$, the number of Y-spaces during the first three time intervals; etc.

Compute the average velocities, v_{av}, for each time interval by using the ratio, s/t, and enter these values in the table. Then record the final velocities, v_f, which is twice the average velocity for a given time interval, or $2v_{av}$.

The accelerations are calculated in three different ways: by the change in v_f in each successive time interval, from the equation $v_f=at$, and from the equation $s=\frac{1}{2}at^2$. Calculate the acceleration for each time interval in each of these three ways and enter the results in the data table. Average the results.

DATA

TRIAL	t (X-axis units)	s (Y-axis units)	v_{av} (s/t)	v_f $(2v_{av})$	Acceleration (Δv_f)	(v_f/t)	$(2s/t^2)$	ave.
1	1							
	2							
	3							
	4							
	5							
	6							
2	1							
	2							
	3							
	4							
	5							
	6							
3	1							
	2							
	3							
	4							
	5							
	6							

QUESTIONS: Your answers should be complete statements.

1. How do the acceleration for the three trials compare with each other?

....1

2. a. Is the acceleration constant for each trial? b. Explain.

a. 2a

b. 2b

3. What are some factors that might affect the accuracy of this experiment?

....

....

....3

4. In what way is this experiment superior to Galileo's method of rolling a ball down an incline to study acceleration?

....

....4

NAME PERIOD DATE INSTRUCTOR'S COMMENTS

7 EXPERIMENT Resolution of Forces

PURPOSE: To show how a single force may be resolved into two component forces.

APPARATUS: Simple crane boom; 2 spring balances, 20-n capacity; slotted weights; weight hanger; twine; protractor. (If the common form of laboratory apparatus is not available, a meter stick may be used for the boom.)

INTRODUCTION: When a lawn mower is pushed, it does not go in the direction it is pushed. It moves in a direction parallel to the surface of the ground. The force which acts along the handle at an angle with the surface of the lawn is resolved into two components. One component acts horizontally and moves the mower along the ground. The other component acts vertically and tends to push the mower into the ground. There are many similar cases where a force that acts on an object in a direction in which it is not free to move is resolved into two component forces.

In Fig. 7-1, three forces act on point *P*. The load F_w pulls directly downward. The tensile force measured by the spring balance *S* pulls along the line *PS*. The boom *PR* exerts a thrust force which acts as the equilibrant of the first two forces. Consequently, the resultant of the load force and the tensile force is equal to, but acts in the opposite direction to, the thrust force.

In this experiment, the thrust force exerted by the boom and the angles at which the three forces act will be measured. Then the equilibrant of the thrust force will be resolved into the load force and the tensile force, and their magnitude determined graphically and mathematically. Finally these values will be compared with those actually used in the experiment.

PROCEDURE: 1. Tie the ring of one spring balance near the top of the upright support rod of the laboratory table. Fasten the crane boom clamp near the bottom of the support rod. Connect the end of the boom by means of twine to the spring balance hook. Add enough weights to the hanger attached to the end of the boom to make the balance *S* read from 12 to 15 n. Then adjust the boom so that it makes a 90° angle with the upright support rod.

Hold a blank sheet of paper against the apparatus so the center of the sheet is behind point *P*. Mark on the paper the position of *P*, and indicate the directions of PF_w, *PR*, and *PS*. Measure the angles *SPR* and RPF_w.

Read the spring balance *S* to determine the force *PS*. Attach a second spring balance at *P* and pull vertically upward until you just counterbalance the downward force exerted by the weights and the weight of the boom. This spring balance reading is the equilibrant of force PF_w. Now, using the same spring balance at *P*, pull outward in the direction of *PR* extended, until the boom just begins to move away from the upright. (The connection at *R* must allow the boom to move horizontally.) This is the thrust force exerted by the boom and is equal in magnitude but opposite in direction to the resultant of *PS* and PF_w. Mark these actual values in their proper places in the data table.

On the paper containing the position of *P* and the directions of the three forces, lay off a vector, to scale, to represent the force *PR*. Using this vector as the diagonal of a parallelogram, and the known directions of PF_w and *PS* as the directions of the sides, complete the parallelogram, $PSRF_w$.

Measure the length of the vectors *PS* and PF_w in your diagram and, from the scale you have used, de-

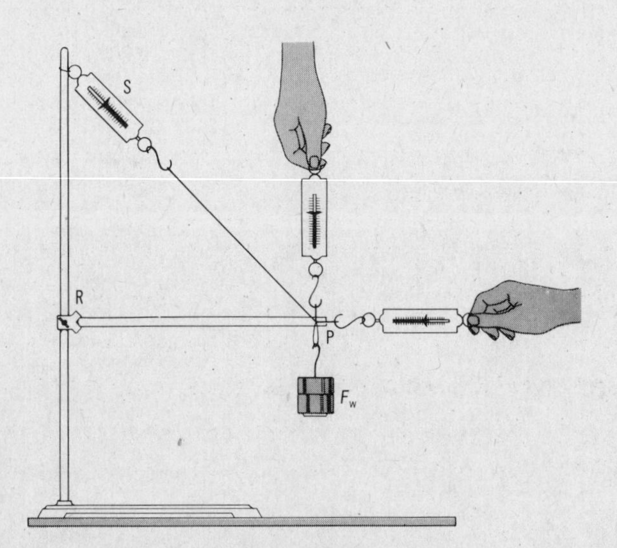

Figure 7-1

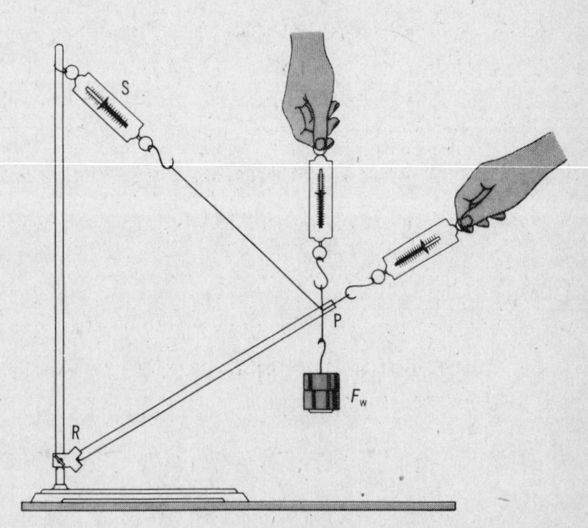

Figure 7-2

termine their graphic values. Determine their values by calculation using the measured values of the vector PR, $\angle SPR$, and $\angle RPF_w$.

2. If time permits, repeat the experiment. This time set the boom at a different angle with the upright support rod as shown in Fig. 7-2.

DATA

TRIAL	PR (n)	$\angle SPR$ (°)	$\angle RPF_w$ (°)	PS (graph) (n)	PS (calc.) (n)	PS (actual) (n)	PF_w (graph) (n)	PF_w (calc.) (n)	PF_w (actual) (n)
1									
2									

QUESTIONS: Your answers should be complete statements.

1. Why could the combined weight of the weights and weight hanger not be used as the force PF_w?

....

....

.... 1

2. In the vector diagram for this experiment, why is the force due to PR shown acting outward from P rather than acting from R toward P?

....

.... 2

3. How does the force along PS change as $\angle SPR$ decreases? What is the practical value of this information?

....

....

.... 3

4. How does the force along PS compare with the total weight supported by the apparatus? Does this help to explain the reason for using a crane boom?

....

....

.... 4

NAME PERIOD DATE INSTRUCTOR'S COMMENTS

8 EXPERIMENT Composition of Forces

PURPOSE: (1) To find the equilibrant of two forces acting at an angle. (2) To show that the equilibrant of two forces is equal to their resultant, but acts in the opposite direction.

APPARATUS: 3 spring balances, 20-n capacity; wooden block; strong twine; small iron ring, 1.0 cm in diameter; ruler; pencil; pencil compass; protractor; 3 iron table clamps, or force board, or composition-of-forces apparatus.

INTRODUCTION: Two different forces often act simultaneously on a body at the same point. The angle between these two forces may be between 0° and 180°. The resultant of these two forces is the single force which could be substituted for them without altering the effect they produce. If the two forces act at an angle, the resultant is equal to the diagonal of the force parallelogram of which the two forces are the sides. The equilibrant of two or more forces is the single force which can produce equilibrium with them. It is equal in magnitude but opposite in direction to their resultant.

PROCEDURE: Fasten three pieces of twine, each from 20 to 25 cm in length, to the iron ring and to the balance hooks as shown in Fig. 8-1. Fasten the balances to the clamps with twine. If a composition-of-forces apparatus is used, follow the manufacturer's directions for attaching the balances. Adjust the length of the twine and the position of the balances so that all balances have a reading within about 3 n of each other. Best results will be obtained if the balances read at least 12 n. The cords must be adjusted

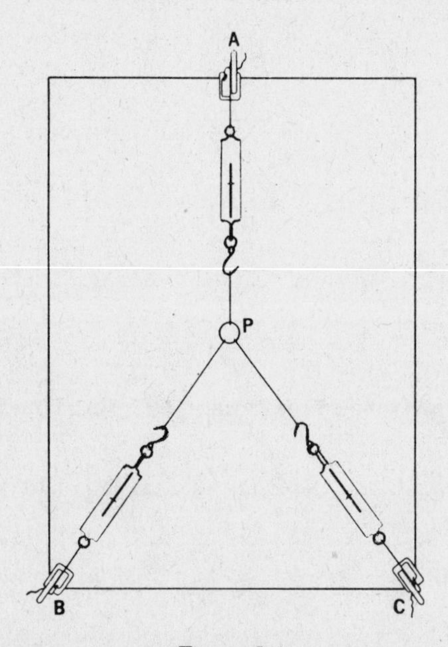

Figure 8-1

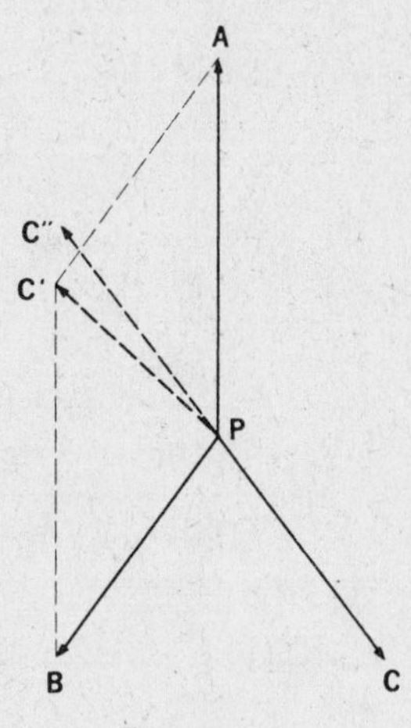

Figure 8-2

so that the pull of each balance is in direct line with the center of the ring.

Place a blank sheet of paper beneath the cords so its center will be directly beneath the center of the ring. Thumb tacks through opposite corners of the paper will keep it from slipping. Three forces are represented by the balances, *A*, *B*, and *C*. Two of them act at an angle on point *P*, represented by the center of the ring. The third force is the equilibrant. Mark the direction of the forces by placing a wooden block on the paper so that its edge just touches the string along the entire length of the block. Be sure it does not displace the string at any point. Draw a line along the edge of the block, just beneath the string, with a fine-pointed pencil. This shows the direction in which the force acts. In a similar manner, locate the direction of the other forces. Read each balance and record its reading along the line which represents the direction of the force.

Remove the paper and extend the lines representing the three forces until they meet. If the work was carefully done, they will meet at point *P*. If they do not meet at a point, consult your instructor to find out whether the error is too great to make your experiment acceptable.

Use some convenient unit (1 cm to represent 2.0 or 2.5 n) and measure off on each line from point *P* a distance corresponding to each of the balance readings. Using the two forces *AP* and *BP* as sides, construct a parallelogram, and draw the diagonal $C'P$. Use solid lines to represent the forces and dashed lines as construction lines. See Fig. 8-2. Determine the graphic magnitude of $C'P$. Extend the third force line, *CP*, beyond *P* a distance equal to *CP* and label this segment $C''P$. Determine the graphic difference in magnitude between $C'P$ and $C''P$. (This is the difference between the theoretical and experimental equilibrants.) Measure the angle between $C'P$ and $C''P$ and record this as graphic direction error.

Using the known values of AP, BP, and $\angle APB$, calculate the magnitude of $C'P$. Also calculate the magnitude of $\angle APC'$. Determine the calculated direction error by obtaining the difference between measured $\angle APC''$ and calculated $\angle APC'$. These calculations are to be shown in smooth form and submitted as part of your experiment report.

If time permits, run two more trials using different magnitudes and directions for AP, BP, and CP from the ones used in the first trial. Find the graphic and calculated errors of magnitude and direction as before. Use AP as the equilibrant in the second trial and BP as the equilibrant in the third trial. Record the errors in the data table.

DATA

TRIAL		*Graphic*	*Calculated*
1	$C'P$		
	$\angle APC'$		
	Magnitude error (between $C'P$ and $C''P$)		
	Direction error ($\angle C'PC''$)		
2	$A'P$		
	$\angle BPA'$		
	Magnitude error (between $A'P$ and $A''P$)		
	Direction error ($\angle A'PA''$)		
3	$B'P$		
	$\angle CPB'$		
	Magnitude error (between $B'P$ and $B''P$)		
	Direction error ($\angle B'PB''$)		

QUESTIONS: Your answers should be complete statements.

1. Could the same apparatus be used with more than three concurrent forces? Explain your answer.

....

....

....

.... 1

2. Would the results of the experiment be changed if the spring balances were attached with the hooks pointing outward? Explain your answer and verify it by experimentation.

....

....

.... 2

3. Is the accuracy of the experiment affected by the length of the twine on the balances? Explain your answer and verify it by experimentation.

....

....

.... 3

NAME PERIOD DATE INSTRUCTOR'S COMMENTS

9 EXPERIMENT Coefficient of Friction

PURPOSE: (1) To find the coefficient of sliding friction. (2) To compare sliding friction with rolling friction.

APPARATUS: Smooth board with hard surface, for use as an inclined plane; wooden block, coated with paraffin; metal car; meter stick; platform balance.

INTRODUCTION: When a block rests on an inclined plane as shown in Fig. 9-1, its weight, F_w concentrated at the center of gravity of the block, acts vertically downward. Since the block cannot move in that direction, the weight of the block is resolved into two component forces. One component, F_p, acts parallel to the plane and tends to slide the block down the plane. The other component, F_N, acts at right angles to the plane and tends to break it, or to make the block stick to the plane. If the slope of the plane is great enough to cause the block to slide at uniform speed, the ratio of the parallel force to the perpendicular force is the *coefficient of friction* between the block and the plane.

The coefficient of friction may also be defined as *the ratio of the force required to slide or roll an object at uniform speed over a horizontal surface to the weight of the object itself.* It can be found experimentally by weighing the object, and then using a spring balance to measure the force needed to slide the object at slow, uniform speed. For example, if a force of 8 n is required to slide a 20-n block over a horizontal surface at a constant rate, the coefficient of friction is 8 n ÷ 20 n or 0.4. The coefficient of friction depends on the materials and the nature of the surfaces.

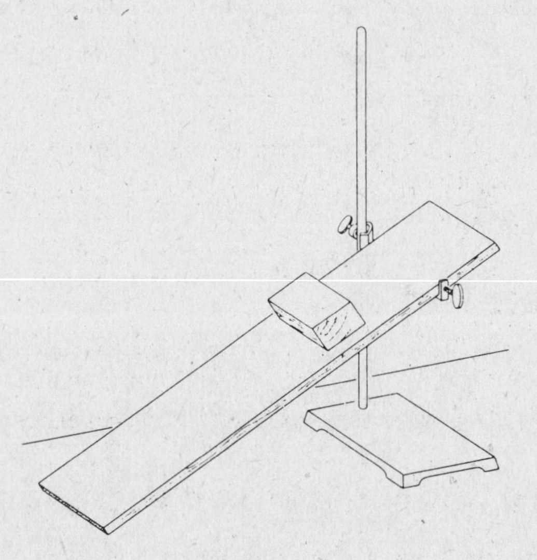

Figure 9-1

PROCEDURE:

1. Sliding friction

Place the plane in the position shown in Fig. 9-1. The slope should be gentle enough so that the wooden block, when placed upon the plane, will not slide down the plane. Now increase the pitch of the plane gradually until a grade is reached at which the block will slide slowly down the plane with uniform speed when it is given a gentle push. Record this angle as θ. At this angle, the force of friction, F_f, is equal in magnitude and opposite in direction to F_p. Using the weight of the block as F_w, construct a vector diagram and find F_p and F_N. Since F_p equals F_f in magnitude in this trial, you can calculate the coefficient of friction, μ.

Use a vector diagram to find the value of F_p when θ is greater than it is in Trial 1. How does F_p compare with F_f in magnitude in this case? What is the magnitude and direction of a force, F_a, that must be applied to the block so that it will slide down the incline at a constant rate? Make a similar study of the case in which θ is less than it is in Trial 1.

Using the same procedure as in Trial 1, find the coefficient of friction when a paraffin-coated block is

DATA

TRIAL	θ	F_w	F_p	F_N	F_f	μ
1						
2						
3						
4						

used. Record your findings and calculations in the data table under Trial 2.

2. Rolling friction

The metal car is used to find the coefficient of rolling friction. First place the plane flat on the table. Then raise one end slightly by putting a pencil under it. Gradually increase the steepness until the car continues to roll slowly down the plane, when pushed gently to start it. Calculate μ as before. Record the data under Trial 3.

Repeat the experiment, but lock the wheels so they will slide. A folded piece of paper, wedged between the wheels and the frame of the car, may be used as a brake. Calculate the coefficient of friction and record the data under Trial 4.

QUESTIONS: Your answers should be complete statements.

1. What information does this experiment furnish concerning the difference between starting and moving friction?

....

.... 1

2. How is the coefficient of friction affected by the orientation of the block? Verify your answer by experimentation.

....

....

....

.... 2

3. Does the coefficient of sliding friction depend on the kinds of material in contact? Support your answer with data from this experiment.

....

....

.... 3

4. Does this experiment supply any evidence concerning the cause of friction? Elaborate.

....

....

....

....

.... 4

5. How is the coefficient of friction related to the angle of the incline?

....

....

....

....

....

....

.... 5

NAME PERIOD DATE INSTRUCTOR'S COMMENTS

10 EXPERIMENT Center of Gravity and Equilibrium

PURPOSE: (1) To observe how the weight of the beam on which the forces act behaves like a force concentrated at the center of gravity of the beam. (2) To determine the conditions for equilibrium of several parallel forces.

APPARATUS: Meter stick and knife-edge support; slotted weights; 5 weight hangers; 5 meter-stick clamps; platform balance. (Twine and interhooking weights may be used in place of the slotted weights, weight hangers, and meter-stick clamps.)

INTRODUCTION: Two conditions must be met in order for equilibrium to be attained with several parallel forces. (1) The sum of the forces acting downward must equal the sum of the forces acting upward. (2) The sum of the clockwise torques must equal the sum of the counterclockwise torques. If the forces are acting on a rigid beam, the weight of the beam, concentrated at its center of gravity, acts as a force which must be considered when producing equilibrium.

PROCEDURE:

1. *Weight of meter stick as a force at its center of gravity*

Place the meter stick on the platform balance and determine its mass. Compute the weight of the stick in newtons. Record these values.

Locate the center of gravity of the meter stick by balancing it on a pencil or other narrow support. Read this location to three significant figures and record.

Support the meter stick at some location other than its center of gravity and bring it into balance with a single mass. (See Fig. 10-1.) Record the value of the mass, its weight, and its distance from the support. Be sure to include the mass of the clamp and hanger when you record the values for the added mass. Also, the positions of the added mass and of the support for the meter stick in Fig. 10-1 are meant to be suggestions. Different positions may have to be used in order to obtain satisfactory data. The masses and distances should be measured as precisely as possible.

Calculate the force required at the center of gravity to produce rotational equilibrium. Compare this force with the measured weight of the meter stick and record the difference as your error.

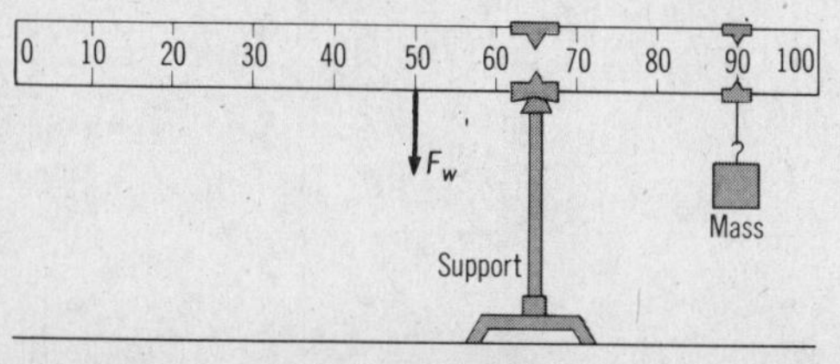

Figure 10-1

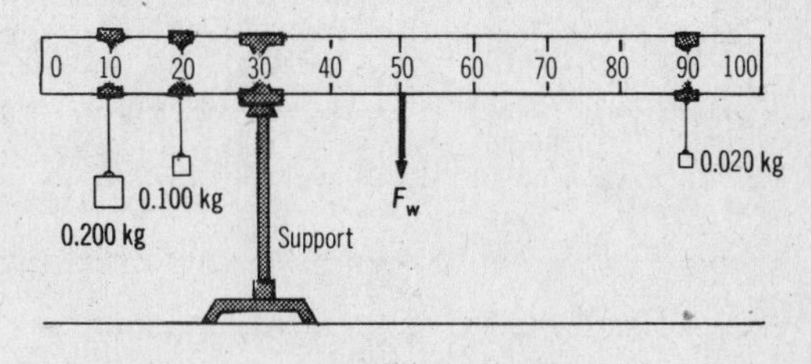

Figure 10-2

DATA

Position of center of gravity	m
Position of meter stick support	m
Mass required for equilibrium	kg
Weight of required mass	n
Position of required mass	m
Length of torque arm for required mass	m
Torque producing equilibrium	mn
Length of torque arm for weight of meter stick	m
Weight of meter stick (experimental)	n
Weight of meter stick (actual)	n
Absolute error	n
Relative error	%

2. *Equilibrium of several parallel forces*

Support the meter stick at a point other than its center of gravity. As a suggestion, the meter stick in Fig. 10-2 is supported at the 0.30-m mark. At the 0.10-m mark, hang a 0.200-kg mass; at the 0.20-m mark, a 0.100-kg mass; and at the 0.90-m mark, a 0.020-kg mass. Find where a 0.050-kg mass must be hung to produce rotational equilibrium. (This mass and its position are not shown in Fig. 10-2.)

After equilibrium has been attained, convert all the forces into force equivalents. Make sure that you include any clamps and weight hangers. Set up a suitable data table and record all torque-producing forces and their respective torque arms. Don't forget the weight of the meter stick and its torque arm in the data. Identify each torque as clockwise or counterclockwise. Enter the sum of the clockwise and counter-clockwise torques and record the difference as the experimental deviation.

QUESTIONS: Your answers should be complete statements.

1. Why isn't the center of gravity of the meter stick necessarily at exactly the 0.50-m mark?

.... 1

2. Are the results affected if the meter stick comes to rest in a position other than horizontal? Explain and verify experimentally.

....

....

....

....

....

.... 2

3. How can the weight of the meter stick be cancelled out in an experiment?

....

....

.... 3

4. How can this experiment be used to find the weight of a rod without using a balance?

....

....

....

....

....

....

.... 4

5. Where is the center of gravity of a ring? How can it be located?

....

....

....

.... 5

NAME PERIOD DATE INSTRUCTOR'S COMMENTS

11 EXPERIMENT Parallel Forces

PURPOSE: (1) To show that the equilibrant of two parallel forces is equal to their sum. (2) To show that the equilibrant of two parallel forces must be between the two forces. (3) To show that the two forces are inversely proportional to the length of the arms upon which they act.

APPARATUS: Meter stick; 2 spring balances, 20-n capacity; 3 meter-stick clamps; weight hanger and slotted weights, or interhooking weights.

INTRODUCTION: Parallel forces are those which act in the same direction or in opposite directions. The resultant of parallel forces in the same direction is equal to the sum of the separate forces. It acts in the same direction they do. It must be applied between them. The equilibrant force is equal to the sum of the separate parallel forces, but acts in the opposite direction. When two parallel forces are held in equilibrium by a third force applied at the appropriate point, the clockwise and counterclockwise torques are equal. It follows then that the two forces are inversely proportional to the lengths of the arms on which they act.

PROCEDURE: Hang the two spring balances from the laboratory table supports so that they are 0.80 m apart. Place three clamps on the meter stick, the one in the middle loop downward, the other two loops upward. For your first trial place one clamp at the 0.10-m mark of the meter stick, another at the 0.90-m mark, and the third clamp C between them at the 0.50-m mark. (Assume that the center of gravity of the meter stick is at or very near the 0.50-m mark.) Now when the meter stick is hung on the hooks of the balances, the two balances will hang parallel to each other.

Take the *initial readings* of both balances with the meter stick in position, and with the third clamp fastened at *C*. Then a weight of 10 n or more, including the weight of the hanger, is applied to clamp *C*. Take the readings of both balances, *A* and *B*, and record them in your data table under the headings marked *final reading*. To find the *true reading*, subtract the initial readings in each case from the final readings. Compute the sum of the true readings.

In the data table, record the total weight F_w (both weights and hanger) as measured by a spring balance, and the distances *AC* and *BC*. Find the difference between the sum of the true balance readings and the total weight F_w. This represents the experimental error in determining the values of the upward and downward forces.

Consider the point *C* a fixed point at the axis of rotation, about which forces tend to turn the meter stick. The true force at *A* acts on the arm *AC* and tends to produce *clockwise rotation.* Calculate the product for each trial and record it in the proper

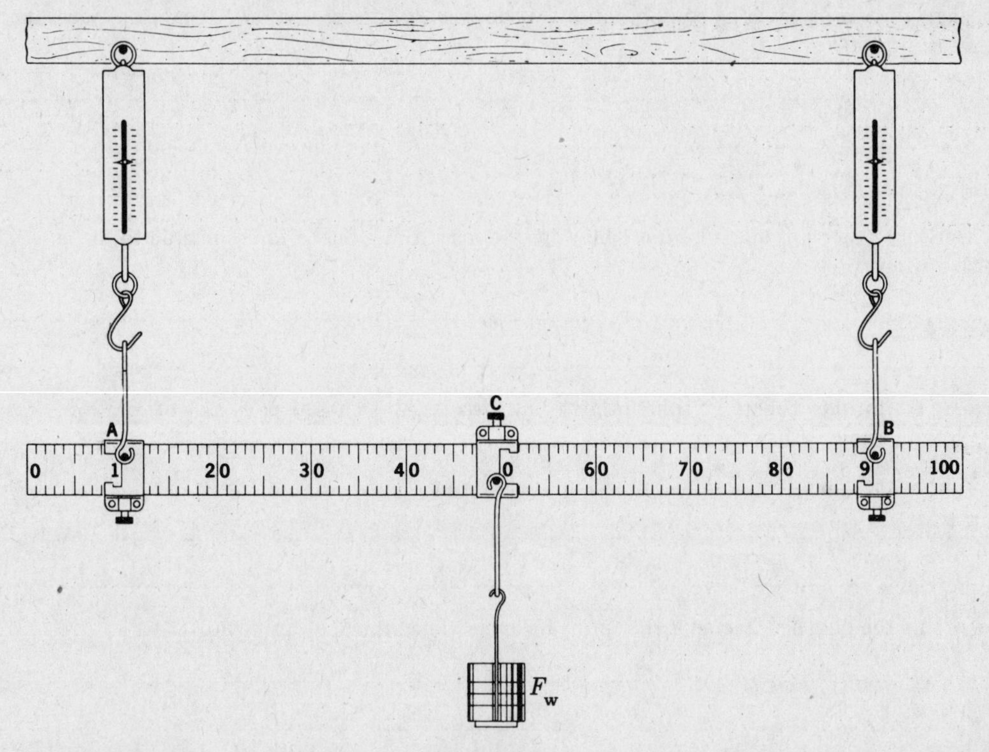

Figure 11-1

torque column. The true force B acts on the arm BC and tends to produce *counterclockwise rotation.* The difference between the two torques is the torque error.

For the second and third trials, move the clamp C to different positions — the 0.25-m and 0.60-m marks, for example.

DATA

TRIAL	Balance A			Balance B		
	Initial reading (n)	*Final reading* (n)	*True reading* (n)	*Initial reading* (n)	*Final reading* (n)	*True reading* (n)
1						
2						
3						

DATA

Sum of true readings (n)	*Total weight* F_w (n)	*Error* (n)	*Distance AC* (m)	*Distance BC* (m)	*Clockwise torque* (mn)	*Counter-clockwise torque* (mn)	*Torque error* (mn)

QUESTIONS: Your answers should be complete statements.

1. If a single force is to act as an equilibrant of two downward forces, what two conditions must the single upward force fulfill?

....

....

.... 1

2. When the meter stick is in equilibrium, how are the magnitudes of the forces related to the lengths of the arms on which they act?

....

.... 2

3. How are the torques affected if some other point (besides C) is chosen as an axis of rotation. Verify your answer with a calculation.

....

....

.... 3

4. How would the data be affected if the spring balances were attached upside down?

....

....

.... 4

NAME PERIOD DATE INSTRUCTOR'S COMMENTS

12 EXPERIMENT Centripetal Force - I

PURPOSE: (1) To study the nature of centripetal force. (2) To measure the relationship among centripetal force, mass, and velocity.

APPARATUS: Glass tube 15 cm long and at least 1 cm in outside diameter; masking tape; nylon fishline; rubber stopper; set of masses; paper clip; clock or watch with second hand, or stopwatch.

INTRODUCTION: An object moving with changing speed in the same direction is undergoing acceleration. If an object moves with constant speed but in changing directions, it is also undergoing acceleration. Both types of acceleration require forces. A change in direction is called centripetal acceleration, and the force producing it is called centripetal force.

The equation relating centripetal force, mass, and velocity is

$$F_c = \frac{mv^2}{r}$$

where F_c is the centripetal force, m is the mass of the moving object, v is its velocity, and r is the radius of the orbit of the object. In this experiment, each of the factors in this formula will be varied as an object is whirled on the end of a string. Centripetal force will be supplied by masses tied to a string that passes through a vertical tube. (See Fig. 12-1.) The effect of gravity on the whirling object is offset by the resulting angle of the string with the horizontal. Thus, r can be taken as the length of the string between the tube and the object (even though the string is not perpendicular to the tube) without introducing a significant error.

Of course, it is possible for an object to have accelerations of speed and direction at the same time. This is the case with the planets, which move around the sun in elliptic orbits. The analysis of this kind of motion requires more advanced mathematical techniques.

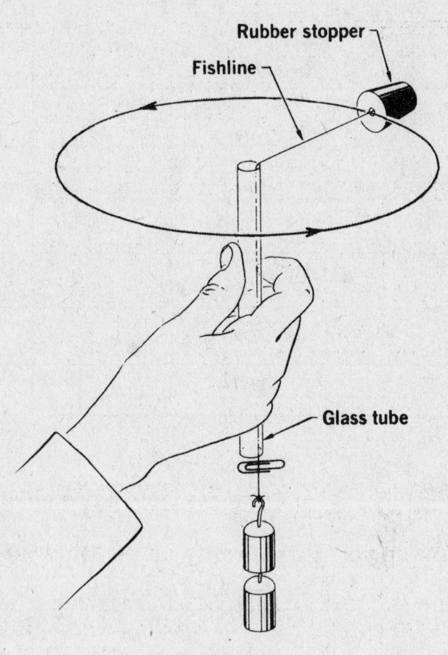

Figure 12-1

PROCEDURE: Cut a 1.5 m length of fishline and fasten one end of it securely to the rubber stopper. Fire polish the ends of the glass tube if they are not smooth. Cover the glass tube with masking tape to prevent breakage during the experiment. Feed the fishline through the glass tube. Fasten a 50-g mass to the free end of the fishline. Adjust the line so that there is about 1.0 m of line between the top of the tube and the stopper. Fasten a paper clip to the fishline just below the bottom of the tube.

Support the 50-g mass with one hand and hold the glass tube in the other. Whirl the rubber stopper by revolving the tube. Slowly release the 50-g mass and adjust the speed of revolution so that the paper clip stays just below the bottom of the tube. Make several trial runs before recording any data.

When you have learned how to keep the rotational velocity and position of the paper clip constant, have a fellow student measure the speed of revolution by measuring the time required for 30 or 40 revolutions and calculating the time of a single revolution from this data Record the centripetal force (weight of the 50-g mass) and velocity (calculated from the radius and period of revolution).

Repeat the procedure, but increase the centripetal force by adding 20 g to the 50-g mass. Record the resulting data. Run a third trial, using a still larger centripetal force.

GRAPH: Plot a graph of the data, using F_c as abscissas and v as ordinates. Plot a second graph of the same data, using F_c as abscissas and v^2 as ordinates.

If time permits, run a series of trials in which (1) the radius is kept constant, but the mass of the whirling object changes and (2) the mass of the object is kept constant, but the radius changes. Record the appropriate data and plot graphs to show the relationships.

DATA

TRIAL	m (kg)	F_c (n)	r (m)	*Period* (sec)	v (m/sec)	v^2 $(m/sec)^2$
1						
2						
3						
4						
5						
6						
7						
8						
9						

QUESTIONS: Your answers should be complete statements.

1. How do your graphs verify the centripetal force formula? Explain the shapes of the two curves.

....

....

....

....

....

.... 1

2. Explain two ways in which the effect of a change in radius in this experiment can be compensated for. Verify by experiment.

....

....

.... 2

3. At what point in the swing of a pendulum does the string exert the greatest centripetal force on the bob? Why?

....

....

....

.... 3

4. Describe the motion of the stopper if the string is cut. Explain.

....

....

....

....

....

.... 4

5. A man with long arms and one with short arms are swinging identical pails of water in vertical circles. Which man must swing his pail faster in order to keep the water from falling out? Why?

....

....

....

....

....

....

.... 5

NAME PERIOD DATE INSTRUCTOR'S COMMENTS

13 EXPERIMENT Centripetal Force - II

PURPOSE: To measure the relationship among centripetal force, mass, and velocity.

APPARATUS: Air table, strobe light, set of pucks designed for the air table, Polaroid camera and camera stand; spring scale; metric ruler; platform balance.

INTRODUCTION: In Experiment 12, the accuracy of the results in the study of centripetal force was affected by friction and by human error in the measurement of time intervals. In this experiment, a special apparatus is used in which an upward flow of air provides a virtually frictionless surface for an object whose curvilinear motion will be studied. Also, human error is minimized by the use of photographs.

PROCEDURE: Set up the air table (as shown in Fig. 13-1), camera stand, and Polaroid camera. (Tables other than the one shown in the figure may involve different setups, and the instructions for the specific table should be carefully studied.)

Turn on the air flow and attach a puck to the center post with the bearing and spring supplied for this purpose. Practice placing the puck in circular motion on the table. Then open the shutter of the Polaroid camera and turn on the strobe light for several short bursts. (Be careful not to overexpose the photo. Three bursts of 3 or 4 flashes each, with about 5 seconds between bursts, will give good results.)

Measure and record the mass of the puck on the platform balance. On the photo, measure the distance between two adjacent images of the puck and calculate its velocity using this distance and the frequency of the strobe light. Make several such calculations and average the results. (Remember to take into consideration the size reduction on the photo.)

Repeat the experiment, using different radii and pucks with different masses.

Figure 13-1

The equation for centripetal force is

$$F_c = \frac{mv^2}{r}$$

where F_c is the centripetal force, m is the mass of the moving puck, v is its velocity, and r is the radius of revolution.

Calculate F_c for each trial, using the data for m, v, and r. Record these as the calculated values for F_c.

With the spring scale, measure the force required to stretch the connecting spring to the length to which it was extended during each trial. Record these values as the measured centripetal forces in the data table.

DATA

TRIAL	m (kg)	v (m/sec)	r (m)	F_c *calculated* (n)	*measured* (n)	*Difference* (n)
1						
2						
3						
4						

QUESTIONS: Your answers should be complete statements.

1. Why is it important that the puck should move in a circular path in this experiment, rather than in an elliptical path?

....

.... 1

2. Which value for F_c do you think is more reliable, the calculated one or the measured one? Why?

....

.... 2

3. Is a circular path more stable than a non-circular one? Verify your answer by experimentation.

.... 3

4. List the possible sources of error in this experiment.

....

.... 4

NAME PERIOD DATE INSTRUCTOR'S COMMENTS

14 EXPERIMENT The Pendulum

PURPOSE: (1) To determine the factors that affect the period of a pendulum. (2) To use a pendulum to determine the local value of the acceleration of gravity.

APPARATUS: Wooden bob; metal bob; pendulum clamp; meter stick; strong linen thread; stop watch; protractor.

INTRODUCTION: A simple pendulum consists of a small dense weight or bob suspended by a nearly weightless cord from a point about which it can vibrate freely. Such a pendulum is shown in Fig. 14-1. The point about which the pendulum swings, *S*, is called the *center of suspension.* As the pendulum swings from *A* to *B* and back to *A* again, it makes a *complete vibration*, or *cycle.* The time required for a cycle is the *period* of the pendulum. The number of cycles a pendulum makes per second is its *frequency.* The *displacement* of the pendulum is its varying distance from *C.* The arc *AC*, representing the maximum displacement of the bob from *C* is the *amplitude* of vibration. The *length* of a simple pendulum is measured from the center of suspension to the center of gravity of the pendulum bob.

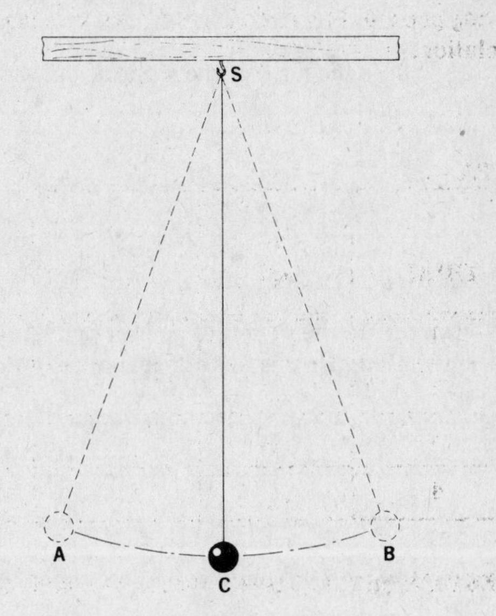

Figure 14-1

PROCEDURE:

1. Factors affecting the period

Conduct a series of trials in which you investigate the effect of the mass and length of a pendulum on its period of vibration. In one set of trials, keep the length constant while using bobs of different materials. In another set, use the same metal bob but change the length of the supporting thread.

In each trial, the amplitude should be between 5° and 10°. To measure the frequency, measure the time required for 20 vibrations to the precision that the stopwatch permits. Obtain three readings for each setup and average the results. Compute the frequency and period from this average and record them in the data table.

2. Value of g

The equation for the period of a pendulum is

$$T = 2\pi\sqrt{\frac{\ell}{g}}$$

where T is the period of the pendulum, ℓ is the length, and g is the acceleration of gravity. Solve this equation for g. Then use the data for each.

GRAPH: Plot a graph of the data, using the periods as abscissas and the lengths as ordinates. Plot a second graph of the same data, using the periods as abscissas and the square roots of the lengths as ordinates.

TRIAL	*Material*	*Length* (m)	*Frequency* (sec^{-1})	*Period* (sec)	*Amplitude* (°)	*g* (m/sec^2)	*Absolute error* (m/sec^2)	*Relative error* (%)
1								
2								
3								
5								
6								

QUESTIONS: Your answers should be complete statements.

1. How is the period of a pendulum affected by very large amplitudes? Verify your answer by experimentation.

....

.... 1

2. How does the mass of the string affect the period of the pendulum?

....

....

....

.... 2

3. How would the period of an iron pendulum be affected by a magnet held under it? Verify this by experimentation. How is this effect similar to the effect of gravity. How is it different?

....

....

....

....

....

....

.... 3

4. How could a pendulum be used as an altitude indicator? What are its drawbacks for this purpose?

....

....

....

....

.... 4

5. Would a simple pendulum work in an orbiting spacecraft?

....

....

.... 5

NAME PERIOD DATE INSTRUCTOR'S COMMENTS

15 EXPERIMENT Efficiency of Machines

PURPOSE: To measure the efficiency of an inclined plane and a pulley system.

APPARATUS: Inclined plane and car; various pulleys; string; set of weights.

INTRODUCTION: Machines are used to make work more convenient by multiplying force at the expense of speed, or vice versa. A machine does not multiply work, however. The work output of a machine is never greater than the work input. This would contradict the law of conservation of energy. In fact, the useful work output is always less than the work input because of the force of friction. The ratio of the useful work output to the work input is called the *efficiency* of the machine. Efficiency is usually expressed as a percentage.

In this experiment, the efficiency of two machines will be measured. (It may be necessary to schedule two laboratory periods for the experiment.) In each case, the work output is the product of the weight of the object being raised, F_w, and the height through which it is raised, h. The work input is the product of the force exerted on the machine, F_a, and the distance through which it acts, s. The efficiency is the ratio of these products, or

$$\text{Efficiency} = \frac{F_w h}{F_a}$$

The efficiency is the percentage ratio of these products, or

$$\text{Efficiency} = \frac{F_w h}{F_a s} \times 100\%$$

PROCEDURE:

1. Inclined plane

Set up the inclined plane as shown in Fig. 15-1 The angle of the plane is arbitrary, but is should be kept constant during this part of the experiment. Place a 200-g mass in the car and find the mass that will allow the car to move up the plane with the same velocity once it has been started. Compute the weights of the masses and record them in the data table. Also record the distances. H is the *vertical* distance through which F_w moves while F_a moves through the distance, s. Make several more trials, each time increasing the mass in the car. Record all necessary data.

2. Pulley

Set up a pulley system like the one shown in Fig. 15-2. Starting with a 500-g mass, find the mass that will make the system move up or down with the same velocity once it has been started. Repeat the procedure with different pulley systems (5 pulleys, 6 pulleys, etc.). Record all data. Be sure to compute the weights of the masses before recording. As with the inclined plane, F_w moves a vertical distance, h, while F_a moves a distance, s.

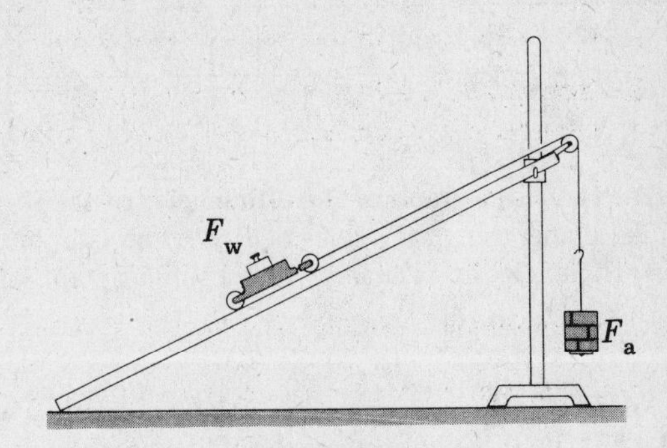

Figure 15-1

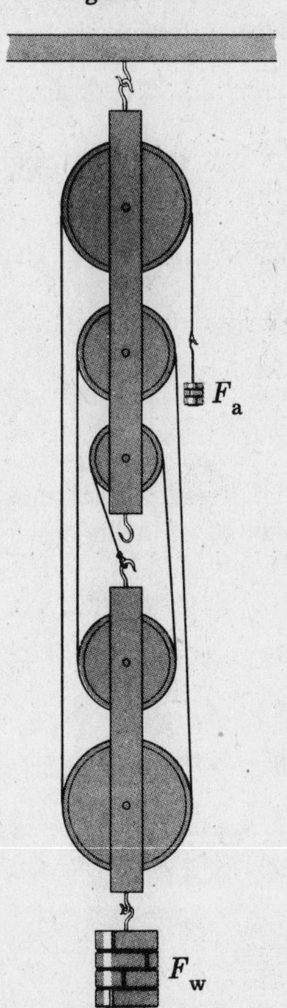

Figure 15-2

CALCULATIONS: Compute the efficiency for each trial in the experiment, using the equation in the Introduction. Record the efficiencies in the table.

DATA

TRIAL	MACHINE	F_w (n)	h (m)	F_a (n)	s (m)	*Work output* (j)	*Work input* (j)	*Efficiency*
1								
2								
3								
4								
5								
6								

GRAPH: Plot a graph of the efficiencies of the inclined plane, using the values of F_w as abscissas and the efficiencies as ordinates. Plot a similar graph of the efficiencies of the pulley systems, using the number of strands supporting the weight as abscissas and the efficiencies as ordinates.

QUESTIONS: Your answers should be complete statements.

1. Is the efficiency formula in harmony with the law of conservation of energy? Explain.

....

....

.... 1

2. Explain the shapes of the curves in the two graphs in this experiment.

....

....

.... 2

3. Compare your results with those of your classmates to see whether there is a relationship between efficiency and the angle of an inclined plane.

....

....

.... 3

4. How could the efficiency in this experiment be increased? Try it.

....

....

.... 4

NAME PERIOD DATE INSTRUCTOR'S COMMENTS

16 EXPERIMENT Conservation of Momentum - I

PURPOSE: (1) To study the quantity of motion described by the product of mass and velocity, which we call momentum. (2) To verify the law of the conservation of momentum.

APPARATUS: Recording timer; timing tape; connecting wire; dry cell or other DC power source; two carts, one with a spring mechanism; platform balance; set of masses.

INTRODUCTION: The product of the mass of a moving object and its velocity is called its *momentum.* A bullet having a small mass and a high velocity may have the same momentum as a truck with a large mass and a very small velocity.

Newton's law of reaction states that every force is accompanied by an equal and opposite force or reaction. Thus, when a rifle is fired, the force on the bullet is accompanied by an equal force or "kick" on the gun. In other words, the momentum of the bullet is equal to the momentum of the gun (and the person holding it).

In this experiment, the gun and bullet will be simulated by two carts with unequal masses. One of the carts contains a spring mechanism to provide the equal and opposite forces for the experiment. The velocities of the carts will be measured with a recording timer. (Experiment 3 contains a more complete description of the timer and its operation.)

The law of the conservation of momentum will also be studied by changing the mass of one of the carts while it is moving. The resulting change in the velocity of the cart will be measured and checked against the equation

$$m_1 v_1 = m_2 v_2$$

PROCEDURE: Set up the carts and timer as shown in Fig. 16-1. Since both tapes pass through the same timer, place two carbon paper discs back to back between the paper tapes. Study the cart with the spring mechanism so that you will know how to cock and release the spring.

Determine the mass of an unloaded cart and record. Load the second cart with a 1-kg mass. Measure the mass of this loaded cart and record. Then cock the spring mechanism and position the carts. Make sure that there are no obstructions in the paths of the carts. Turn on the timer and trip the spring mechanism. When the carts have stopped moving, turn off the timer and remove the tapes. Label each one to correspond to the cart to which it was attached.

Examine the tapes and locate the region where the spacing between the dots is uniform. This region will begin a short distance from the starting point of the tapes. Count off the same number of intervals (at least 10) on each tape and measure the total distance. Record as s_a and s_b in the data table.

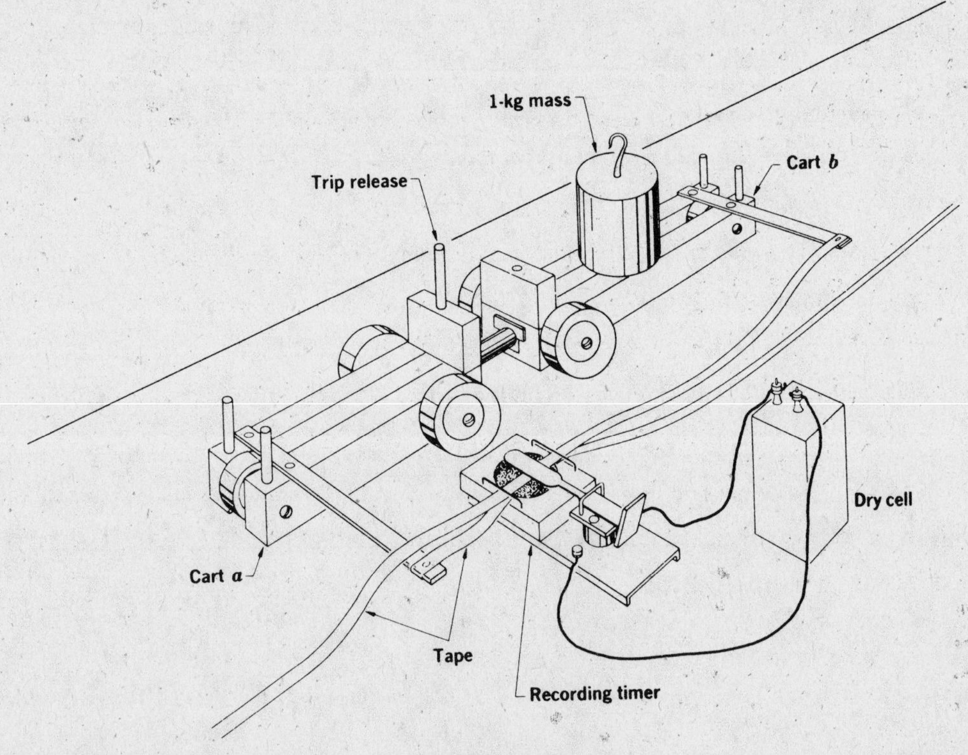

Figure 16-1

Determine the period of the timer as outlined in Experiment 3, or obtain the period from your instructor. Use the period to calculate the velocity of each cart and record the answer. Calculate the momentum of each cart. Compare the two momentums and record the difference.

Repeat the experiment two more times, using different combinations of masses. Record all data and calculate the momentums as before.

If time permits, change the mass of a cart en route and measure the resulting change in velocity. A good way to do this is to use an empty cart and place a 1-kg mass on it shortly after the spring is tripped. Measure the velocity of the cart before and after the mass was added and calculate the momentum for each case. Record your data.

Another way to vary the mass of the cart is to remove a 1-kg mass while the cart is in motion. Tie a thread to the mass before the cart starts to move in order to remove this mass with a minimum of disturbance. Try to think of other variations of this experiment and compare the results with the one described.

DATA

	Distance		*Mass*		*Velocity*		*Momentum*		*Difference*
TRIAL	s_a (m)	s_b (m)	m_a (kg)	m_b (kg)	v_a (m/sec)	v_b (m/sec)	$m_a v_a$ (kg m/sec)	$m_b v_b$	(kg m/sec)
1									
2									
3									
4									

QUESTIONS: Your answers should be complete statements.

1. What role does friction play in this experiment? How do the paper tapes reveal this role?

....

....

....

....

....

.... 1

2. What role does inertia play? How do the tapes show this?

....

....

....

.... 2

3. Show that equal and opposite forces are exerted on the carts.

....

....

....

.... 3

NAME PERIOD DATE INSTRUCTOR'S COMMENTS

17 EXPERIMENT Conservation of Momentum - II

PURPOSE: To apply the law of conservation of momentum to elastic, two-dimensional collisions.

APPARATUS: Air table; strobe light; set of pucks designed for the air table; Polaroid camera and camera stand.

INTRODUCTION: In Experiment 16, the law of the conservation of momentum was seen to hold true for objects moving along the same straight line. This is an example of motion in one dimension. In this experiment, the same law will be applied to motion in two dimensions. A moving object will be made to collide with a stationary one in such a way that both objects will move away from the collision in different directions. Vector diagrams will then be used to compare the momenta of the objects before and after the collision.

The reduction of friction is an important factor in conservation of momentum experiments, since friction is a dissipative and not a conservative force. In Experiment 16, this was accomplished by the use of wheels on the moving objects. In this experiment, a special apparatus is used in which an upward flow of air provides a virtually frictionless surface for the moving objects. The round pucks used in the experiment are also designed to make the collisions as elastic as possible.

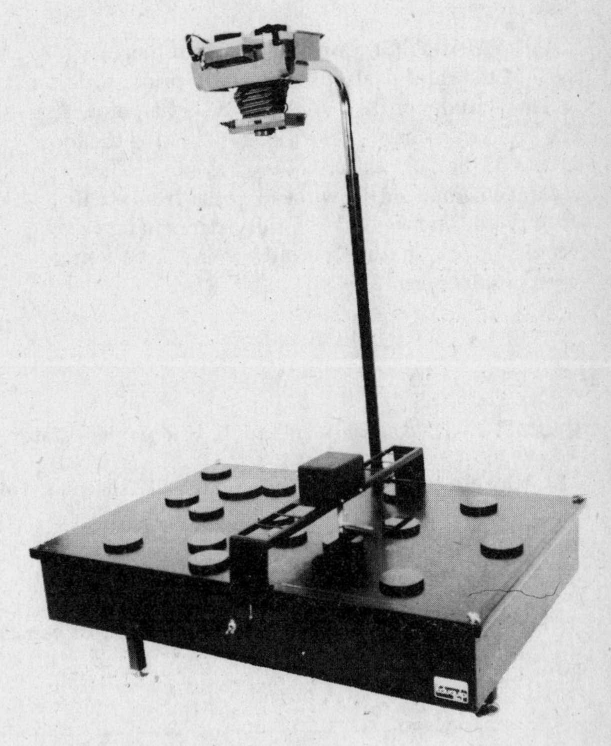

Courtesy of Eduquip, Inc.

Figure 17-1

PROCEDURE: Set up the air table, camera stand, and Polaroid camera as shown in Fig. 17-1. (Tables other than the one shown in the figure may involve different setups, and the instructions for the specific table should be carefully studied.)

Turn on the air flow and place a puck in the center of the table. Propel a second puck toward the stationary one from the side of the table. Try to hit the stationary puck slightly off-center, so that both pucks will keep moving after the collision. Practice this procedure until you can control the angle between the paths of the two pucks after impact with a fair degree of precision.

Now turn on the strobe light and open the shutter of the Polaroid camera. Repeat the collision. Close the camera shutter and turn off the strobe when one of the pucks reaches the side of the air table. Develop the film to see whether you have obtained a good multiple exposure of the collision. The photo should show the positions of the pucks before and after impact as a series of white dots. These were made as the strobe light bounced from the small reflectors in the centers of the pucks.

Repeat the procedure, but this time double the mass of the incident puck by stacking two similar pucks on top of each other and taping them together. For the third trial, triple the mass of the incident puck and double the mass of the stationary one, or use some other variation of the masses.

Record the masses of the incident and target pucks, m_i and m_t, for each trial in the data table in terms of unit pucks. In each photo, measure the distance between dots before and after collision. Record these distances in the data table as the speed of each puck before and after collision, v_i, v_i', v_t, and v_t', in terms of mm/dot. Then calculate the momentum of each puck before and after collision, m_iv_i, m_iv_i', and m_tv_t'.

DATA

TRIAL	m_i (units)	m_t (units)	v_i (mm)	v_t (mm)	v_i' (mm)	v_t' (mm)	m_iv_i (u mm)	m_tv_t (u mm)	m_iv_i' (u mm)	m_tv_t' (u mm)
1										
2										
3										

DATA ANALYSIS: Construct a vector diagram for each of the three trials, using an appropriate scale for the magnitudes of the momenta. Select a point, P, as the point of collision. See Fig. 17-2. Using the momenta of the pucks after collision as sides, construct a parallelogram and draw the diagonal from P. Record the difference between this diagonal and the vector for m_iv_i, both in magnitude and direction, as your absolute error.

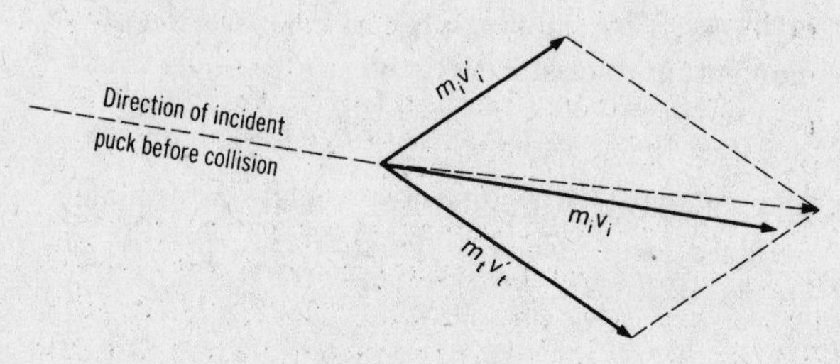

Figure 17-2

QUESTIONS: Your answers should be complete statements.

1. Why are the dots in the photos equally spaced before and after the collisions in this experiment?

....

.... 1

2. How do the vector diagrams verify the law of conservation of momentum?

....

....

....

....

.... 2

3. What are some sources of error in this experiment?

....

....

.... 3

4. How would the results of this experiment be affected if the pucks were less elastic?

....

.... 4

NAME PERIOD DATE INSTRUCTOR'S COMMENTS

18 EXPERIMENT Hooke's Law

PURPOSE: To show that the relative distortion of matter is directly proportional to the force per unit area applied to it, provided the limit of perfect elasticity is not exceeded.

APPARATUS: Hooke's law apparatus; set of masses.

INTRODUCTION: The force per unit area tending to distort the body on which it acts is called *stress;* the relative amount of distortion produced is called *strain.* As we use an ordinary spring balance, we observe the the relative amount the spring is stretched (strain) is directly proportional to the weight applied (stress). If the spring is perfectly elastic, it will return to its original form when the stress is removed. These facts are summarized in Hooke's law, which states that within the limits of perfect elasticity, strain is directly proportional to stress. While our experiment is concerned only with elasticity of extension, Hooke's law also applies to compression, bending, and twisting.

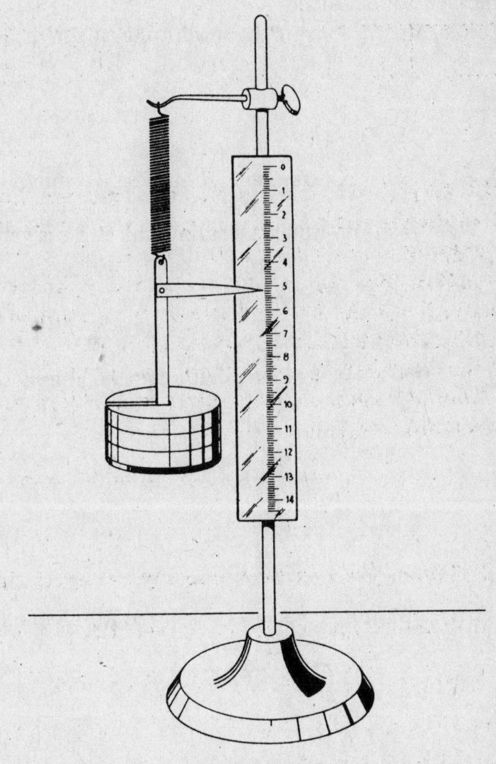

Figure 18-1

PROCEDURE: The mirror scale of the Hooke's law apparatus shown in Fig. 18-1 makes it possible to take readings with great accuracy. It consists of a long narrow mirror placed beside a graduated scale. In making a reading, the eye should be on a line with a point on the spring and the image of that point in the mirror. The scale is read at the point where the line touches the mirror edge of the scale.

Adjust the position of the scale or the weight hanger so that the first reading with the scale pan alone falls at the 0.0-cm mark on the scale. Then place successively larger masses on the hanger and measure the amount of elongation from the scale. The size of the

DATA

TRIAL	*Mass* (g)	*Weight* (n)	*Elongation* (cm)	*Elongation* (m)	*Elongation per newton* (m/n)
1					
2					
3					
4					
5					
6					
7					
8					
9					
10					

masses to be used will depend on the stiffness of the spring. Consult your instructor to learn what masses to use. For example, successive readings may be taken using first a 100-g mass, a 200-g mass, and masses of 300 g, 400 g, 500 g, 600 g, 700 g, and 800 g in order. Make sure that the weight limit of the apparatus is not exceeded.

For each trial, convert the magnitude of the mass used to its corresponding weight in newtons, using the local value for the acceleration of gravity. Convert the elongations to meters. Calculate the elongation per newton.

GRAPH: Plot a graph of the data obtained in this experiment, using the weights as abscissas and the total elongations as ordinates.

QUESTIONS: Your answers should be complete statements.

1. What relationship exists among the various values of the elongation per newton?

....

.... 1

2. Describe an experiment that might be conducted to show that Hooke's law also applies to bending.

....

....

....

.... 2

3. Why are spring balances considered unreliable for very precise work?

....

....

....

.... 3

4. In what way is a Hooke's law apparatus similar to a pendulum?

....

....

....

.... 4

5. Would this experiment yield the same results on the moon? Explain.

....

....

....

.... 5

NAME PERIOD DATE INSTRUCTOR'S COMMENTS

19 EXPERIMENT Elastic Potential Energy

PURPOSE: To verify the law of conservation of potential energy in an oscillating spring.

APPARATUS: Ring stand; three 4-inch rings; spring; spring support; set of masses; ruler.

INTRODUCTION: In a swinging pendulum, there is a continuous interchange between potential energy and kinetic energy. The potential energy is at a maximum when the bob is at the highest point in its swing. The kinetic energy is at a maximum when the bob is at its lowest point and is moving with maximum velocity. Except for the effect of friction, the potential energy at the top of the swing is equal to the kinetic energy at the bottom, in accordance with the law of conservation of energy.

The apparatus in Fig. 19-1 is similar to a pendulum in some respects. An oscillating mass falls under the influence of gravity, as in the pendulum. However, the oscillation is entirely vertical rather than in an arc. Furthermore, the mass oscillates about a midpoint where the tension in the spring is equal to the force of gravitation on the mass. At the top of the oscillation, the mass has maximum gravitational potential energy. At the bottom of the oscillation, the spring has maximum elastic potential energy. In this experiment, we will compare these two forms of potential energy as a test of the law of conservation of energy.

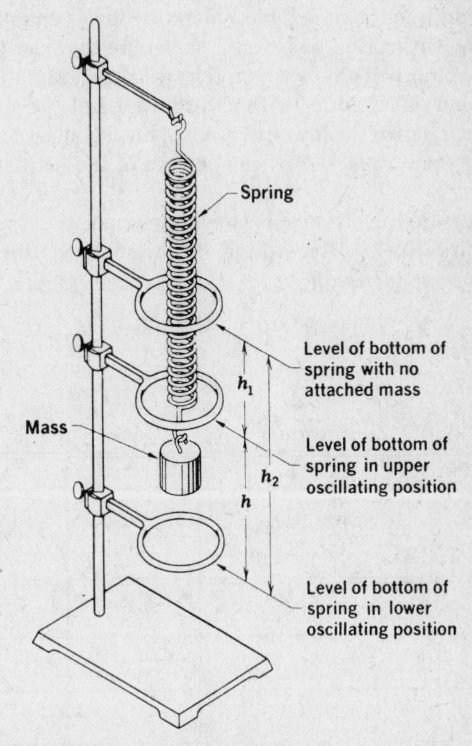

Figure 19-1

PROCEDURE:

1. Elastic constant of the spring

To calculate the potential energy stored in a stretched spring, it is first necessary to find the force required to stretch the spring by a given amount. This is called the *elastic constant* of the spring. It is expressed in units of newtons per meter, n/m. The procedure for finding the elastic constant is similar to that used to verify Hooke's Law in Experiment 18.

Hang the spring from the spring support and mark the position of the bottom of the spring with a ring as in Fig. 19-1. Find the zero reading for the spring by hanging the largest mass on it that the spring will hold without beginning to stretch. Now add about 50 g and again mark the position of the bottom of the spring with a ring. Convert the mass units to equivalent force units and record it in the data table along with the elongation produced. Repeat the procedure with additional masses until you reach the capacity of the spring as specified by your instructor. Record the force and elongation for each trial and find the average value of the elastic constant. In each

DATA Zero reading ________ g.

TRIAL	*Mass* (kg)	*Force* (n)	*Elongation* (m)	*Elastic constant* (n/m)
1				
2				
3				
4				
5				
6				

case, remember to subtract the zero reading from the total mass before recording it in the data table.

2. Conservation of potential energy

Hang a mass from the spring so that it will stretch to about half of the maximum elongation recorded in Part 1. (Subtract the zero reading before recording this mass in the data table.) Mark the lower end of the spring with a ring, as before. Now raise the mass until it is about halfway between the two rings. Mark the bottom of the spring with a third ring. Let the mass fall and move the lower ring down until it marks the lowest point reached by the bottom of the oscillating spring.

Measure the distance between the upper two rings. Record it as h_1. Record the distance between the upper and lower ring as h_2.

Run several more trials, varying both the mass and the amplitude of oscillation.

CALCULATIONS: Compute the elastic potential energy of the spring for each trial with the equation

$$E_{p(elastic)} = ½k(h_2^2 - h_1^2)$$

where k is the elastic constant of the spring. Compute the gravitational potential energy for each trial with the equation

$$E_{p(gravitational)} = mgh$$

where $h = h_2 - h_1$. Compare the two values for potential energy and record the difference as your experimental error.

DATA Zero reading ________ g.

TRIAL	*Mass* (kg)	s_2 (m)	s_1 (m)	h (m)	E_p *elastic* (j)	E_p *grav.* (j)	*Deviation* (j)
1							
2							
3							
4							
5							
6							

QUESTIONS: Your answers should be complete statements.

1. When is the kinetic energy of the oscillating spring at a maximum? How can the experiment be modified to verify your answer?

....

....

....

....

....

....

.... 1

2. Using your textbook, explain the formula for elastic potential energy.

....

....

....

....

....

....

....

.... 2

3. What happens to the energy of the oscillating spring as the oscillations slow down and the mass comes to rest?

....

....

.... 3

4. How would the data change if this experiment were conducted on a high mountain? Would the experiment work in an orbiting space vehicle? Explain.

....

....

....

....

.... 4

NAME PERIOD DATE INSTRUCTOR'S COMMENTS

20 EXPERIMENT Specific Heat

PURPOSE: To determine the specific heat of one or more metals.

APPARATUS: Calorimeter; thermometer, Celsius -10° to 110°, graduated to 0.2°; stirring rod; steam boiler; tripod, for steam boiler; burner and rubber tubing; set of masses; platform balance; ice cubes; metal block (lead, aluminum, brass, and copper are satisfactory); magnifier; strong twine; barometer.

INTRODUCTION: The method of mixtures is used in determining the specific heat of a metal. A certain mass of water in a calorimeter is warmed a measured number of degrees by a solid of known mass as it cools through a measured number of degrees.

The heat gained or lost by a substance when it undergoes a change in temperature is calculated by

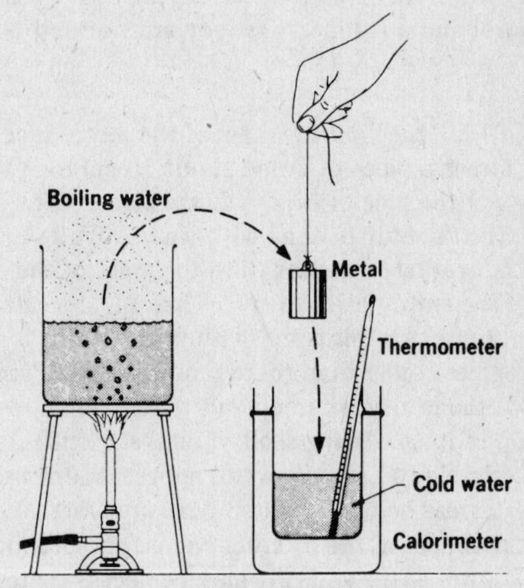

Figure 20-1

DATA

	TRIAL 1	TRIAL 2
1. Kind of metal	....	....
2. Mass of metal	....................g	....................g
3. Mass of calorimeter cup and stirrer	....................g	....................g
4. Mass of calorimeter cup, stirrer, and water	....................g	....................g
5. Mass of water	....................g	....................g
6. Specific heat of calorimeter	....................cal/gC°	....................cal/gC°
7. Temperature of solid, initial (temperature of boiling water)	....................°C	....................°C
8. Temperature of water and calorimeter, initial	....................°C	....................°C
9. Temperature of solid, water, and calorimeter, final	....................°C	....................°C
10. Temperature change of water and calorimeter	....................C°	....................C°
11. Calories gained by calorimeter cup and stirrer	....................cal	....................cal
12. Calories gained by water	....................cal	....................cal
13. Total calories gained	....................cal	....................cal
14. Calories lost by solid	....................cal	....................cal
15. Temperature change of solid	....................C°	....................C°
16. Specific heat of solid	....................cal/gC°	....................cal/gC°
17. Accepted value for specific heat of solid	....................cal/gC°	....................cal/gC°
18. Absolute error	....................cal/gC°	....................cal/gC°
19. Relative error	....................%	....................%

multiplying the mass of the substance by its change in temperature and by its specific heat. According to the law of heat exchange, the total amount of heat lost by a hot object equals the total amount of heat gained by the cold object with which it comes in contact. Consequently, in this experiment, the total heat lost by the solid on cooling equals the heat gained by the water and calorimeter as they are warmed.

PROCEDURE: Find the mass of the metal block. Then attach a piece of twine about 30 cm long to it and lower the block into the steam boiler. The boiler should be about half filled with water. While the water is heating to boiling, find the mass of the inner cup of the empty calorimeter. Then fill the calorimeter about two-thirds full with water which is several degrees colder than room temperature. Find the mass of the calorimeter cup and water. Then replace the cup in its insulating shell. For best results, it is also desirable to have the water approximately as many degrees below room temperature when starting the experiment as the mixture will be above room temperature at the end. (It may be necessary to use ice cubes to get the initial temperature low enough.) The temperature change of the water should be at least 10 C°.

Take the temperature of the boiling water with the thermometer. Since the solid is being heated in the boiling water, its original temperature is the temperature of the boiling water. Then stir the water in the calorimeter and take its temperature. When making these temperature readings, you should estimate to the nearest 0.1° by using a hand magnifier. Make sure the thermometer bulb is completely immersed in the liquid, and keep your line of sight at right angles to the stem of the thermometer. Considerable error may result from poorly taken temperature readings.

Lift the solid by means of the twine and hold it just above the water, *in the steam*, long enough to let the water which adheres to the solid evaporate. Then quickly transfer the solid to the calorimeter of cold water as shown in Fig. 20-1, and replace the cover. Agitate the solid in the calorimeter and then stir the water. The temperature of the mixture should be carefully read to the nearest tenth of a degree when it has risen to its highest point.

If your instructor so directs, make a second trial with a different metal.

QUESTIONS: Your answers should be complete statements.

1. Why is it desirable to have the water a few degrees colder than room temperature when the initial temperature is taken?

....

....

....

.... 1

2. Why is the mass of the outer shell of the calorimeter and the insulating ring not incuded in the data for this experiment?

....

....

....

.... 2

3. What does this experiment show about the specific heat of water?

....

.... 3

4. How does the heat conductivity of the metals used in this experiment affect the accuracy of the results?

....

....

....

.... 4

5. **Why should the hot metal be dry before it is introduced into cold water?**

....

....

....

.... 5

NAME PERIOD DATE INSTRUCTOR'S COMMENTS

21 EXPERIMENT Coefficient of Linear Expansion

PURPOSE: To measure the coefficient of linear expansion of one or more metals.

APPARATUS: Coefficient of linear expansion apparatus; meter stick; steam boiler; tripod, for steam boiler; rubber tubing, for steam delivery; thermometer; burner and rubber tubing; 100-ml beaker, for catching condensate; rods or tubes of aluminum, brass, copper, and steel.

INTRODUCTION: Nearly all substances expand when heated and contract when cooled. The increase in length of a unit length of a solid when heated through a temperature change of one degree is called the *coefficient of linear expansion.* Such an increase is difficult to measure directly. Generally a rod about 1.0 m long is heated through a temperature change of about 80 C° and the expansion measured by a device which multiplies the small amount of expansion actually occurring.

With the type of apparatus shown in Fig. 21-1, the rod, with one end fixed, is placed in a steam jacket. When steam passes through the jacket, the rod expands and the free end pushes against the short arm of a bent lever. This forces the long arm of the lever over a graduated scale.

In other types of coefficient of linear expansion apparatus the amount of expansion is measured directly by means of a micrometer screw, or by an axle which rotates through a measurable angle as the expanding rod or tube moves over it.

PROCEDURE: The general method to be followed in this experiment, regardless of the type of apparatus used, is to cause a metal rod or tube of known initial length and known initial temperature to undergo a known increase in temperature and measure the resulting amount of expansion. One end of the metal rod or tube is kept stationary. The other end is free to expand under the heating effect of the steam used, and its movement is measured by some suitable device.

Inspect the apparatus to determine how the amount of expansion is to be measured. In the bent lever type, the amount of expansion is the product of the change in scale reading (usually in millimeters) and the ratio of the length of the short lever arm to the length of the long lever arm. In the micrometer type, the expansion is the difference between two micrometer readings—one taken with the rod at room temperature, the other taken with the rod at steam temperature. (Make sure you "back off" the micrometer screw during the expansion.) In the rotation method, the expansion is the product of the number of degrees of rotation and the circumference of the axle divided by exactly 360°.

Make determinations with as many different metals as your instructor directs.

Record all necessary data in an appropriate table in the space provided. Compare the experiment coefficient of linear expansion with the accepted value for the metals used. (Appendix B, Table 17.) Compute the relative error for each trial.

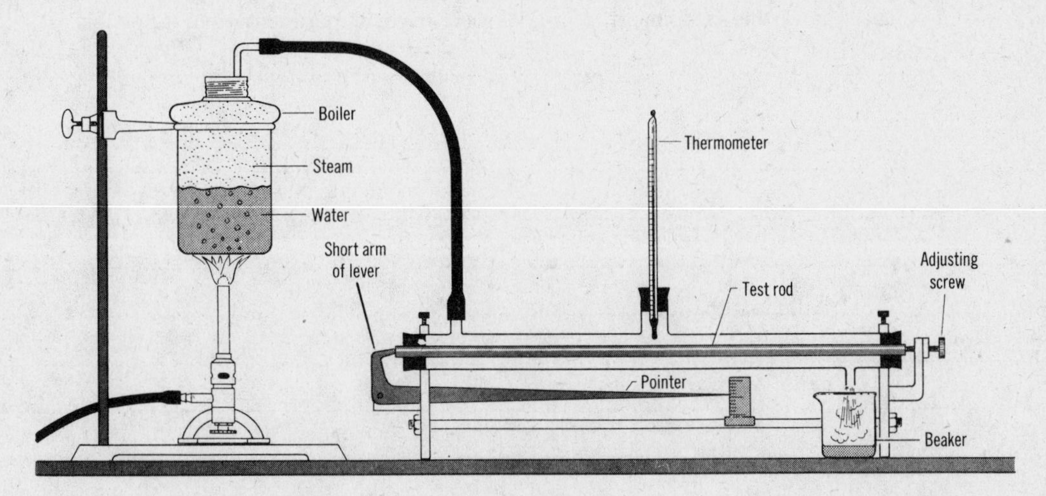

Figure 21-1

DATA

QUESTIONS: Your answers should be complete statements.

1. On the basis of this experiment, do you think that the linear expansion of the metal rods was the same for each degree rise in temperature? Justify your answer.

1

2. A metal sheet has a round hole in it. When the sheet is heated, does the hole get bigger or smaller? Justify your answer on the basis of this experiment. Verify the answer by experimentation, if possible.

2

3. Can you think of a way to modify this experiment so that you could see linear expansion directly?

....

....

....

....

.... 3

4. List three factors that determine the precision of this experiment.

....

....

....

....

....

....

....

....

....

....

....

.... 4

NAME PERIOD DATE INSTRUCTOR'S COMMENTS

22 EXPERIMENT Charles' Law

PURPOSE: (1) To learn how the volume of a gas is affected by a change in temperature. (2) To verify Charles' law.

APPARATUS: Tall glass cylinder, 30 cm high; boiler; tripod for boiler; burner and rubber tubing; thermometer; magnifier; glass tube, globule of mercury, and meter stick, or Charles' law tube; ice.

INTRODUCTION: Charles' law states that the volume of a dry gas is directly proportional to the Kelvin temperature, provided the pressure remains constant. In this experiment the volume occupied by a gas at the ice-point and at the steam-point will be measured. Using the volume of gas measured at the ice-point, the Kelvin temperatures at the ice-point and at the steam-point, and the Charles' law equation, $V/V' = T_K/T_K'$, the volume the gas should occupy at the steam-point will be calculated. This calculated value will then be compared with the actual volume of the gas at the steam-point to verify Charles' law. During the course of the experiment it is assumed that the atmospheric pressure remains constant.

In this experiment, the length of a column of gas trapped in a glass tube is measured, rather than a volume of gas directly. The length of a column of gas of uniform cross-sectional area is directly proportional to the volume of the gas, and can be substituted for the volume in a gas law equation.

The apparatus used in this experiment consists of a glass tube, not more than 1.5 mm in diameter, and about 350 mm long. A tiny globule of mercury is introduced into the tube (not more than enough to make a column 5 mm in length) and one end of the tube is sealed so that the enclosed air column is about 250 mm long. The glass tube should then be fastened with rubber bands to a section of meter stick. See Fig. 22-1. When the tube is heated, the air expands and pushes the mercury globule up the tube. The volume of air trapped in the tube is proportional to the length of the column of air as read from the meter stick.

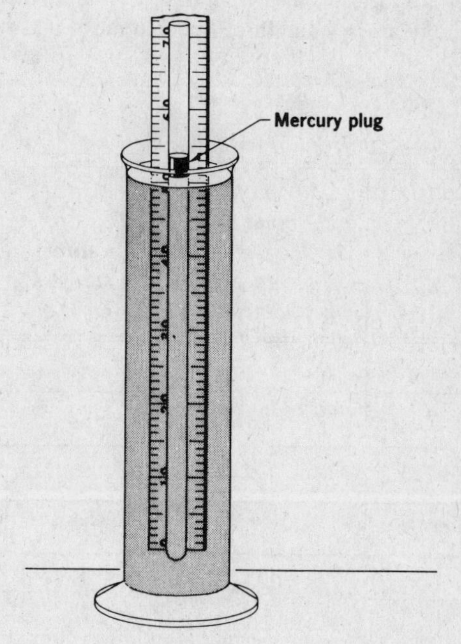

Figure 22-1

PROCEDURE: Record in the data table to the nearest tenth of a centimeter the meter stick reading of the sealed end of the air column. Fill the cylinder with ice and water and support the glass tube and metric scale in the ice-water mixture so it stands vertically. The surface of the water should stand at about the same level as the globule of mercury in the glass tube. Find the temperature of the ice-water mixture to the nearest tenth of a degree. It should be approximately zero degrees Celsius. After the glass tube has stood in the ice-water mixture for about three minutes, read the position of the top of the air column to the nearest tenth of a centimeter. Allow the glass tube to remain in the ice-water mixture for one additional minute. Reread the position of the top of the air column to the nearest tenth of a centimeter. If there is no change in the length of the air column, record the reading in the data table.

Remove the glass tube and meter stick from the ice-water mixture, and suspend them in the chimney of the steam boiler so they will be surrounded by the steam arising from the boiling water. After being suspended in the steam for about three minutes, read the position of the top of the air column to the nearest tenth of a centimeter. Also read the temperature of the steam surrounding the tube to the nearest tenth of a degree. After one additional minute make confirmatory readings using the same procedure as was used at the ice-point.

Make two more trials, determining the position of the top of the air column and the temperature in both the ice-water mixture and in the steam for each trial. Record the data.

CALCULATIONS: For each trial, determine the length of the air column at the temperature of the ice-water mixture and at the temperature of the steam. Determine the average length of the air column at the temperature of the ice-water mixture and the average length of the air column at the temperature of the steam. Also determine the average temperature of the ice-water mixture and the average temperature of the steam in degrees Celsius.

Convert the average Celsius temperatures of the ice-water mixture and the steam to Kelvin temperatures. Using the average length of the column of air in the ice-water mixture as V, the average Kelvin temperature of the ice-water mixture as T_K, and the average Kelvin temperature of the steam as T_K', calculate the theoretical length of the air column at the temperature of the steam V', by means of the formula $V/V' = T_K/T_K'$. Find the error between your experimental value and this calculated value. Assuming the calculated value to be correct, find the percentage error.

DATA *Bottom of air column* ________cm

TRIAL	*Top of air column in ice-water mixture* (cm)	*Top of air column in steam* (cm)	*Length of air column in ice-water mixture* (cm)	*Length of air column in steam* (cm)	*Temperature ice-water mixture* (°C)	*Temperature steam* (°C)
1						
2						
3						
Average						

	Temperature ice-water mixture (°K)	*Temperature steam* (°K)	*Length of air column in steam (calculated)* (cm)	*Absolute error* (cm)	*Relative error* (%)
Average					

QUESTIONS: Your answers should be complete statements.

1. (a) How could this experiment be modified to determine whether air expands uniformly when it is heated?
(b) How could this experiment be modified to determine whether other gases obey Charles' law?

a. ....

....

.... 1a

b. ....

.... 1b

2. Does the expansion of the glass tube affect the accuracy of this experiment? Explain.

....

....

....

....

....

....

.... 2

3. Devise a method of determining the length of the air column at an intermediate temperature such as 50° C and verify Charles' law for expansion from the ice point to this temperature.

....

....

....

....

....

....

....

....

....

....3

NAME PERIOD DATE INSTRUCTOR'S COMMENTS

23 EXPERIMENT Boyle's Law

PURPOSE: To learn how the volume of a given mass of gas varies with the pressure exerted on it.

APPARATUS: J-tube apparatus; mercury; medicine dropper.

INTRODUCTION: Boyle's law states that if the temperature remains constant, the volume of a dry gas varies inversely with the pressure exerted on it. This means that as the pressure on a gas is increased, the volume of the gas will become smaller at the same rate. Or, as the pressure on the gas is decreased, the volume will increase. If we use p to represent the pressure and V to represent the volume, Boyle's law can be stated algebraically as pV = a constant.

In this experiment we shall use a modification of the method actually used by Robert Boyle in deriving this pressure-volume relationship of gases which now bears his name.

PRECAUTION: In this experiment you will be using mercury. In order to protect any hand jewelry you wear from the action of mercury, remove these articles and place them in your pocket or purse. Be careful not to spill any mercury, and do not handle it. After the experiment is completed, wash your hands thoroughly with soap and water before replacing your jewelry.

PROCEDURE: Using the medicine dropper, put enough mercury into the J-tube to fill the curved portion and reach the graduations on the meter stick mounted on the wooden support. Tip the tube sideways enough to let air flow past the mercury and make the mercury columns the same height in both tubes as in *A* of Fig. 23-1. Record the meter stick reading at the top of the short arm and the height of the mercury in both arms of the tube. The difference between the meter stick reading at the top of the short arm and the height of the mercury in the short arm indicates the *original volume* of air taken. (The volumes of cylinders of the same diameter are directly proportional to their lengths.) The barometer reading is the *original pressure.*

Now add 10 to 15 cm of mercury to the long arm of the tube. Take readings of the height of the mercury level in each arm of the tube. Calculate the new volume of air by subtracting the height of the mercury in the short arm of the tube from the height of the tube itself. Calculate the new pressure by adding the barometer reading to the difference between the height of the mercury levels in the two columns. In the same manner, take three or four more readings, and record the results. In each case calculate the pV product.

When you have obtained all the necessary data, follow your instructor's directions for removing the mercury from the J-tube. Be especially careful when handling a J-tube which contains mercury.

GRAPH: Plot a graph of your data, using the pressures as abscissas and the volumes as ordinates.

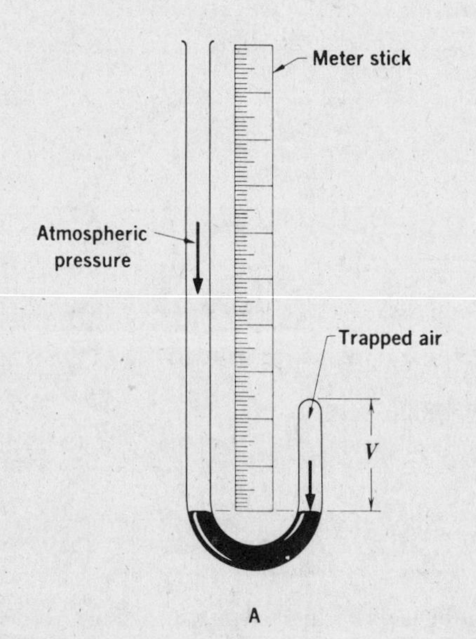

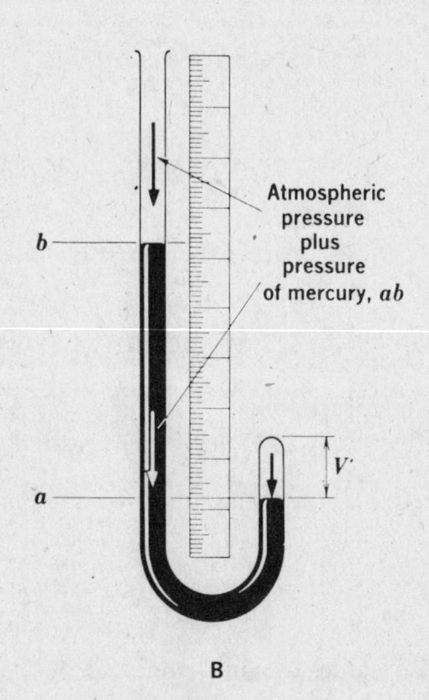

Figure 23-1

DATA

Height of short arm ________ cm   Barometer ________ cm

TRIAL	*Mercury level in short arm* (cm)	*Volume of gas V* (cm)	*Mercury level in long arm* (cm)	*Difference in mercury levels* (cm)	*Pressure on gas p* (cm)	pV
1						
2						
3						
4						
5						
6						
7						
8						

QUESTIONS: Your answers should be complete statements.

1. Describe the relationship between the pressure exerted on a gas and the volume occupied by the gas.

....

....

....1

2. (a) How would the presence of air bubbles in the mercury affect the results of this experiment? (b) How would the presence of water in the mercury affect the data?

a. ....

....

....2a

b. ....

....

....2b

3. In what way is a Boyle's law tube like a barometer?

....

....

....

....3

4. What affect does an increase in pressure have on the density of a gas?

....

....

....4

5. What determines the upper limit of the pressure that can be exerted in a Boyle's law tube?

....

....5

NAME PERIOD DATE INSTRUCTOR'S COMMENTS

24 EXPERIMENT Heat of Fusion

PURPOSE: To determine the heat of fusion of ice.

APPARATUS: Calorimeter; platform balance; set of masses; thermometer, Celsius -10° to 110°, graduated to 0.2°; stirring rod; pan for ice; towel; ice cubes; magnifier.

INTRODUCTION: The number of calories needed to melt one gram of any substance at its normal melting point without any temperature change is called its *heat of fusion.* In this experiment the heat of fusion of ice will be determined by using the method of mixtures. The temperature drop of a given amount of rather hot water in a calorimeter when a certain quantity of ice is added will be measured. The heat lost by the water does two things; (1) it melts the ice; (2) it warms the water formed by the melting ice from zero up to the final temperature.

PROCEDURE: First find the mass of the empty inner cup of the calorimeter. Then fill it about half full of water having a temperature of about 35° C, and find the mass of the calorimeter cup and water. Then replace the cup in its insulating shell. Next stir the water in the calorimeter and read its temperature to the nearest tenth of a degree, using the magnifier. Wipe two ice cubes, or an equivalent amount of ice chunks, with a towel to remove adhering water. Then put them in the calorimeter carefully so that there is no splashing. Stir rather rapidly until all the ice is melted, and read the temperature to the nearest tenth of a degree. Remove the calorimeter cup from its shell and find the combined mass of the cup and its contents. If your instructor directs, make a second trial.

DATA

	TRIAL 1	TRIAL 2
1. Mass of calorimeter cup and stirrer	....................g	....................g
2. Mass of calorimeter cup, stirrer, and water	....................g	....................g
3. Mass of water	....................g	....................g
4. Mass of calorimeter cup, stirrer, and water (after ice is melted)	....................g	....................g
5. Mass of ice	....................g	....................g
6. Specific heat of calorimeter	....................cal/gC°	....................cal/gC°
7. Temperature of water and calorimeter, initial	....................°C	....................°C
8. Temperature of water and calorimeter, final	....................°C	....................°C
9. Temperature change of water and calorimeter	....................C°	....................C°
10. Calories lost by calorimeter	....................cal	....................cal
11. Calories lost by water	....................cal	....................cal
12. Total calories lost	....................cal	....................cal
13. Calories used to warm water formed by melted ice	....................cal	....................cal
14. Calories used to melt ice	....................cal	....................cal
15. Heat of fusion of ice	....................cal/g	....................cal/g
16. Accepted value for heat of fusion of ice	....................cal/g	....................cal/g
17. Absolute error	....................cal/g	....................cal/g
18. Relative error	....................%	....................%

QUESTIONS: Your answers should be complete statements.

1. Since heat of fusion does not result in a temperature change, where does the energy go?

....

....

.... 1

2. How can heat of fusion be used to guard against frost damage?

....

....

....

....

.... 2

3. What source of error is present in this experiment that was not present in previous heat experiments?

....

....

....

....

.... 3

NAME PERIOD DATE INSTRUCTOR'S COMMENTS

25 EXPERIMENT Heat of Vaporization

PURPOSE: To determine the heat of vaporization of water.

APPARATUS: Steam boiler, fitted with rubber tubing, a water trap, and a piece of glass tubing about 20 cm long; tripod, for steam boiler; calorimeter; platform balance; set of masses; thermometer, Celsius -10° to 110°, graduated to 0.2°; stirring rod; a few chunks of ice are desirable but not essential; burner and rubber tubing; asbestos paper; magnifier; asbestos board; wooden dip stick.

INTRODUCTION: The heat of vaporization is the amount of heat required to vaporize one gram of a liquid substance at its normal boiling point without any change in its temperature. In this experiment a determination of the heat of vaporization of water will be attempted. This is rather difficult to do accurately, so the work must be done carefully to avoid possible sources of error. A known mass of steam at a measured temperature will be passed into a known mass of cold water at a measured temperature. By noting the temperature rise of the water, the number of calories yielded by the steam as it condenses can be calculated. The heat lost by condensation equals the heat absorbed during vaporization.

PROCEDURE: Set up a steam boiler and partly fill it with water (Fig. 25-1). Connect the water trap and tubing as shown, and light the burner. While the water is heating, find the mass of the inner cup of the calorimeter. Then fill it about two-thirds full of water at about 5° C and determine the combined mass of the cup and water. Replace the cup in its insulating shell. Stir the water thoroughly. Take its temperature, estimating to the nearest tenth of a degree, *just before* the glass tube is placed in the calorimeter and steam flows in. Place the asbestos board as a heat shield between the boiler and the calorimeter.

Pass a steady current of steam into the water, stirring continuously, until the temperature rises to about 35° C. Then remove the glass tube from the calorimeter and stir the water thoroughly. Read the thermometer to the nearest tenth of a degree when the temperature of the stirred water reaches its highest point. Remove the calorimeter cup from its outer shell. Find the mass of the cup and contents. The increase in mass equals the mass of steam introduced.

For best results, about 15 g of steam should be introduced into the cold water. This can be determined by using a wooden dip stick and marking it at the height to which 15 more ml of water will raise the surface of the water in the calorimeter cup.

The temperature of the steam may be read directly from the steam boiler, or it may be computed from the barometer reading. In making your calculations you must remember that the steam furnished heat upon condensing, and also that it formed an equal mass of water at about 100° C. The water thus formed also furnished heat as it cooled from the boiling point down to the final temperature.

Record all necessary data in the data table. If your instructor directs, make a second trial.

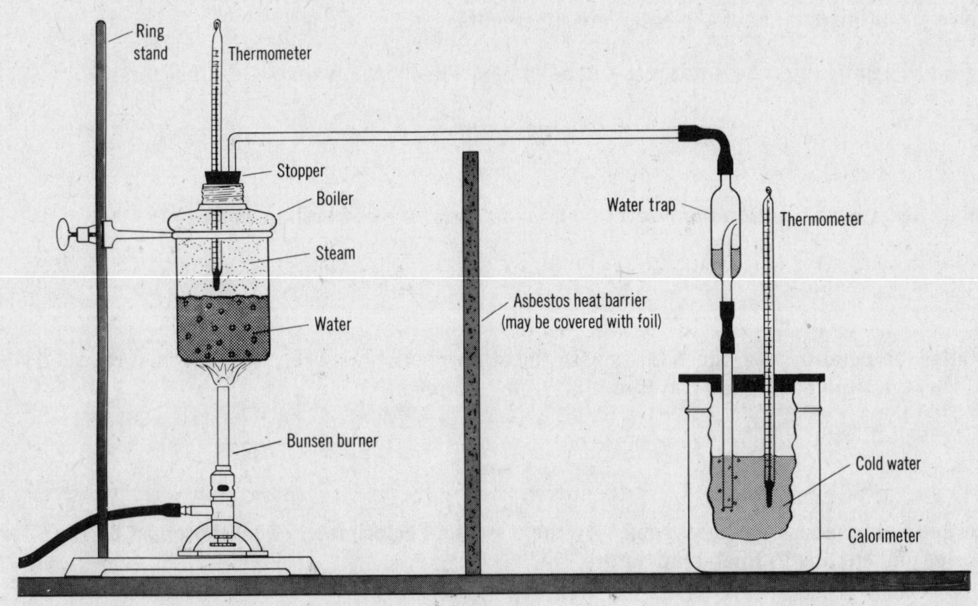

Figure 25-1

DATA

	TRIAL 1	TRIAL 2
1. Mass of calorimeter cup and stirrer	..................g	..................g
2. Mass calorimeter cup, stirrer, and water	..................g	..................g
3. Mass of water	..................g	..................g
4. Specific heat of calorimeter	..................cal/gC°	..................cal/gC°
5. Temperature of water and calorimeter, initial	..................°C	..................°C
6. Temperature of steam	..................°C	..................°C
7. Temperature of water and calorimeter, final	..................°C	..................°C
8. Mass of water, calorimeter cup, stirrer, and added steam	..................g	..................g
9. Mass of steam	..................g	..................g
10. Temperature change of water and calorimeter	..................C°	..................C°
11. Temperature change of water from condensed steam	..................C°	..................C°
12. Calories gained by calorimeter	..................cal	..................cal
13. Calories gained by water	..................cal	..................cal
14. Total calories gained	..................cal	..................cal
15. Calories lost by water formed from steam in cooling to final temperature	..................cal	..................cal
16. Calories lost by steam in condensing	..................cal	..................cal
17. Heat of vaporization of water	..................cal/g	..................cal/g
18. Accepted value for heat of vaporization of water	..................cal/g	..................cal/g
19. Absolute error	..................cal/g	..................cal/g
20. Relative error	..................%	..................%

QUESTIONS: Your answers should be complete statements.

1. Since heat of vaporization does not result in a temperature change, where does the energy go?

....

.... 1

2. Why does steam burn a person more severely than an equal mass of boiling water?

....

.... 2

3. The boiling temperature of a liquid varies with the atmospheric pressure. Does the heat required for vaporization vary with the boiling temperature of a liquid? Explain.

....

.... 3

4. Why should the number of calories gained by the water and calorimeter equal those lost by the steam in condensing and cooling to final temperature?

....

.... 4

NAME PERIOD DATE INSTRUCTOR'S COMMENTS

26 EXPERIMENT Pulses on a Coil Spring

PURPOSE: (1) To observe the behavior of transverse and longitudinal pulses on a stretched coil spring. (2) To verify the effects of free-end and fixed-end terminations of a medium on the traveling pulse.

APPARATUS: Two long coil springs, a Slinky and a heavy metal coil matched to give approximately equal amplitudes of reflected and transmitted pulses; light string; meter stick or metric tape; stopwatch; chalk.

INTRODUCTION: The Slinky is a large "soft" spring, edge wound of flat wire, which transmits a pulse slowly enough to be observed with relative ease. When a heavy metal spring is attached to the end of the Slinky, the energy of a traveling pulse is partly reflected at the junction and partly transmitted across the junction. The inertia of the heavy spring is greater than that of the Slinky. Thus, at the junction of the two springs, the Slinky has a relatively fixed-ended termination while the heavy spring has a relatively free-ended termination.

SUGGESTION: At least four students should work as a group with one set of apparatus as there are several assignments to be carried out during each experiment. These assignments should be rotated frequently within the group to provide all members with a full measure of participation.

PROCEDURE:

1. Transverse pulse

Place a large, light spring (called a *Slinky*) on a smooth floor and stretch it to a length of approximately 8 m. (*Suggestion:* Mark one end position on the floor with chalk and mark other positions 6 m, 8 m, 10 m, and 12 m away. Determine the best stretch distance for your coil by experiment, being careful not to exceed the stretch limit of the spring being used.) Station group partners at each end of the Slinky to hold it securely in position while stretched. CAUTION: *Avoid releasing an end of the stretched spring; the untangling process can be very difficult.*

With the ends of the stretched Slinky held rigidly in place, form a pulse by grasping a loop near one end of the spring and displacing it to one side by a quick back-and-forth motion of the hand. Practice this motion until a pulse can be formed that travels down only one side of the spring. Why is the pulse called a *transverse pulse*? Make a statement about a transverse pulse relating the motion of the separate coils of the spring to the path traversed by the pulse.

Observation: ....

....

....

....

....

The coil spring is the medium through which the pulse travels. Send a short pulse down the spring. Observe the shape of the pulse as it moves along the spring. How does the shape change? Can you suggest a reason? Upon what does the initial amplitude of the pulse depend? Does the speed of the pulse appear to change with its shape? Generate single pulses of different amplitudes. Does the pulse speed appear to depend on the size of the pulse?

Observation: ....

....

....

....

....

Measure the length of the stretched Slinky and the travel time of a pulse generated at one end. Assuming that the pulse speed remains unchanged after reflection, the time required for the pulse to make a few excursions back and forth along the spring could be measured. Can you devise a way to check this assumption? Compute the speed of the traveling pulse.

Change the tension in the Slinky and determine the pulse speed as before. Is the speed of propagation through the spring affected by the tension? Does the stretched Slinky, under different tensions, represent the same, or different transmitting media?

Observation: ....

....

....

....

....

Generate single pulses simultaneously from opposite ends of the stretched Slinky. Observe the way they meet and pass through each other. The superposing of the two pulses is called *interference.* Send two pulses of approximately equal amplitudes toward each other on the same side of the spring. Describe the interference as they pass through each other. How does the pulse amplitude during interference compare with the individual amplitudes before and after superposition?

Repeat the experiment, but with the two pulses traveling on opposite sides of the spring. Compare the interference with that of the previous trial. Observe the region of interference closely and see if there is a point on the spring that does not move appreciably during interference. What conclusions can you draw about the displacement of the medium at a point where two pulses interfere?

Observation: ....

....

....

....

....

With the far end of the Slinky held firmly in place (fixed-end termination), send a single pulse down one side of the spring. Observe the *reflected* pulse. Compare its amplitude with that of the transmitted pulse just before reflection. What is its phase relative to the transmitted pulse?

Attach a light string about 2 m long to the far end of the Slinky and maintain the tension on the spring by holding the end of the string. This approximates a *free-end* termination for the Slinky. Send a pulse down one side of the spring as before and observe the pulse reflected from the "free" end. Compare reflection from the "free" end of the spring with reflection from the fixed end.

Observation: ....

....

....

....

....

Investigate the transfer of a pulse from one medium to another by attaching a heavy metal spring of about 2 cm diameter to the end of the Slinky. Would you expect the pulse speed to remain the same in both sections of the stretched combination? To the extent that the second spring produces an impedance mismatch at its junction with the Slinky, the pulse energy arriving at the junction will be partly reflected and partly transmitted.

Send a single pulse down the stretched combination, first from one end and then the other. Observe what happens when each pulse reaches the junction of the two springs. (Try not to be distracted by an end reflection if the junction is near the end of one spring.) How does the speed of propagation in the heavy spring compare with that in the Slinky?

Send a pulse down one side of the Slinky toward the junction and observe the relative phase of the pulse reflected at the junction. Similarly determine the relative phase of the pulse transferred across the junction to the heavy spring. Repeat these phase observations for a pulse sent down the heavy spring toward the junction. If the difference in pulse speeds in the two springs were greater, how would you expect the relative portions of the pulse energy reflected and transmitted at the junction to be affected? Summarize your observations of the reflection of pulses from the junction of two media in which they have different speeds.

Observation: ....

....

....

....

....

2. Longitudinal pulse

With the heavy metal spring removed, stretch the Slinky as before and ensure that the far end is held firmly against the floor by a partner. While holding your end of the spring securely, pull about 0.5 m of stretched spring together toward the end with your free hand and release it. CAUTION: *Do not let go of the end of the Slinky.* Observe the pulse that travels back and forth through the spring. Why is it called a longitudinal pulse? Make a statement about a longitudinal pulse relating the motion of the separate coils of the spring to the path traversed by the pulse.

Observation: ....

....

....

....

....

NAME PERIOD DATE INSTRUCTOR'S COMMENTS

27 EXPERIMENT Wave Properties

PURPOSE: (1) To learn how to produce and project water waves with a ripple tank. (2) To study reflection, refraction, diffraction, and interference of water waves.

APPARATUS: Ripple tank, similar to the one shown in Fig. 27-1; tank attachments to produce straight and circular waves; hand stroboscope; cork; stopwatch or watch with second hand; straight barrier; rubber hose, at least 2 cm in diameter; glass plate; protractor; paraffin blocks.

INTRODUCTION: A very familiar type of wave motion is the motion of water waves. When you drop a stone into a body of calm water, the disturbance produced by the stone spreads out from the point of disturbance in a rhythmic circular pattern. Energy is thereby transmitted through the water in a two-dimensional pattern.

Many waves, including sound and light waves, are three-dimensional, but many of their properties can be adequately simulated and studied by means of two-dimensional water waves. The ripple tank has been designed for this purpose. Essentially, it consists of a thin layer of water supported by a sheet of glass. Waves are produced in the water by a vibrating bar or by one or more vibrating prongs. A light source above the water is used to project images of the water waves on a screen below the ripple tank. The waves act as lenses which reproduce the crests and troughs of the waves as light and dark images on the screen.

SUGGESTION: At least two sessions should be provided for this experiment.

PROCEDURE:

1. Producing straight and circular waves

Fill the ripple tank to a depth of about 0.5 cm. Check the depth of all four corners of the tank to make sure that it is level. Dip the edge of a ruler into the tank and observe the wave that is produced. Turn on the projection light and adjust it until sharp images are produced on the screen. Dip the end of a pencil into the tank and study the resulting wave. Compare it with the wave produced by the ruler. Draw a sketch of the results in Block A.

Attach the device for producing straight waves and start the vibrating motor. Adjust the position of the vibrator and the motor speed until distinct periodic waves are produced that are parallel to the side of the tank toward which they move. Place a cork on the water and observe its motion.

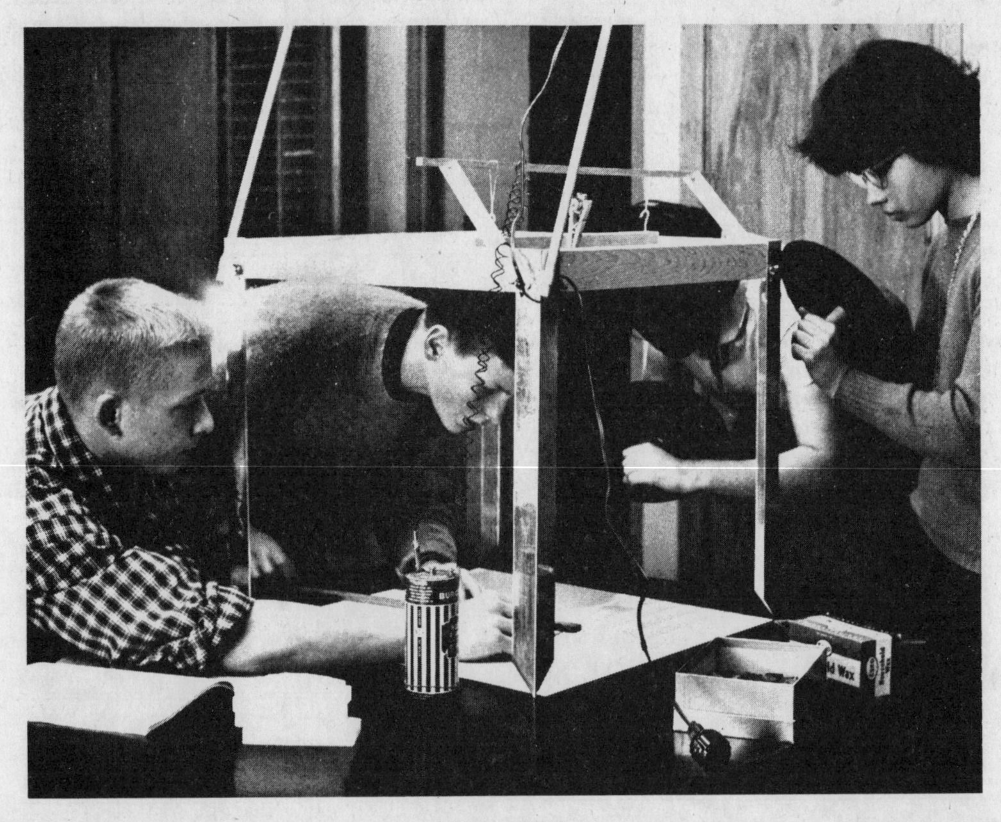

Figure 27-1

A

Observation: ....

....

....

....

....

2. Wavelength, frequency, velocity

A ripple tank projects magnified images of water waves. Therefore, in making measurements on the screen, it is necessary to take this magnification into account. To measure the magnification, place an object of known length (such as a pencil) in the tank and measure its length on the screen. Record the magnification factor in the data table.

Set up periodic straight waves in the tank and "stop" them with the hand stroboscope. Your instructor will show you the proper way to use this instrument. With the help of a fellow student, measure the distance between two crests on the screen. Correct this reading with the magnification factor and record the actual wavelength in the data table.

Another way to measure wavelength is to place a straight barrier in the path of the waves, perpendicular to the direction of wave motion. Produce standing waves by adjusting the position of the barrier or the frequency of the generator. Recall the spacing of adjacent nodes or adjacent crests in a standing wave pattern when measuring the wavelength this way.

Measure the frequency of the wave by holding a piece of paper against the vibrating motor counting the number of times the vibrator strikes the paper in a given period of time. (It may be necessary to count the vibrations in groups of three or more.) Record the frequency.

With the stopwatch, measure the time required for a wave to cover a known distance on the screen. Make several trials and record the average. Correct for magnification. Record the speed.

Calculate the speed from the wavelength and frequency. Compare it with the speed as measured directly. Record the difference as your error.

3. Reflection

Place a straight barrier diagonally in the ripple tank. Produce periodic straight waves and observe them as they strike the barrier. Draw a sketch of the resulting wave pattern in Block B.

B

Adjust the barrier so that the waves strike it at an angle of 45°. Measure the angle of the reflected waves. (Angles of incidence and reflection are measured as the angles between the direction of wave motion and a perpendicular to the surface from which they are reflected.) Slowly turn the barrier and study the effect on the angles of incidence and reflection. Record your conclusions in a general statement.

Conclusion: ....

....

....

....

....

Replace the straight barrier with a parabolic one. A thick rubber hose serves well for this purpose. In Block C sketch the pattern that results. Mark the spot on the screen where the reflected waves are focused. Turn off the vibrator and dip your pencil into the tank at the point in the tank that corresponds to

DATA Magnification Factor: ________

TRIAL	*Wavelength* (m)	*Frequency* (hz)	*Speed* *measured* (m/sec)	*Speed* *calculated* (m/sec)	*Absolute error* (m/sec)
1					
2					
3					

the mark on the screen. Record your observations and conclusions concerning reflection from a curved surface.

C

Observations and conclusions: ....

....

....

....

....

4. Refraction

The velocity of water waves depends on the depth of the water. Friction has a greater effect on the propagation of the wave in shallow water than in deep water. Consequently, waves slow up when they pass from deep to shallow water. If the waves strike the boundary between deep and shallow water at an angle, the wavefront will change direction at the boundary. This bending of waves is called *refraction.*

To produce refraction, place a glass plate in the tank and adjust the water level so that the plate is covered with a very thin layer of water. Position the plate so that the forward edge is at an angle to the approaching wavefronts. Turn on the vibrator and sketch the resulting wave pattern in Block D. Measure the angle of incidence and angle of refraction with a protractor. Change the orientation of the glass plate and measure the angles again. Look up the sines of these angles and calculate the index of refraction using the equation

$$n = \frac{\sin i}{\sin r}$$

where n is the index of refraction, and i and r are the

D

angles of incidence and refraction, respectively. Record these values.

Measure the speed of the incident and refracted waves with the method described in Step 2. Calculate the index of refraction using the formula

$$n = \frac{\text{speed in deep water}}{\text{speed in shallow water}}$$

Record this value and compare it with the one obtained from the angles. Record the difference as your error.

5. Diffraction

Place a paraffin block on edge in the tank so that it extends about half way across the tank and is parallel to the advancing straight waves. Start the vibrator. Observe the waves as they pass the edge of the paraffin block. Sketch the wave pattern in Block E.

Vary the frequency of the vibrator and describe any changes in the wave pattern.

Place another paraffin block in the tank, leaving an opening between the blocks about 3 cm long. Adjust the frequency of the vibrator to produce waves about 1.5 cm long. Sketch the resulting wave pattern in Block F. Vary the opening between the blocks as

E

DATA

TRIAL	i (°)	r (°)	$\sin i$	$\sin r$	n	*Speed* Deep (m/sec)	*Speed* Shallow (m/sec)	n	*Absolute error*
1									
2									
3									

well as the frequency of vibration and describe the effect on the wave pattern.

F

Observation: ....

....

....

....

....

The bending of waves as they pass obstacles or small openings is called *diffraction.*

6. Interference

Arrange two prongs and the vibrator in such a way that circular waves will be produced a short distance apart. Turn on the vibrator and sketch the resulting wave patterns in Block G.

G

When two waves with the same frequency and wavelength interact, it is called *interference.* See how the interference pattern is affected when the *frequency* and distance between wave sources are changed. Sketch the result in Block H.

H

QUESTIONS: Your answers should be complete statements.

1. What does the motion of the cork show about the transmission of energy in wave motion?

....

....

.... 1

2. Is the magnification factor involved in measuring wave frequency? Elaborate.

....

.... 2

3. What is the general relationship between wavelength and diffraction?

.... 3

4. List similarities and differences between water waves and sound waves.

....

....

....

.... 4

5. How does this experiment explain the fact that sound travels around corners?

....

....

.... 5

NAME PERIOD DATE INSTRUCTOR'S COMMENTS

28 EXPERIMENT Resonance; The Speed of Sound

PURPOSE: (1) To determine the best resonant length of a closed tube for sounds of known frequency. (2) To determine the speed of sound from the wavelengths and frequencies of several sounds.

APPARATUS: Tuning forks of known frequency. C-fork, 261.6 cps; E-fork, 329.6 cps; G-fork, 392 cps; and C′-fork, 532.2 cps are all satisfactory. (Physical pitch forks of 256 cps to 512 cps may also be used); glass tube, 2.5 cm to 4.0 cm in diameter and at least 40 cm long; tall cylinder; thermometer; meter stick; tuning fork hammer or large rubber stopper.

INTRODUCTION: Resonance, or sympathetic vibration, occurs when the natural vibration rates of two objects are the same. The air column in a closed glass tube produces its best resonance when it is approximately one-fourth as long as the wavelength of the sound which it reinforces. For example, a tube one meter long produces resonance with a sound wave which is about four meters long. A small correction in wavelength must be made for the internal diameter of the tube. The wavelength of the sound may be calculated from the resonant length of the tube by the equation $\lambda = 4(l + 0.4d)$, where λ is the wavelength, l is the length of the closed tube, and d is the diameter of the tube.

In this experiment the best resonant length of a closed tube will be determined. From this length and the diameter of the tube, the wavelength of the sound will be calculated. The speed of sound will then be calculated from the equation $v = f\lambda$, where v is the speed of sound, f is the frequency, and λ is the wavelength.

SUGGESTION: This experiment can be performed successfully by pairs of students, providing *they do not talk* while other members of the class are finding the best resonant length of the tube. However, to avoid the confusion which arises when several students in the laboratory are attempting to find the best resonant length, some instructors may prefer to perform this experiment with the class working as a group.

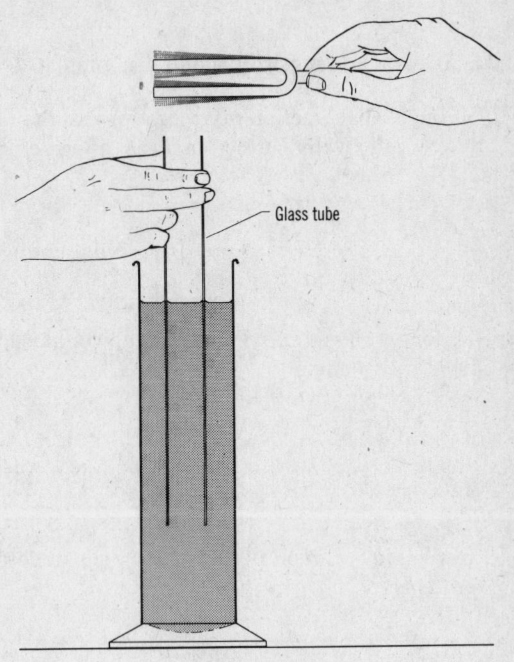

Figure 28-1

PROCEDURE: Hold the tube vertically in a cylinder nearly full of water, as shown in Fig. 28-1. Make a tuning fork vibrate by holding it by the shank and striking it with the tuning fork hammer or striking it on a large rubber stopper. *Do not strike the tuning fork on the table top or other hard surface.* As you hold the vibrating tuning fork over the open end of the tube, find that position where resonance produces the loudest sound by moving the glass tube slowly up and down in the water of the cylinder. Then hold the glass tube firmly while another student measures the length of the tube from its top to the water surface inside the tube. This length should be recorded to the nearest thousandth of a meter. Carefully measure the internal diameter of the tube to the nearest thousandth of a meter also. Use the thermometer to

DATA

TRIAL	*Length of tube* (m)	*Diameter of tube* (m)	*Wavelength* (m)	*Frequency* (hz)	*Temperature* (°C)	*Speed of sound* *experimental* (m/sec)	*Speed of sound* *accepted* (m/sec)	*Relative error* (%)
1								
2								
3								

determine the temperature of the air inside the tube in degress Celsius. A second and a third trial should be taken using forks of different frequencies.

For each fork, compute the speed of sound from the resonant length of the tube and the known frequency of the fork. Compare this value with the accepted value for the speed of sound at the temperature inside the tube. Compute the relative error.

QUESTIONS: Your answers should be complete statements.

1. Through what fraction of a vibration has the prong of a tuning fork moved while the sound wave traveled down to the water surface and was reflected back up to the fork again?

....

....

.... 1

2. If a longer glass tube were available, would it be possible to find another position where resonance is produced? Explain.

....

....

.... 2

3. How could you modify this experiment to determine whether the water vapor inside the tube affects the results?

....

.... 3

4. How does the amplitude of vibration affect the data? Verify your answer by experimentation.

....

....

.... 4

5. How could you modify the experiment to find the resonant length of an open pipe?

....

....

.... 5

NAME PERIOD DATE INSTRUCTOR'S COMMENTS

29 EXPERIMENT Frequency of a Tuning Fork

PURPOSE: To determine the frequency of a tuning fork.

APPARATUS: Pendulum and tuning fork apparatus, or vibrograph; glass plates; stopwatch; whiting suspended in alcohol.

INTRODUCTION: In this experiment, a stylus attached to one prong of a vibrating tuning fork traces a curved path on a coated glass plate which is pulled steadily beneath the stylus. At the same time, a small pendulum to which a needle is attached traces a second curved path. See Fig. 29-1. If the length of the pendulum is known, the time required for it to make a complete vibration can be calculated from the pendulum equation $T = 2\pi\sqrt{l/g}$. Or, it may be just as simple to count the number of vibrations the pendulum makes per second. Next the number of vibrations the fork makes while the pendulum is making a complete vibration is counted. Then the frequency of the fork can be calculated.

SUGGESTION: One set of apparatus is sufficient for an entire class since it takes only a very short time to trace the vibrations after the glass is prepared. A little practice is necessary to enable the student to draw the glass at the proper speed. There should be one glass plate for each student.

PROCEDURE: Prepare a glass plate with a coating of whiting suspended in alcohol. As the alcohol evaporates, the whiting is left as an even coating on the plate. Lay the plate on the apparatus base and adjust the tuning fork so the stylus will just touch the plate. The stylus of the fork should be rather close to the needle of the pendulum. Adjust the pendulum so that the point of its needle will just touch the glass plate as the pendulum vibrates through an arc of small amplitude.

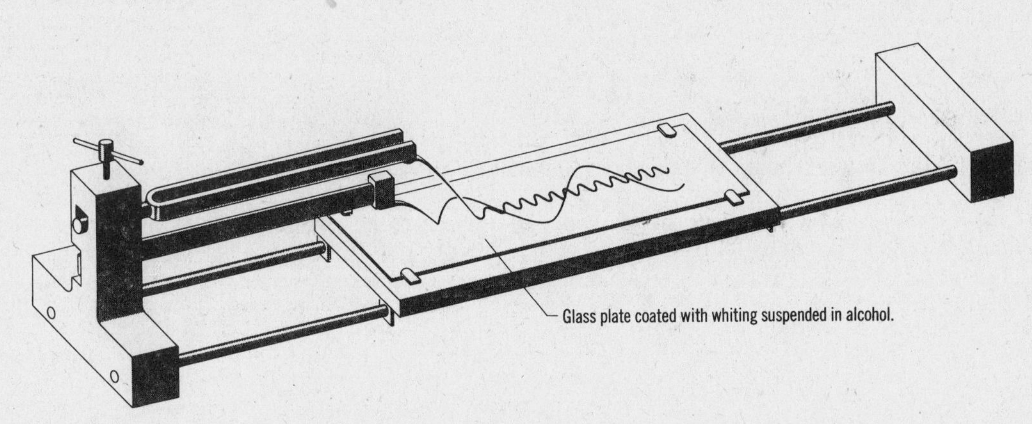

Figure 29-1

DATA

1. Frequency of pendulum	..........hz
2. Period of pendulum (time required for a complete vibration)	..........sec
3. Number of complete vibrations fork makes while pendulum makes one vibration	....
4. Frequency of tuning fork	..........hz

Start the pendulum swinging, and set the fork into vibration by pinching the two prongs together, then releasing them suddenly. Draw the glass plate beneath the fork and pendulum at such a rate that the pendulum will make at least two complete vibrations while the plate is beneath the fork and the pendulum. Make the fine adjustments necessary and enough practice runs to insure getting a successful recording.

After all adjustments have been made, determine the frequency of the pendulum using a stopwatch and the method learned in earlier experiments. Place a freshly coated glass plate on the base, set both the pendulum and the fork in motion, and make a record of their tracings.

Count the number of vibrations made by the fork while the pendulum was making a complete vibration. Then calculate the time required for the pendulum to make a complete vibration, and also the number of vibrations the fork makes in one second.

ALTERNATE PROCEDURE: Stretch a No. 6 piano wire across two triangular blocks on a table or an inclined plane board, and tune it in unison with a C tuning fork by adding weights to the hanger. Measure the length l of the string, from one block to the other. Then use another block beneath the string, and shorten it until it is in unison with a second fork of unknown frequency higher pitched than the C fork. Measure the length l' of the wire or string. Using the appropriate law of strings, determine the frequency of the second fork. Record your information in the data table below.

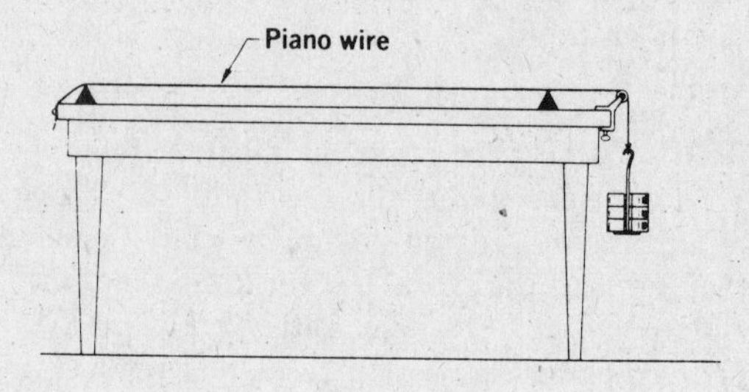

Figure 29-2

DATA

1. Length (l) of string when tuned in unison with C fork	..........m
2. Length (l') of string when tuned in unison with unknown fork	..........m
3. Frequency of fork of unknown vibration rate	..........hz

QUESTIONS: Your answers should be complete statements.

1. How do the following factors affect the frequency of a tuning fork? (a) amplitude of vibration; (b) mass; (c) temperature; (d) position.

....

....

....

....

....

....

....

.... 1

2. Does the pressure of the stylus affect the results of this experiment? Check your answer by experimentation.

....

....

....2

NAME PERIOD DATE INSTRUCTOR'S COMMENTS

30 EXPERIMENT Photometry

PURPOSE: (1) To measure the intensity of sources of light by photometric means. (2) To determine the efficiency of several light sources.

APPARATUS: Meter stick; meter stick supports; Joly, Bunsen, or photoelectric photometer; galvanometer and hook-up wire (for use with photoelectric photometer only); lamp sockets with cord and plug; calibrated (standard) 40 w lamp, or new 40 w lamp; 25 w, 40 w, 60 w, and other incandescent lamps as indicated by the instructor.

INTRODUCTION: The candle is the unit in which the intensity of a source of light is measured. While today the candle is defined in terms of the intensity of the light emitted from a source at the temperature of solidifying platinum, originally the candle was defined as the amount of light emitted by a standard sperm candle.

In order to determine the intensity of a light source, we compare it with a standard light source. This is usually done by means of a photometer. The photometer is placed with respect to the light sources so that the two sides of the photometer screen are equally illuminated. Under this condition of equal illumination, the intensities of the sources are directly proportional to the squares of their respective distances from the photometer. If the intensity of the unknown lamp is I_2, the intensity of the standard is I_1, the distance of the unknown lamp from the photometer is s_2, and the distance of the standard from the photometer is s_1, then $I_2 = I_1 \times s_2^2/s_1^2$.

In this experiment we shall determine the intensity of several incandescent lamps by comparison with a standard lamp, or a new 40 watt lamp which is assumed to have an intensity of 32 candles. The efficiency of the lamps in candles/watt will then be found.

SUGGESTION: A good way to isolate each setup in the laboratory is to make portable hoods out of cardboard or plywood that are large enough to cover the meter stick and accessory apparatus. The hoods should be painted black on the inside. A hole should be cut in the front of the hood so that photometer observations can be made without lifting the hood. In the absence of such hoods, the laboratory should be darkened and the instruments positioned in such a way as to keep the errors due to extraneous light as low as possible.

PROCEDURE:

(Bunsen or Joly photometer)

Make a preliminary examination of your photometer head under varying conditions of illumination. If a grease-spot type is used, observe effects by transmission and by reflection of light. Arrange the meter stick, photometer, standard light source, and unknown light source as illustrated in Fig. 30-1. Place the two light sources at opposite ends of the meter stick and adjust the position of the photometer until it is equally illuminated from both sides. With a Bunsen photometer, this condition is reached when the grease-spot is made as nearly invisible as possible. With a Joly photometer, the translucent blocks must appear equally illuminated to the same depth. Knowing the positions of the photometer and both lamps, record the distance of each lamp from the photometer head.

Repeat this procedure with the other light sources specified by your instructor.

ALTERNATE PROCEDURE:

(Photoelectric photometer)

Connect the galvanometer to the photoelectric photometer terminals and place the photometer on the meter stick facing a standard lamp mounted on one end of the meter stick. Move the photometer as necessary to give an approximately midscale deflection on the galvanometer, then adjust this position for a precisely known output reading on the galvanometer scale. *Remember this scale reading.* Knowing the positions of the photometer and standard lamp, record the distance s_1 of the standard source (from the photometer head).

Substitute an unknown lamp for the standard lamp, keeping the same socket position, and move the photometer head to a position which gives precisely the same output reading on the galvanometer as for the standard lamp. Record the distance s_2 of the unknown source (from the photometer head).

Repeat this procedure with the other light sources specified by your instructor.

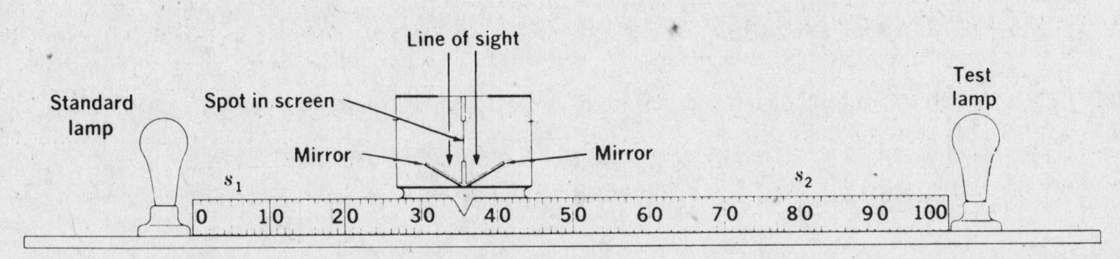

Figure 30-1

Note: Because of variations in galvanometer sensitivities and photometer output currents, the instructor may wish to modify the "midscale deflection" standard suggested above.

CALCULATIONS: For each trial the distances of the unknown source and standard source from the photometer are known. Assuming that your photometer functions according to the inverse square law, calculate the intensity of the unknown source for each trial. Compute the efficiency of each lamp in cd/watt.

OPTIONAL: Using three or four identical light sources, see if you can get data that shows that the squares of the distances from the sources to the photometer are proportional to the intensities of the source. Can you show that the photometer does function according to the inverse square law?

DATA

TRIAL	*Power rating of unknown source* (w)	*Intensity of standard source* I_1 (cd)	*Distance of standard source* s_1 (cm)	*Distance of unknown source* s_2 (cm)	*Intensity of unknown source* I_2 (cd)	*Efficiency of unknown source* (w)
1						
2						
3						

QUESTIONS: Your answers should be complete statements.

1. How does the efficiency of an old incandescent lamp compare with that of a new one? Give several possible reasons for the difference. Why is platinum used as a standard of luminous intensity?

....

....

....

.... 1

2. Could this experiment be used to compare the intensities of light sources having different colors? Justify your answer. Try to verify it experimentally.

....

....

....

....

....

.... 2

3. Is the intensity of an incandescent lamp equal in all directions? Explain and verify.

....

....

.... 3

4. Is the intensity of a lamp changed by placing a mirror behind it? Explain.

....

.... 4

5. In terms of efficiency, do you think it is better to use one 100-watt lamp or two 50-watt lamps? Explain.

....

.... 5

NAME PERIOD DATE INSTRUCTOR'S COMMENTS

31 EXPERIMENT Plane Mirrors

PURPOSE: (1) To verify the laws of reflection. (2) To show how images are formed by plane mirrors.

APPARATUS: Plane mirror; rectangular wooden block; ruler; protractor; pins; drawing paper; rubber bands or cellulose tape.

INTRODUCTION: The laws of reflection state: (1) the incident ray, the reflected ray, and the normal to the reflecting surface lie in the same plane and (2) the angle of incidence is equal to the angle of reflection. This experiment is designed to enable you to verify both of these laws.

The image of an object in a plane mirror is a virtual image. It is the same size as the object, erect, reversed right and left, and as far behind the mirror as the object is in front of the mirror. By a process of construction, we shall see how the image of an object in a plane mirror is formed. From this constructed image, we can verify the characteristics of images formed by plane mirrors.

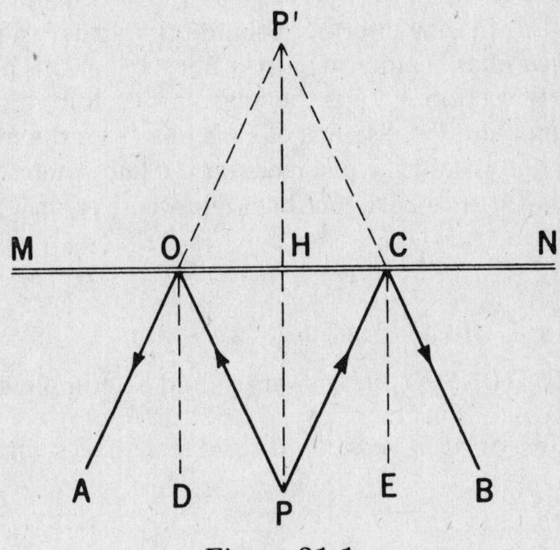

Figure 31-1

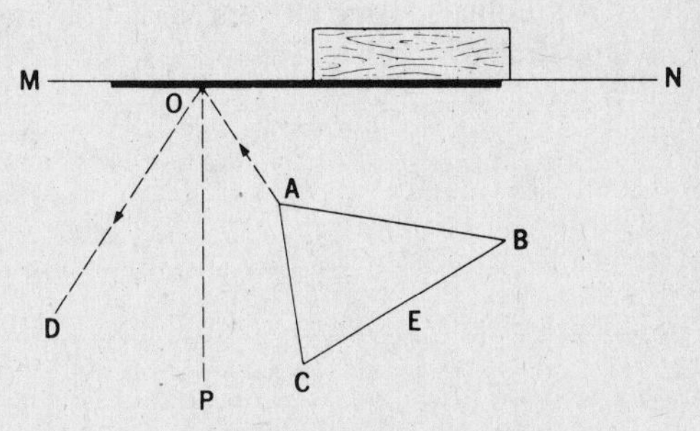

Figure 31-2

PROCEDURE:

1. Laws of reflection

Draw a line *MN* across the middle of a sheet of unlined paper. Place the mirror on the line *MN* so that the edge of its reflecting surface coincides with *MN*. The mirror must stand vertically and may be fastened to the rectangular wooden block by means of a rubber band or a piece of cellulose tape. Place a pin at some point *P*, as indicated in Fig. 31-1, about 3 or 4 cm from the front of the mirror. Lay a straight edge or ruler on the paper, at *A* for example, far enough from the point *P* so the angle *AOD* will be 30° or more. Sight along the edge of the ruler at the *image* of the pin *P* as you see it in the mirror. Then draw a line along the edge of the ruler, using a sharp-pointed pencil. In the same manner, locate a second sight line for the pin *P*, but from an entirely different angle, from *B* for example. Then remove the mirror and extend the two sight lines until they meet at the point *P'*. *All lines extending behind MN should be dashed lines.* Draw the lines *PO* and *PC*, which *represent incident rays of light from the pin to the mirror.* Draw the lines *OD* and *CE* perpendicular to the mirror line *MN*. Measure the distances *HP* and *HP'* to the nearest millimeter. By using a protractor, measure the angles of incidence *POD* and *PCE*, and the angles of reflection *AOD* and *BCE*.

2. Formation of images

As before, draw a line *MN* across the middle of a sheet of unlined paper and place the mirror on the line. In front of the mirror and not less than 4 cm from it, draw a scalene triangle. The sides of the triangle should be between 4 cm and 7 cm long. See Fig. 31-2. Place a pin at one of the vertices of the triangle, at *A* for example, and locate two sight lines for the image of the pin. Label both of these sight lines *A*. Without moving the mirror, proceed to lo-

DATA

Length of line *PH* ....mm	Angle *POD* ....°	Angle *PCE* ....°
Length of line *P'H* ....mm	Angle *AOD* ....°	Angle *BCE* ....°
Error ....mm	Error ....°	Error ....°

cate two sight lines for the image of the pin at *B*. Also locate two sight lines for the image of the pin at *C*. If necessary, the mirror may be moved from side to side along the line *MN* to permit a better view, but its reflecting surface must always coincide with the line *MN*.

Remove the mirror and extend the two sight lines for *A* until they intersect behind the mirror, using *dashed lines* behind the mirror line. Label this point of intersection A'; it is the image of *A*. Join *A* and A'; measure the distance of each one from the mirror line, and record the distances on the lines themselves. The difference between them represents error.

In the same manner extend the sight lines for *B* until they meet at B' to form the image of *B*. Join *B* and B', and measure their distances from the mirror line. Extend too, the sight lines for *C* until they meet at C'. Measure the distances of *C* and C' from the mirror line. Connect A', B', and C' with dashed lines to form the triangle $A'B'C'$. Measure the sides of the triangle *ABC*, and write the measurements along the respective sides. Measure also the sides of the image of the triangle $A'B'C'$, and write their lengths along the respective sides. Submit the completed drawing, properly labeled and identified, as part of your laboratory report.

QUESTIONS: Your answers should be complete statements.

1. Why is it impossible to form real images with a plane mirror?

....

....

.... 1

2. Why are ordinary plane mirrors coated on the back instead of the front? When would such a back coating be undesirable?

....

....

....

....

.... 2

3. To take a clear picture of an image in a plane mirror, should the camera be focused on the image or on the surface of the mirror? Try to verify your answer by experimentation.

....

.... 3

4. Why must the mirror in this experiment by perpendicular to the table for best results?

....

....

.... 4

5. What error is introduced by the thickness of the mirror?

....

....

.... 5

NAME PERIOD DATE INSTRUCTOR'S COMMENTS

32 EXPERIMENT Concave Mirrors

PURPOSE: (1) To study the images formed by concave mirrors. (2) To study the images formed by convex mirrors.

APPARATUS: Spherical mirrors, 4.0 cm diameter; and with a focal length of approximately 25 cm, concave and convex; meter sticks; image screen of white cardboard, ground glass, or plastic, approximately 25 cm square; candle or clear 25-w aquarium lamp.

INTRODUCTION: Concave mirrors cause parallel rays of light to converge. Different types and sizes of images may be formed by concave mirrors depending upon the distance of the object from the mirror. In this experiment we shall study the formation of these images.

Convex mirrors cause parallel rays of light to diverge. Only one type of image is produced by a convex mirror. Convex mirrors, used as rear view mirrors on buses and trucks, furnish the driver with a wide field of vision, but a misleading impression of distance.

PROCEDURE:

1. *Concave mirror*

a. Determine the focal length of the concave mirror by projecting the image of the sun on a small screen, and carefully measuring the image distance from the vertex of the mirror. If a direct view of the sun is not obtainable from the laboratory, the sharply defined image of a distant object outside the laboratory window will give a close approximation of the focal length of the mirror. The laboratory should be darkened as much as possible with one window being uncovered sufficiently to admit light from the distant object to reach the mirror. Record the focal length in the data table.

b. Set up the meter sticks and concave mirror as shown in Fig. 32-1. The apex of the V should be slightly below the center of the mirror. Mount the candle on one meter stick as far away from the mirror as possible. Measure this distance and record it as s_o. Mount the image screen on the other meter stick and move it back and forth until a sharp image of the candle is obtained. (It may be necessary to adjust the position of the mirror and to change the angle between the meter sticks in order to locate the image properly on the screen.) Measure the distance between the mirror and image screen and record it as s_i.

Measure the height of the candle flame as accurately as possible. Record it as h_o. Measure the height of the image of the flame and record it as h_i.

c. Interchange the candle and image screen. Make any necessary adjustments in the position of the image screen so as to obtain a sharply defined image. Measure and record s_o, s_i, h_o and h_i.

d. Find the position of the candle and screen for which h_o and h_i are equal.

e. Move the candle so that s_o is equal to the focal length of the mirror. Try to locate the image. Record your observations.

Observation: ....

....

....

f. Position the candle between the focal length and the mirror. Try to locate the image. What do you observe when you look into the mirror?

Observation: ....

....

....

2. *Convex mirror*

Turn the mirror around so that the convex surface is toward the meter sticks. Place the candle at the far end of one meter stick. Describe the image in the mirror. Move the candle closer to the mirror and record the corresponding changes in the image.

Observation: ....

....

....

....

CALCULATIONS: Compute the value of $\frac{1}{s_o}$, $\frac{1}{s_i}$, and $\frac{1}{f}$. Add $\frac{1}{s_o}$ and $\frac{1}{s_i}$ and compare it with $\frac{1}{f}$. Record the difference as your absolute error.

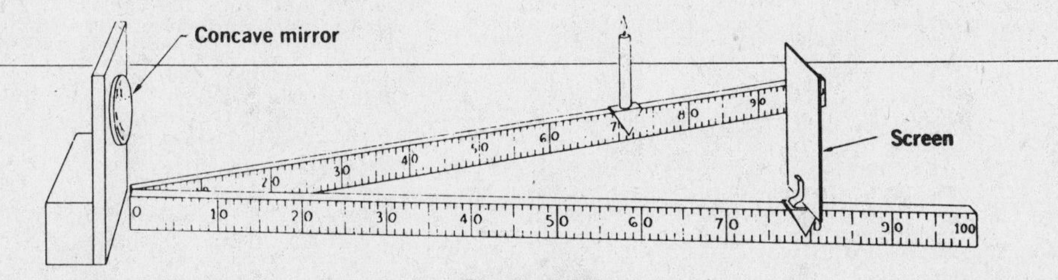

Figure 32-1

DATA *Focal length of concave mirror* ________ cm

TRIAL	s_o (cm)	s_i (cm)	h_o (cm)	h_i (cm)	$\frac{1}{s_o}$	$\frac{1}{s_i}$	$\frac{1}{s_o} + \frac{1}{s_i}$	$\frac{1}{f}$	*Abs. error*
1									
2									
3									
4 (Convex)									

QUESTIONS: Your answers should be complete statements.

1. What is the relationship between the data for Parts *b* and *c*?

....

.... 1

2. What is the relationship between the focal length of the mirror and the location of the object and image in Part *d*?

....

.... 2

3. What type of image is formed in Part *f*? Why is it possible to see the image when looking into the mirror, whereas it is not possible to form it on the screen?

....

....

.... 3

4. List a practical application for each part of this experiment.

....

....

....

....

....

....

.... 4

NAME PERIOD DATE INSTRUCTOR'S COMMENTS

33 EXPERIMENT Index of Refraction of Glass

PURPOSE: To determine the index of refraction of glass by means of refracted light rays.

APPARATUS: Glass plate, about 7 cm square and 9 mm thick, or a glass cube 5 cm on each side; glass prism, equilateral, with faces about 7.5 cm long and 9 mm thick; ruler; pencil; compass; pins; protractor; drawing paper.

INTRODUCTION: The index of refraction of a substance is defined as the ratio of the speed of light in a vacuum to its speed in that substance. Since the speed of light in air is only slightly different from the speed in a vacuum, a negligible error is introduced when we measure the index of refraction by permitting light to travel from air into another medium.

Willebrord Snell provided a simple direct method of measuring the index of refraction by defining it in terms of functions of the angle of incidence and the angle of refraction. The mathematical relationship, known as Snell's law, is

$$n = \frac{\sin i}{\sin r}$$

where n is the index of refraction, i the angle of incidence, and r the angle of refraction. In Fig. 33-1,

$$\sin i = \frac{AC}{AO} \text{ and } \sin r = \frac{DB}{OB}$$

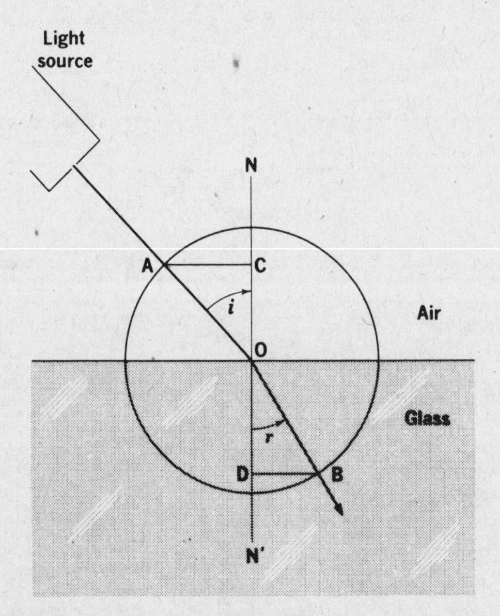

Figure 33-1

Since AO and OB are radii of the same circle, they are equal. Therefore,

$$n = \frac{\sin i}{\sin r} = \frac{AC}{DB}$$

The index of refraction of glass varies with its composition and with the wavelength of the light incident on the glass. Ordinary crown glass, when illuminated by white light, has a refractive index of 1.52; medium flint glass has an index of 1.63. Your experimental results should be precise enough to enable you to identify the kind of glass used in this experiment.

PROCEDURE:

1. *Index of refraction of glass plate*

Place the glass plate or cube on the center of a sheet of blank paper and outline it with a sharp-pointed pencil. About 1 cm from the lower left-hand corner of the plate place a pin A as close to the glass as possible, as in Fig. 33-2. At B, about 1 cm from the upper right-hand corner of the glass plate, stick a second pin, as close to the plate as possible. At C, not less than 7 cm from B, stick a third pin *so that A is in line with B and C as seen through the glass plate.* Keep the eye you sight with near the level of the table top. *(Do not align the pins with B and C as seen above the glass plate.)*

Remove the glass plate, and join the points, A, B, and C to represent the path of the light traveling *from* the pin C through the air to B and through the glass to A. Identify the *incident ray* and the *refracted ray*.

From B construct the normals, NB and BN' to the line DE. Using as large a radius as possible, describe a circle with point B as the center which intersects BA, BN', BC, and BN.

From the point of intersection of the circle with the incident ray BC draw a line x perpendicular to BN. Measure the length of x to the nearest 0.01 cm and record it in your data table. From the point of intersection of the circle with the refracted ray BA draw a line y perpendicular to the normal BO. Measure the length of y to the nearest 0.01 cm and record its value. Compute the index of refraction and record it in the data table.

By means of a protractor, measure the angle of incidence i and the angle of refraction r. From these data, find the index of refraction of the glass plate.

2. *Index of refraction of glass prism*

Arrange the glass prism on a separate sheet of blank paper, and using information gained in the previous trial, proceed to determine the index of refraction of the prism. Record all necessary data.

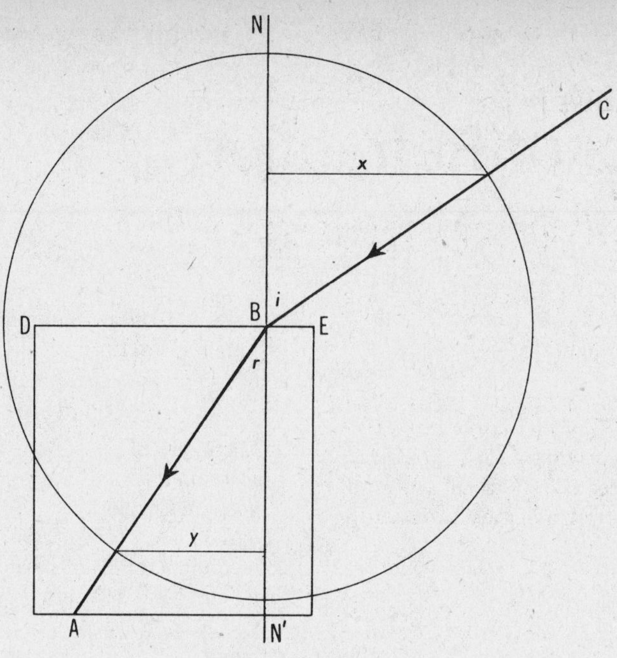

Figure 33-2

DATA

TRIAL	x (cm)	y (cm)	*Index of refraction* $\frac{x}{y}$	$\angle i$ (°)	$\angle r$ (°)	*Index of refraction* $\frac{\sin i}{\sin r}$	*Kind of glass*
■							
▲							

QUESTIONS: Your answers should be complete statements.

1. What is the size of the angle of refraction if the angle of incidence is 0°?

....

.... 1

2. What is meant by the critical angle?

....

....

.... 2

3. From the values of the index of refraction, calculate the critical angle for the glass prism.

....

....

.... 3

4. How could you verify the critical angles of the plate and prism experimentally?

....

....

....

....

....

.... 4

NAME PERIOD DATE INSTRUCTOR'S COMMENTS

34 EXPERIMENT Index of Refraction by a Microscope

PURPOSE: To measure the index of refraction of glass and water using a compound microscope.

APPARATUS: Microscope with 16 mm (10x) and 32 mm (4x) objectives (single 16 mm divisible objective will do); vernier calipers, metric; white paper; white index cards; rubber cement; glass plate, 6-10 mm thick; evaporating dish; magnifier; distilled water; china-marking pencils.

INTRODUCTION: The index of refraction of a substance is defined as the ratio of the speed of light in a vacuum to its speed in that substance. Since the speed of light in air is only slightly different from its speed in a vacuum, a negligible error is introduced if the index of refraction is measured by comparing the speed in air to the speed in the second medium.

Because of refraction, objects appear to be closer when viewed through a dense medium such as glass or water, than when viewed through air. This phenomenon can be used to find the index of refraction n of the medium by means of the equation

$$n = \frac{t}{t - d}$$

where t is the thickness of a transparent medium and d is the apparent shortening of the perpendicular line of sight through the medium.

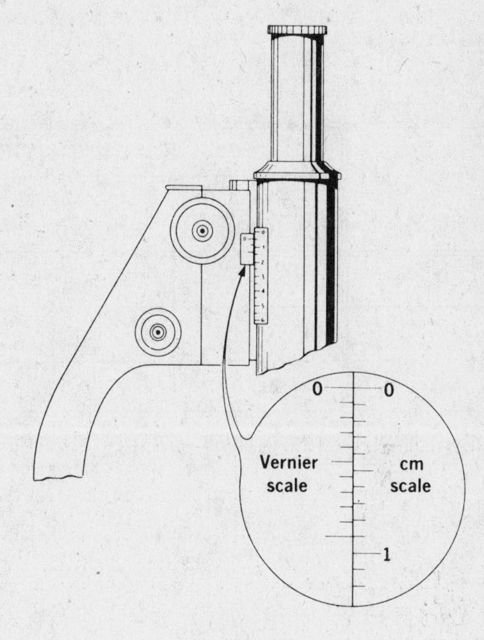

Figure 34-1

PROCEDURE:

1. Index of refraction of glass

The microscope may be used to measure distances normal to the stage by attaching a vernier scale to the side of the rack as shown in Fig. 34-1. Using a very sharp pencil, *carefully* construct a centimeter scale about 6 cm long with 1 mm subdivisions on the edge of a piece of white paper and a vernier scale on another piece of white paper. The vernier should consist of 10 equal divisions in a space of 9 mm. Figure 1-4 provides a suitable pattern for making these scales. Mount the scales on the microscope as shown in Fig. 34-1 so that the zero index of the fixed vernier scale is at or below the zero index of the centimeter scale when the body tube is fully lowered.

Determine whether the 16 mm objective provides the *working distance* required for the thickness of the glass plate. This may be done by making a pencil mark on a white index card, mounting the card on the stage, and focusing on it with the 16 mm objective in place. Without shifting the position of the card, place the glass plate over the mark and *cautiously* lower the body tube. If the mark can again be brought into sharp focus, the working distance of the lens is adequate and it may be used in this part of the experiment.

CAUTION: *The objective must not be forced down in contact with the glass plate.* If this second focus is not achieved, the 32 mm objective must be used; or if the 16 mm objective is divisible, the front lens section may be removed to provide the equivalent lens.

With an adequate working distance assured, again mount the card and bring the pencil mark into sharp focus. Record the body-tube position as the *focus in air* in Part 1 of the data table, using a magnifier to estimate the vernier reading to the nearest 0.001 cm. With the glass plate over the card, again bring the mark into sharp focus. Read the body-tube position as before and record as the *focus in glass.* The distance the objective has been raised is the distance d. Now focus the microscope on the top surface of the glass plate. Read the body-tube position on the vernier scale as before and record as the *surface position.* The total distance the objective has been raised is the thickness of the glass plate t.

Make two additional marks on the card and determine d and t for each mark, recording the vernier-scale readings as before. Make three trials and compute the average value of n.

2. Index of refraction of water

Make a mark in the bottom of a clean evaporating dish (a sharpened china-marking pencil may be used) and bring it into focus under the microscope using the 32 mm objective. Record the body-tube position as the *focus in air* in your data table. Add distilled water to the dish until it is approximately two-thirds full, focus on the mark and then on the surface, recording the required data for each operation. If difficulty is experienced in focusing on the water surface, a little chalk dust sprinkled on the surface by tapping a blackboard eraser over the dish may help. Repeat these measurements for two other depths of water and record the required data. Make three trials and compute the average value of *n*.

DATA Part 1

TRIAL	FOCUS POSITION *In air* (cm)	*In glass* (cm)	*Surface* (cm)	*d* (cm)	*t* (cm)	*n*
1						
2						
3						
Average value of n for the glass plate						

DATA Part 2

TRIAL	FOCUS POSITION *In air* (cm)	*In water* (cm)	*Surface* (cm)	*d* (cm)	*t* (cm)	*n*
1						
2						
3						
Average value of n for water						

QUESTIONS: Your answers should be complete statements. Problem solutions should be set down in the space provided below.

1. Show that the expression for the index of refraction employed in this experiment can be derived from Snell's law.

1

2. A pail 35.0 cm deep is filled with water to within 10.0 cm of the rim. What is the apparent depth of the water?

2

NAME PERIOD DATE INSTRUCTOR'S COMMENTS

35 EXPERIMENT Converging Lenses

PURPOSE: (1) To find the focal length of a converging lens. (2) To study the image-forming characteristics of a converging lens.

APPARATUS: Object screen; electric lamp and object box; meter stick; supports for meter stick; lens holder; screen holder; cardboard screen with metric scale; double convex lens, preferably 10-cm or 15-cm focal length; cobalt glass filter.

INTRODUCTION: Converging lenses can produce both real and virtual images; diverging lenses can produce only virtual images. In this experiment we shall study image formation by a converging lens. A spherical lens made of crown glass has a focal length very nearly equal to its radius of curvature, a fact commonly used in lens diagrams.

When an object is illuminated, each point on its surface acts as a source of diverging rays. When some of these rays from a point on the object are incident on a converging lens properly placed, they converge at a point on the opposite side of the lens forming an image of the object point. Collectively these image points form an image of the object. Since the rays of light converge to form the image, it is a *real image* and can be formed on a screen.

If a lens is less than its focal distance from an object, the refracted rays do not converge and no real image is formed. Instead a *virtual image* can be seen by the eye looking through the lens in the direction of the object, but cannot be formed on a screen.

PROCEDURE:

1. Focal length.

Case 1. Support a converging lens and a cardboard screen on a meter stick as shown in Fig. 35-1. Place the lens at the 50-cm mark and the screen near the 70-cm mark and point the meter stick at the sun. Then, looking through a cobalt glass filter, move the screen along the meter stick until a position is found where the image of the sun on the screen is as small as possible. The distance from the lens to the screen is the *focal length* of the lens. Record the focal length of the lens in the data table to the degree of precision that your metric scale will allow.

If the sun is not visible from the laboratory window, a house or a tree several hundred meters distant may be used as the object. (The cobalt glass filter is unnecessary in this case.) The image may be found most satisfactorily by darkening the laboratory and, with your back to the open window, aiming the meter stick back over your shoulder. Adjust the position of the screen until a sharply defined image is observed. Record the distance between the lens and the screen in your rough notes.

Move the lens to the 60-cm position and repeat the procedure using a different distant object. Average the two results and record as the approximate focal length of the lens.

2. Formation of images.

Case 2. At one end of the meter stick place the illuminated object screen. Place the lens far enough from the object so it will be at a distance greater than twice the focal length of the lens. See Fig. 35-2. Move the cardboard screen along the meter stick until the image that is formed upon it is as well defined as possible. Read and record in the data table the position of the object, of the lens, and of the screen. Measure, too, the height of the object, h_o, and the height of the image, h_i. Record all these measurements to the degree of precision permitted by the metric scale you are using.

Case 3. For the second trial, place the lens at a position which is distant from the object just twice the focal length of the lens. Then adjust the screen until the best-defined image is obtained. Read all positions and measure the heights of object and image and record in the data table.

Case 4. For the third trial, place the lens so that its distance from the object is more than once and less than twice the focal length of the lens. Adjust the

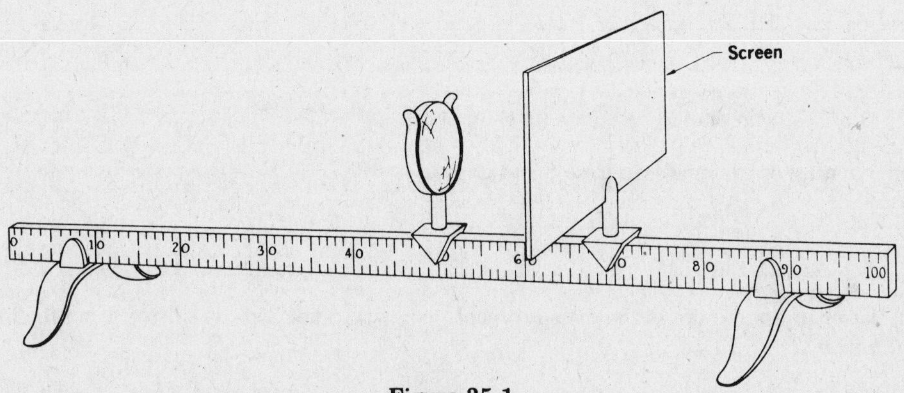

Figure 35-1

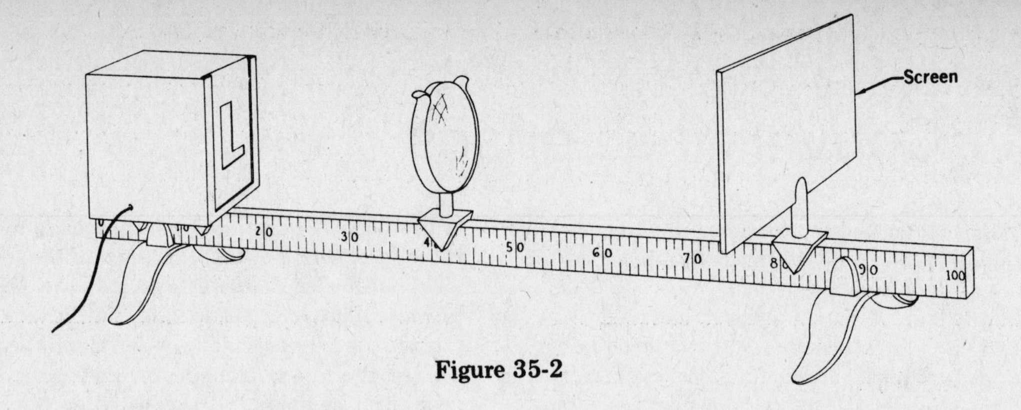

Figure 35-2

screen to secure the best-defined image, and record all measurements as in the preceding cases.

Case 5. Place the lens so that it is exactly one focal length away from the object. Try to form an image on the screen.

Observation: ....

....

....

....

Case 6. Place the lens so that it is less than one focal length away from the object. Try to form an image on the screen. Remove the screen, and placing your eye close to the lens, look through the lens at the object. What kind of image may be produced by a converging lens using Case 6?

....

....

....

3. Optional

If you wish to do some additional work with lenses, try to devise a method of locating the virtual images formed by converging and diverging lenses. Investigate the *parallax method* of locating images and see if this suggests a way of determining the image distance of a virtual image.

CALCULATIONS: For each of the three trials recorded, calculate the object distance s_o and the image distance s_i. Next calculate the reciprocals of these distances. Add the values of these reciprocals and compare the sum with the reciprocal of the focal length, $1/f$. In each case also divide the object distance by the image distance and the object size by the image size and compare these two results. Record these values.

DATA *Focal length of lens* ________ cm

Position of object (cm)	*Position of lens* (cm)	*Position of image* (cm)	s_o (cm)	s_i (cm)	$\frac{1}{s_o}$	$\frac{1}{s_i}$	$\frac{1}{s_o} + \frac{1}{s_i}$	$\frac{1}{f}$	h_o (cm)	h_i (cm)	$\frac{s_o}{s_i}$	$\frac{h_o}{h_i}$

QUESTIONS: Your answers should be complete statements.

1. Explain the nature of the images in Case 5 and Case 6.

....

.... 1

2. Why is it better to use the sun as the distant object than it is to use a house or tree when finding the focal length of the lens?

.... 2

3. Give a practical application of each of the six cases of image formation by convex lenses.

....

....

....

....

....

.... 3

NAME PERIOD DATE INSTRUCTOR'S COMMENTS

36 EXPERIMENT Focal Length of a Lens

PURPOSE: To measure the focal length of diverging lenses.

APPARATUS: Object screen; electric lamp and object box; meter stick and supports; double convex lens, 5-cm focal length; double concave lenses, 10-cm and 15-cm focal lengths; two lens holders; cardboard screen; screen holder.

INTRODUCTION: The relation between the object distance s_o, the image distance s_i, and the focal length f of a lens is given by the fundamental lens equation

$$\frac{1}{s_o} + \frac{1}{s_i} = \frac{1}{f}$$

One standard practice in optics is to use the optical center of the lens as the origin and to measure distances from right to left as negative and distances from left to right as positive, and to let the incident rays travel from left to right. Consequently, s_o is positive for *real* objects and negative for *virtual* objects; s_i is positive for *real* images and negative for *virtual* images; and f is positive for converging lenses and negative for diverging lenses.

In order to find the focal length of a diverging lens we shall use a converging lens ahead of it to converge the rays before they reach the diverging lens. If a converging lens forms a real image at I, Fig. 36-1, the introduction of the diverging lens between this lens and the real image will cause the image to be formed farther away at I'. By taking the distance from the diverging lens to the position of I as the object distance s_c and to the position of I' as the image distance s_i, the lens formula may be used to determine the focal length f of the diverging lens.

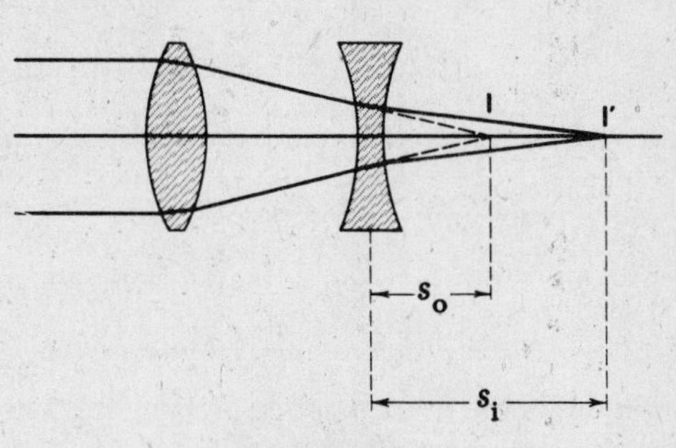

Figure 36-1

PROCEDURE: 1. Mount the converging lens on the optical bench between the object box and the cardboard screen and adjust the lens and screen until a sharp image is formed. Determine at this point whether a reduced or enlarged image is preferable. This image locates I of Fig. 36-1. Observe the edges of the image closely for color fringes while slowly moving the screen through the position of sharpest focus. Insert a diverging lens between the converging lens and the screen and again move the screen until a sharp image is formed. This is the position I'.

Record the values of s_o and s_i in the data table to the nearest 0.01 cm and calculate the focal length. Be sure to use + and - signs correctly when recording these distances.

Make at least three additional determinations of f by varying s_o and s_i. Record all data as before and find the average focal length of the lens.

2. Substitute a second diverging lens and determine its focal length as before, finding the average of at least four trials.

DATA

FIRST LENS				SECOND LENS			
TRIAL	s_o (cm)	s_o (cm)	f (cm)	TRIAL	s_o (cm)	s_i (cm)	f (cm)
1				1			
2				2			
3				3			
4				4			
		Average				*Average*	

QUESTIONS: Your answers should be complete statements.

1. a. What variation in color of the edge of the image occurs when the screen is moved through the point of sharpest focus? b. Can you suggest a possible explanation?

a. ....

....

....1a

b. ....

....

....

....1b

2. Why does the usual practice of assigning a negative value to the image distance s_i of a diverging lens not apply in this experiment?

....

....

....

....

....

....2

3. What measurement was assigned a negative value in this experiment? Why?

....

....

....3

NAME PERIOD DATE INSTRUCTOR'S COMMENTS

37 EXPERIMENT Lens Magnification

PURPOSE: To determine experimentally the magnification of short-focal-length lenses commonly used as simple magnifiers.

APPARATUS: Three converging lenses of different focal lengths ranging from 5 cm to 15 cm; ring stand, with 2 iron rings, 1 in and 2 in O.D.; object screen; microscope slide 2.5 × 7.5 cm with 1 cm^2 of millimeter cross-section paper mounted; image screen; white Bristol board with millimeter scale; meter stick; lens holder; screen holder.

SUGGESTIONS: The object screens may be made up in advance of the laboratory period and saved for use year after year; or each student may be supplied with the raw materials from which he makes his own object screen. In either case, the object square of cross-section paper should be cut slightly larger than 1 cm^2 so the graduation lines at each edge of the object piece will be perpendicular to the edge.

INTRODUCTION: A converging lens of short focal length is frequently used to magnify small objects and may be in the form of a reading glass, a simple magnifier, or the eyepiece of a compound microscope or refracting telescope. The lens is held slightly less than one focal length away from the object and the eye is placed close to the lens on the side opposite the object. This is a practical application of the principle of Case 6 for converging lenses; the image is virtual, erect, enlarged, and appears to be on the same side of the lens as the object.

The linear magnification M of a lens is simply the ratio of the image size h_i to the object size h_o.

$$M = \frac{h_i}{h_o} \qquad (1)$$

From $h_o/h_i = s_o/s_i$, it is apparent that the lens magnification may be expressed in terms of the image distance s_i and object distance s_o.

$$M = \frac{s_i}{s_o} \qquad (2)$$

The normal eye can focus on objects as close as 25 cm, this distance being known as the *distance for most distinct vision.* Thus, if a simple magnifier is placed so that the image distance is 25 cm, the maximum detail of the object will be revealed by the image. Because the object is very near the principal focus, the magnification of a simple magnifier is *approximately* equal to the ratio of the distance for most distinct vision to the focal length f of the lens.

$$M = \frac{25\text{ cm}}{f}\ \text{(approx.)} \qquad (3)$$

To arrive at a more precise expression for the magnification of a simple magnifier than the approximation given above, consider the lens equation:

$$\frac{1}{s_o} + \frac{1}{s_i} = \frac{1}{f} \qquad (4)$$

Multiplying by s_i and rearranging terms, Equation 4 becomes

$$\frac{s_i}{s_o} = \frac{s_i}{f} - \frac{s_i}{s_i} = \frac{s_i}{f} - 1$$

Substituting in Equation 2:

$$M = \frac{s_i}{f} - 1 \qquad (5)$$

Since the image formed at the distance for most distinct vision is virtual, $s_i = -25$ cm,

$$M = \frac{25\text{ cm}}{f} + 1 \qquad (6)$$

where f is the focal length expressed in centimeters.

PROCEDURE: Determine the focal lengths of the three lenses to the nearest 0.01 cm by the method used in Experiment 35 and record each measurement in the data table. Magnification data will be taken for the three lenses in the order in which you have listed them in the data table. Take the necessary precautions to insure that the focal-length identity of each lens is retained.

Arrange the support stand, lens support, and object support as shown in Fig. 37-1. Arrange the image

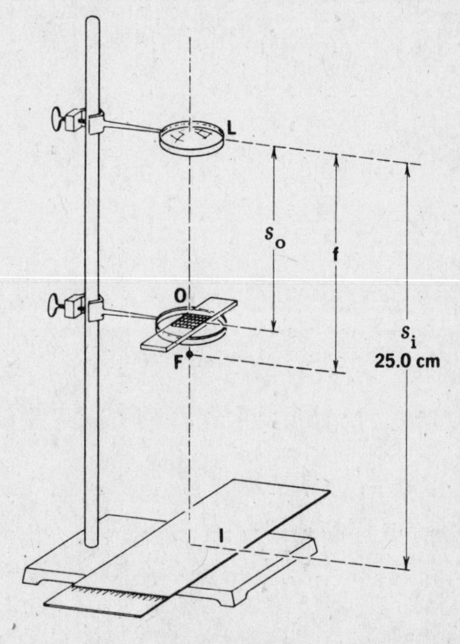

Figure 37-1

screen on the base of the stand so the metric scale is centered under the two ring supports.

Mount the first lens and adjust the lens support for a value of s_i of 25.0 cm measured from the lens plane to the image screen. Record s_i in the data table for all three trials since this distance will remain constant for all lenses. Arrange the object screen below the lens so the object is centered on the principal axis of the lens.

With one eye close to the lens and sighting through the lens along the principal axis and the other eye focused on the image screen, vary the object position until a sharp image of the aperture appears to fall on the scale of the image screen.

Determine whether a final adjustment of the object position is required by moving the eye laterally back and forth across the lens (parallax). Readjust the object position, if necessary, to eliminate any relative motion (parallax) between the virtual image and the lines of the metric scale as the eye is moved.

Position the object screen and the image screen as necessary to locate one edge of the centimeter square object on an appropriate mark on the metric scale.

Carefully read the image size h_i on the scale to the nearest 0.01 cm and record. Assume the cross-section graduations of the object to be accurate to 0.01 cm in recording the object size (one dimension). Measure the object distance s_o to the nearest 0.01 cm and record.

Repeat the procedure with the second lens and record the required data in the appropriate columns of the data table.

Repeat using the third lens and record the required data as before.

CALCULATIONS: From the tabulated data compute the magnification of each lens by three methods using equations 1, 2, and 6. Determine the average value for the magnification of each lens. Record all results in the appropriate columns of the data table.

DIAGRAM: Using a well-sharpened pencil, straight edge, and compass, carefully construct a ray diagram for a simple magnifier. Show an object, an image, and all necessary construction lines. Label fully and attach to your report.

DATA

TRIAL	f (cm)	s_i (cm)	s_o (cm)	h_i (cm)	h_o (cm)	MAGNIFICATION $\frac{s_i}{s_o}$	$\frac{h_i}{h_o}$	$\frac{25.0\ cm}{f} + 1$	Ave.
1									
2									
3									

NAME PERIOD DATE INSTRUCTOR'S COMMENTS

38 EXPERIMENT The Compound Microscope

PURPOSE: To construct the lens system of a compound microscope and determine experimentally its magnification.

APPARATUS: Optical bench consisting of a meter stick, 2 supports, light source, 3 screen holders, and 3 lens holders; 2 converging lenses, 5-cm focal length; object screen, black Bristol board with 4-mm diam. aperture covered with wire gauze (see suggestion below); first-image screen, white Bristol board 10 X 12.5 cm with millimeter scale; second-image screen, white Bristol board 12.5 X 15 cm with 3.5-cm diam. aperture, and metric scale (see suggestion below); steel metric rule graduated in 0.5 mm.

SUGGESTION: The object screens and second-image screens should be prepared in advance of the laboratory period. Screens constructed according to the following directions are quite satisfactory and may be retained for use year after year.

Object screen: Cut brass sheet, B and S No. 20, or iron sheet, B and S No. 28, into pieces 3 X 5 cm. Carefully drill a hole approximately 4 mm in diameter through the center of each sheet. (Use a No. 23 drill and dress with a flat file.) Mount the metal plate on a standard object screen made of black Bristol board with a triangular wire-gauze aperture to provide a small, round, wire-gauze aperture.

Second-image screen: The aperture location given below accommodates a lens system having a principal axis 5.5 cm or 7 cm above a meter-stick optical bench and is suitable for 3.75-cm and 5-cm diameter lenses when mounted in their respective standard lens supports on a meter stick. If in doubt, determine the height of the lens center above the optical bench and adjust the aperture center location accordingly.

Cut white Bristol board into 12.5 X 16 cm rectangles and locate the aperture center 8 cm from either 12.5-cm edge and 7 cm from one 16-cm edge. Draw a line from edge to edge through the aperture center parallel to the two 16-cm sides. Draw a second line through the aperture center perpendicular to the first extending it about 1 cm each way. Turn the screen so that the aperture center is at the proper height for your optical bench when measured from the *bottom* of the screen before cementing the paper scale.

Cut a metric paper scale (printed for horizontal use) to slightly in excess of 15 cm. Locate the 7.5-cm graduation mark precisely on the aperture center with the millimeter graduations centered on the edge-to-edge line and bond the scale to the screen with a rubber-base cement. Trim the screen to 15 cm by the scale graduations. Cut out an aperture of approximately 3.5-cm diameter. (A 1⅜ in Greenlee chassis punch is an excellent tool for this purpose.)

INTRODUCTION: The lens system of the compound microscope consists of two high quality, short focal-length, converging lenses. One lens, the *objective*, is located slightly more than its focal length from the object O and produces a real, inverted, and enlarged image I in front of the second lens, the *eyepiece* or *ocular*. This real image becomes the *object* for the eyepiece located slightly less than its focal length away. Thus, the eyepiece is used as a simple magnifier to form a virtual, erect, and enlarged second image I'.

The linear magnification M of a lens is equal to the ratio of the image size h_i to the object size h_o and to the ratio of the image distance s_i to the object distance s_o. Thus, the magnification M_o of the microscope objective is

$$M_o = \frac{h_i}{h_o} = \frac{s_i}{s_o} \tag{1}$$

where h_i is the diameter of the first image I, h_o the diameter of the object aperture O, s_i the distance from the objective lens L_o to the first-image screen I, and s_o the distance from the objective lens L_o to the object screen O.

Similarly, the magnification M_e of the eyepiece is

$$M_e = \frac{h_i'}{h_i} = \frac{s_i'}{s_o'} \tag{2}$$

where h_i' is the diameter of the second image I', s_i' the distance from the eyepiece lens L_e to the second-image screen I' and s_o' the distance from the eyepiece lens L_e to the first-image screen I.

The overall magnification M of the compound lens system is, of course, equal to the ratio of the second-image size to the object size. From Equation 1,

$$M = \frac{h_i'}{h_o} \tag{3}$$

Substituting in Equation 2:

$$h_i = M_o h_o$$

$$M = \frac{h_i'}{M_o h_o}$$

Solving for h_o:

$$h_o = \frac{h_i'}{M_o M_e}$$

Substituting in Equation 3:

$$M = M_o M_e \tag{4}$$

In this experiment we shall determine experimentally the separate magnifications of the objective and

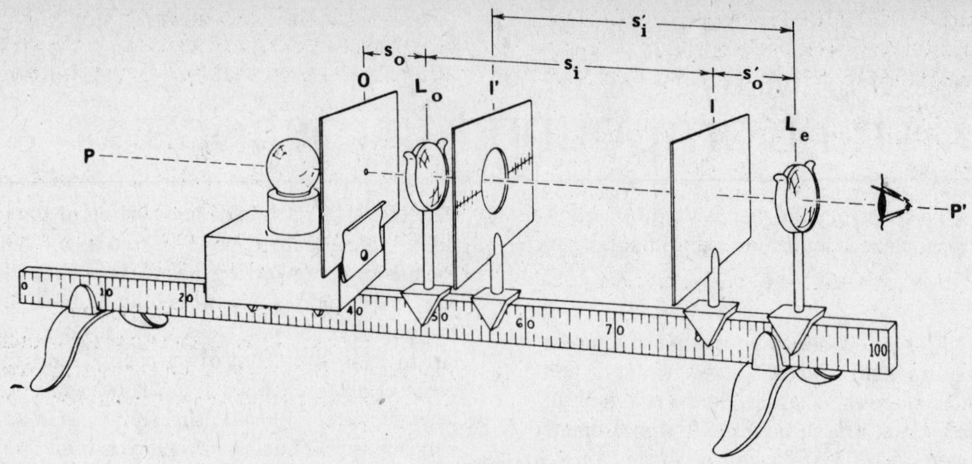

Figure 38-1

eyepiece lenses using both size and distance ratios. We shall then determine the magnification of the lens system using both the size ratio and the product of the individual lens magnifications.

PROCEDURE: If the focal lengths of the two lenses are not known, they should be determined before proceeding with the experiment. Using one short focal-length lens as a simple magnifier and a steel metric rule graduated in 0.5 mm, carefully measure the diameter of the object-screen aperture estimating to the nearest 0.01 cm. Record as s_o in the data table.

Set up the optical bench as shown in Fig. 38-1. Locate one short focal-length lens near the right end of the bench to serve as the eyepiece L_e and adjust the position of the first-image screen I approximately one focal length away. Locate the second image I' approximately 25 cm from L_e, the remaining short focal-length lens (as the objective L_o) about 5 cm from I', and the object screen slightly more than a focal length away from L_o. The object-screen aperture and its luminous source, the two lenses, and the second image-screen aperture must be centered on a common principal axis PP'.

Illuminate the object (the laboratory should be darkened) and adjust the positions of the first-image screen I and the objective L_o to give a sharply defined real image, keeping the image distance roughly 5 times the object distance. Shift the position of the object aperture slightly with respect to the principal axis if necessary to center the image on the first-image screen in front of the eyepiece lens. A final focus can be attained most readily by slight adjustments of the objective lens.

Shift the first-image screen I in its holder (being careful not to change its location on the optical bench) so the image falls on the millimeter scale. Read the diameter of the image estimating to the nearest 0.01 cm and record as h_i in the data table. Similarly record the object distance s_o and the first image distance s_i to the nearest 0.01 cm.

With one eye close to the eyepiece lens, view the back of the first-image screen and adjust the eyepiece slightly to bring it into sharp focus. Remove the screen I from its holder and reset screen I' if necessary to place it approximately 25 cm from L_e. Again with one eye close to the eyepiece and looking along the principal axis, view the virtual second image of the object.

Focus the other eye on the second-image screen I' and adjust the object screen slightly in its holder to superimpose the virtual image symmetrically about the aperture of the second-image screen. A little practice in this use of both eyes will develop skill in viewing the image superimposed on the metric scale of the screen. Adjust the position of the eyepiece slightly to yield the best definition of the wire-gauge

DATA

OBJECTIVE				EYEPIECE			M_o			M_e			M		
h_o (cm)	h_i (cm)	s_o (cm)	s_i (cm)	h_i' (cm)	s_o' (cm)	s_i' (cm)	$\frac{h_i}{h_o}$	$\frac{s_i}{s_o}$	Ave.	$\frac{h_i'}{h_i}$	$\frac{s_i'}{s_o'}$	Ave.	$\frac{h_i'}{h_o}$	M_oM_e	Ave.

image on the screen. With the image properly positioned on the metric scale, read the image diameter to the nearest 0.01 cm, and record as h_i'. Similarly record s_o' and s_i'.

As a second trial, increase the object distance s_o about 0.5 cm by moving the objective L_o 0.5 cm to the left and the object O 1 cm to the left, repeat the entire procedure, and record the required data.

If time permits, complete a third trial by increasing s_c as before. Record data as in previous trials.

CALCULATIONS:

1. Magnification of the objective

Compute the magnification of M_o of the objective lens for each trial from both size and distance data according to Equation 1, determine the average magnification for each trial, and record the results in the data table.

2. Magnification of the eyepiece

Compute the magnification M_e of the eyepiece lens for each trial using the data according to Equation 2. Determine the average eyepiece magnification for each trial and record the results.

3. Magnification of the lens system

By Equations 3 and 4, compute the overall magnification M of the compound lens system for each trial, determine the average value M for each trial, and record.

QUESTIONS: Your answers should be complete statements. Problem solutions should be set down in the space provided below.

1. Considering the focal length of the lens and the related object and image distances, the principle of which case for converging lenses is used (a) for the objective lens of a microscope, (b) for the eyepiece lens?

   a. ....

   ....1a

   b. ....

   ....1b

2. Assuming that the focal length of the eyepiece f_e is very small compared with the length of the microscope tube l (the distance between objective and eyepiece lenses) and that the object to be magnified is very near to the principal focus of the objective lens of focal length f_o, show algebraically that the following expression for the total magnification is approximately correct:

$$\frac{l \times 25\text{ cm}}{f_o \times f_e}$$

2

3. A converging lens of focal length 2.5 cm is used as an eyepiece lens to form a virtual image at the distance of most distinct vision. What is its magnification?

3

4. The lens of Problem 3 is used as the eyepiece of a compound microscope, the objective of which is a converging lens of 0.75 cm focal length. The real image is formed 12 cm from the objective. When the eye is held close to the eyepiece the virtual image is viewed at the distance of most distinct vision. What is the magnification of the instrument?

4

NAME PERIOD DATE INSTRUCTOR'S COMMENTS

39 EXPERIMENT The Refracting Telescope

PURPOSE: To construct the lens system of a refracting telescope and determine experimentally its magnification.

APPARATUS: Optical bench consisting of a meter stick, 2 supports, 1 screen holder, and 3 lens holders; light box, with large-object screen; image screen, white Bristol board 10 × 12.5 cm; 1 long focal-length converging lens, f = 25 to 35 cm; 2 short focal-length converging lenses, f = 3 to 10 cm; 1 short focal-length diverging lens, f = 10 to 15 cm; telescope magnification scale, in centimeter divisions, mounted; measuring tape, 15 m.

SUGGESTION: 1. Suitable magnification scales may be made from 20-cm paper scales printed for horizontal use. Make the centimeter graduations heavy black lines and extend them to within 1.5 cm of the lower edge of the paper strip. Below these heavy lines print the numerals 1 cm high. Cement the paper scale to a suitable backing and mount on a ring stand with a utility clamp.

2. If time is short, the direct measurement of magnification with the magnification scale can be omitted from the procedure.

INTRODUCTION: Refracting telescopes are of three general types: (1) *celestial* or astronomical, (2) *terrestrial*, and (3) *Galilean* telescope or opera glass. The essential elements of the lens system of the refracting telescope are the same as those of the microscope; an objective forms a real image of the object to be magnified, and an eyepiece, using this image as an object, forms an enlarged virtual image.

A telescope does not magnify objects in the sense that a microscope does. In a microscope, the image is actually larger than the object. In a telescope, however, the image is much smaller than the object, but since the image is much closer to the eye than the object is, the image *appears* to be larger than the object. Hence, the magnification of a telescope is an *apparent* magnification rather than a real one.

The astronomical telescope in its simplest form consists of a pair of converging lenses, an objective of long focal length f_o, and an eyepiece of short focal length f_e. When the object to be viewed is far away, a real first image is formed at the principal focus of the objective. The eyepiece, located approximately a focal length f_e away, forms an enlarged second image. The apparent magnification of the astronomical telescope *for distant objects* is

$$M = \frac{f_o}{f_e} \qquad (1)$$

The telescope length l (separation of the lenses) for *distant objects* is

$$l = f_o + f_e \qquad (2)$$

When the telescope is focused on a nearby object (s_o being a measurable distance), the first image is formed beyond the principal focus and the tube length must be *increased.* Equations 1 and 2 would lead to erroneous results, if applied to such a case. The tube length is the sum of the distances the first image is located from the objective (s_i) and from the eyepiece (s_o'), this image serving as the object for the eyepiece lens.

$$l = s_i + s_o' \qquad (3)$$

The total magnification of the telescope is, of course, equal to the product of the magnification of the separate lenses and to the ratio of the size of the second image h_i' to the apparent size of the object h_o.

$$M = M_o M_e = \frac{h_i'}{h_o} \qquad (4)$$

The terrestrial telescope is similar to the astronomical version except for the inclusion of an inverting lens between the objective and the eyepiece to reinvert the real image. This enables the observer to view the virtual image in the same erect position as the object.

The Galilean telescope or opera glass consists of a conventional objective and a diverging (negative) eyepiece lens. The objective alone would form a real image of a distant object at its principal focus. The diverging eyepiece lens is placed ahead of this focal point so that its principal focus coincides with that of the objective. The real image then becomes a *virtual object* for the eyepiece and an enlarged virtual image is formed. Since this image is erect, the instrument may be used as a terrestrial telescope.

The apparent magnification of the Galilean telescope *for distant objects* is given by the expression

$$M = \frac{f_o}{f_e} \qquad (5)$$

Since f_e is negative for a diverging lens, the magnification is positive. The telescope lengh l for *distant ob-*

jects is the *algebraic* sum of the focal lengths of the two lenses (Equation 2).

PROCEDURE:

1. Lens focal lengths

Determine the focal length of each of the three converging lenses by the method used in Experiment 35. Use the method of Experiment 36 to determine the focal length of the diverging lens. Devise a suitable means for maintaining the focal length identity of each lens in use throughout the experiment.

2. Astronomical telescope

Construct a simple astronomical telescope on the optical bench using two converging lenses, a long focal-length objective and a short focal-length eyepiece. Locate the eyepiece near the right end of the bench and determine, from the focal lengths of the lenses used, the approximate location of the objective. Focus as sharply as possible on a distant object as before and record the lens positions in the first data table. Compare the tube length l with the sum of the focal lengths and compute the magnification M from Equation 1.

3. Galilean telescope

Construct a Galilean telescope on the optical bench using the long focal-length converging lens as the objective and the diverging (negative) lens as the eyepiece. Mount the eyepiece lens near the right end of the bench and, with the objective lens located approximately a focal length f_o away, aim the telescope at a distant object through an open window. Considering the fact that the eyepiece is a negative lens, should the lens separation be increased or decreased to bring the object into focus? Establish a rough focus by moving the objective lens and then adjust the eyepiece for the sharpest image possible. Record the objective and eyepiece positions as L_o and L_e respectively in the data table. Compare the tube length l with the algebraic sum of the focal lengths. Compute the magnification M from Equation 5.

Mount an image screen on the optical bench approximately one focal length in front of the eyepiece and adjust its position as necessary to form a sharply focused first image of an illuminated object located a few meters in front of the objective (the laboratory should be darkened). Record the object distance s_o and the image distance s_i in the second data table.

Adjust the eyepiece position as necessary to focus on the back of the image screen; then remove the screen from its holder and adjust the eyepiece for sharpest detail of the virtual image of the illuminated object. Has the tube length l been increased or decreased as compared to that for the distant object? Taking the first image as the object for the eyepiece, record the object distance s_o' and the second-image distance s_i' (the distance of most distinct vision). Compute the separate magnification of the objective and eyepiece, the total magnification of the telescope, and record as M_o, M_e, and M respectively.

Replace the illuminated object with a magnification scale, being careful to maintain the same object distance. Observe the scale through the telescope with one eye and directly with the other eye. Make slight adjustments of the eyepiece if necessary so the scale seen through the lens and the scale seen by the unaided eye appear equally distant. Determine the magnification of the telescope by comparing the relative sizes of the scales as seen through the telescope and by the unaided eye. Record as the ratio of the second-image size s_i' to the object size s_o.

If time permits, increase the object distance as much as possible and collect data as you have previously done in this experiment.

OPTIONAL: Convert the astronomical telescope to a terrestrial instrument using a single inverting lens. Determine a position for the inverting lens which will leave the total magnification of the instrument unchanged. Plan a suitable table for recording data and results. Construct a ray diagram of the lens system showing all significant positions and distances.

DIAGRAMS: On a separate sheet of paper, construct fully labeled ray diagrams of the astronomical and Galilean telescopes. Submit the diagrams with your report.

DATA *(Distant object)*

TELESCOPE	f_o (cm)	f_e (cm)	L_o (cm)	L_e (cm)	l: $L_e - L_o$ (cm)	l: $f_o + f_e$ (cm)	M
Astronomical							
Galilean							

DATA *(Near object)*

TRIAL	f_o (cm)	f_i (cm)	s_o (cm)	s_i (cm)	s_o' (cm)	s_i' (cm)	l $s_i + s_o'$ (cm)	M_o $\frac{s_i}{s_o}$	M_e $\frac{s_i'}{s_o'}$	M $M_o M_e$	M $\frac{h_i'}{h_o}$
1											
2											

QUESTIONS: Your answers should be complete statements. Problem solutions should be set down in the space provided below.

1. Compare the distances between the lenses of a refracting telescope when it is used for observing nearby and distant objects. Explain.

....

....

....

.... 1

2. Two telescopes, one an astronomical and the other a Galilean, use identical objectives and a magnification of 3X for each. Show that the tube length of the astronomical telescope is twice that of the Galilean instrument.

2

NAME PERIOD DATE INSTRUCTOR'S COMMENTS

40 EXPERIMENT Color

PURPOSE: (1) To learn what determines the colors of transparent and opaque objects. (2) To learn what principle governs the combining of colored lights. (3) To learn what principle governs the combining of colored pigments.

APPARATUS: Glass slides: red, yellow, green, blue; colored construction paper: red, orange, yellow, green, blue, violet; poster paints: red, yellow, green, blue; glass plates; glass rods.

SUGGESTION: A commercial color projection apparatus is ideal for this experiment. Good results may be obtained using three slide projectors, or even three strong flashlights in a well-darkened room. Special color glass slides provide truer colors, but good results may be obtained by mounting several thicknesses of colored cellophane or spotlight gelatin between ordinary projection slide glasses.

PROCEDURE:

1. *Color of objects*

a. Transparent objects. Look through the red, yellow, green, and blue glass slides in turn at a piece of white paper placed on your laboratory desk. How does the color of the white paper viewed through the slide compare with the color of the slide?

Observation: ....

....

....

Now look through pairs of glass plates at the piece of white paper. Use the red and green plates, and red and blue plates, and the yellow and blue plates, for example.

Observation: ....

....

....

....

b. Opaque objects. In a darkened room, use a flashlight or a projector and the colored glass plates to shine beams of light of various colors on sheets of colored paper. In the table record the observed color of the various sheets of paper under different colored beams of light.

2. *Combining colors*

Using a color apparatus, or three projectors, or three flashlights and colored slides in a well-darkened room, observe the colors produced by the overlapping of colored beams of light.

3. *Combining pigments*

Use a glass rod to mix a drop of each of the following poster paint colors on a glass plate and observe the color produced: yellow and blue; red and green; red and blue; blue and green; red, blue, and green. Record your observations.

DATA

COLOR OF INCIDENT LIGHT	COLOR OF OBJECT IN WHITE LIGHT					
	Red	*Orange*	*Yellow*	*Green*	*Blue*	*Violet*
Red						
Yellow						
Green						
Blue						

DATA

COLORS COMBINED	COLOR OBSERVED
Yellow and blue	
Red and green	
Red and blue	
Blue and green	
Red, blue, and green	

DATA

COLORS MIXED	COLOR OBSERVED
Yellow and blue	
Red and green	
Red and blue	
Blue and green	
Red, blue, and green	

QUESTIONS: Your answers should be complete statements.

1. Upon what factors does the color of a transparent object depend?

....

....

.... 1

2. What factors determine the color of an opaque object?

....

....

.... 2

3. What determines the color observed when two beams of colored light are superposed?

....

....

.... 3

4. What name is given to any two colored beams of light that form white light when they are superposed?

....

.... 4

5. What name is given to the three colors that form white light when superposed?

....

.... 5

6. What determines the color when two or more pigments are mixed?

....

....

....

.... 6

7. What are the three primary pigments?

.... 7

8. What is the relationship between primary colors and primary pigments?

....

.... 8

NAME PERIOD DATE INSTRUCTOR'S COMMENTS

41 EXPERIMENT Diffraction and Interference

PURPOSE: (1) To produce diffraction and interference patterns. (2) To study the wave nature of light.

APPARATUS: Straight filament clear glass electric lamp (showcase lamp will serve); lamp base and receptacle for vertical mounting of lamp: diffraction slits, single and double, of various widths and spacings; color filters of various colors (colored cellophane or spotlight gelatin will serve).

SUGGESTION: If diffraction slits are not available, satisfactory slits can be made by the students themselves. Clear glass microscope slides may be sprayed with black lacquer, then scored crosswise with a razor blade using a straight edge as a guide. With some practice using steady pressure, successful slits may be made. Several single slits and several double slits should be made on the slide so the most satisfactory ones can be used in the experiment. To make double slits, clamp two double edged razor blades together. Strips of paper may be used as wedges for slits of varying separation.

A narrow slit 0.2 to 0.5 mm wide cut in a square of stiff black Bristol board and mounted in front of an ordinary frosted lamp will provide a suitable line source of light if a straight filament lamp is not available.

INTRODUCTION: According to Huygens' principle, every point on a light wave front may be considered as a new source of light, but in general, all parts of each new wavelet are cancelled by destructive interference except that moving in the same direction as the original wave front. Thus the whole wave front appears to move as a unit.

If a series of wave fronts strike a barrier having two narrow slit openings, as shown in Fig. 41-1, each opening acts as a new source and new wavelets travel out in phase with each other. In certain regions they will reinforce each other producing bright bands. In alternate regions they will interfere producing dark bands. White light, since it is a polychromatic source, yields indistinct bands.

PROCEDURE:

1. Diffraction

Darken the laboratory and set up a line source of light in a vertical position. Hold a single slit close to your eye so that it is oriented vertically (parallel to the line source) and view the line source through it from a distance of one or two meters. Repeat using other single slits of different width. Make sketches of the best diffraction patterns you observe through a narrow slit and a wide slit. Briefly describe the difference between the two patterns.

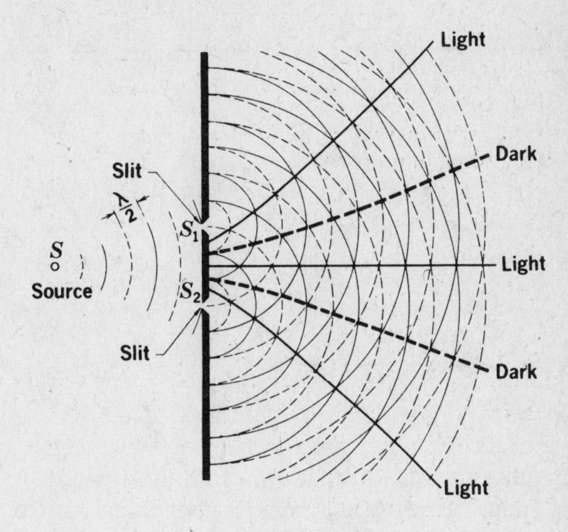

Figure 41-1

Observation: ....

....

....

....

....

Narrow slit Wide slit

2. Interference

Observe the line source through double slits. Compare the pattern seen through pairs of slits of different spacing. Sketch your best pattern observed through very closely spaced slits and your best pattern observed through the more widely spaced slits. Observe a single-slit pattern again and carefully note the similarities and differences.

Observation: ....

....

....

....

Narrow spacing Wide spacing

Examine a double-slit pattern carefully to see if color fringes are evident. Suggest an explanation for any color effects observed. Place a color filter in front of the white-light source. Record the result.

Observation: ....

....

....

Change to another color. Again record the result.

Observation: ....

....

....

Select two color filters from the end regions of the spectrum, red and blue for example, and observe the interference pattern with first one and then the other in front of the line source. Describe the changes you observe.

Observation: ....

....

....

Cover half the source with the red filter and the other half with the blue filter. From the pattern you observe estimate the ratio of the wavelength of blue light to the wavelength of red light.

Observation: ....

....

....

QUESTIONS: Your answers should be complete statements.

1. How does the diffraction pattern through a single slit change as a slit is made narrower?

....

....

....

.... 1

2. How do the interference patterns compare when formed by two narrow slits very closely spaced and by two narrow slits more widely spaced?

....

....

....

.... 2

3. How do the interference patterns of red light and blue light compare when formed by the same pair of slits?

....

....

.... 3

4. What do you think would happen to the interference pattern from a white-light line source if you were able to score many extremely narrow slits very close together, perhaps 2000-3000/cm, on your viewing plate?

....

....

....

.... 4

NAME PERIOD DATE INSTRUCTOR'S COMMENTS

42 EXPERIMENT Wavelength by Diffraction

PURPOSE: To measure the wavelength of light by means of a diffraction grating.

APPARATUS: Optical bench consisting of a meter stick, 2 supports, grating holder and support, metric scale and slit, for meter stick mounting; bentpin riders, for meter stick; Bunsen burner; evaporating dish; diffraction grating, transmission replica, of the order of 4×10^3 lines/cm; incandescent light source; sodium chloride; mercury vapor discharge tube and power supply if available.

INTRODUCTION: According to Huygens' principle, every point on a light wave front may be considered as a new source of light. In general, all parts of each new wavelet are cancelled due to destructive interference except that part traveling in the same direction as the original wave front.

When a diffraction grating of the transmission type is placed in the path of a plane wave front only alternate parts of it pass through. The ruled lines are opaque to the light and the uniform spacings between the lines, being transparent to the light, provide a large number of fine transmission slits very close together. The new wavelets originating at the slits interfere in such a way that several new wave fronts are set up, one traveling in the original direction and the others at various angles from this direction depending on the wavelength.

If a narrow slit is illuminated by white (polychromatic) light and viewed through a transmission grating, a white image of the slit will be seen directly in line with the slit opening and pairs of continuous spectra will be seen equally spaced on opposite sides of the slit opening. When the slit is illuminated by monochromatic light, successive pairs of slit images

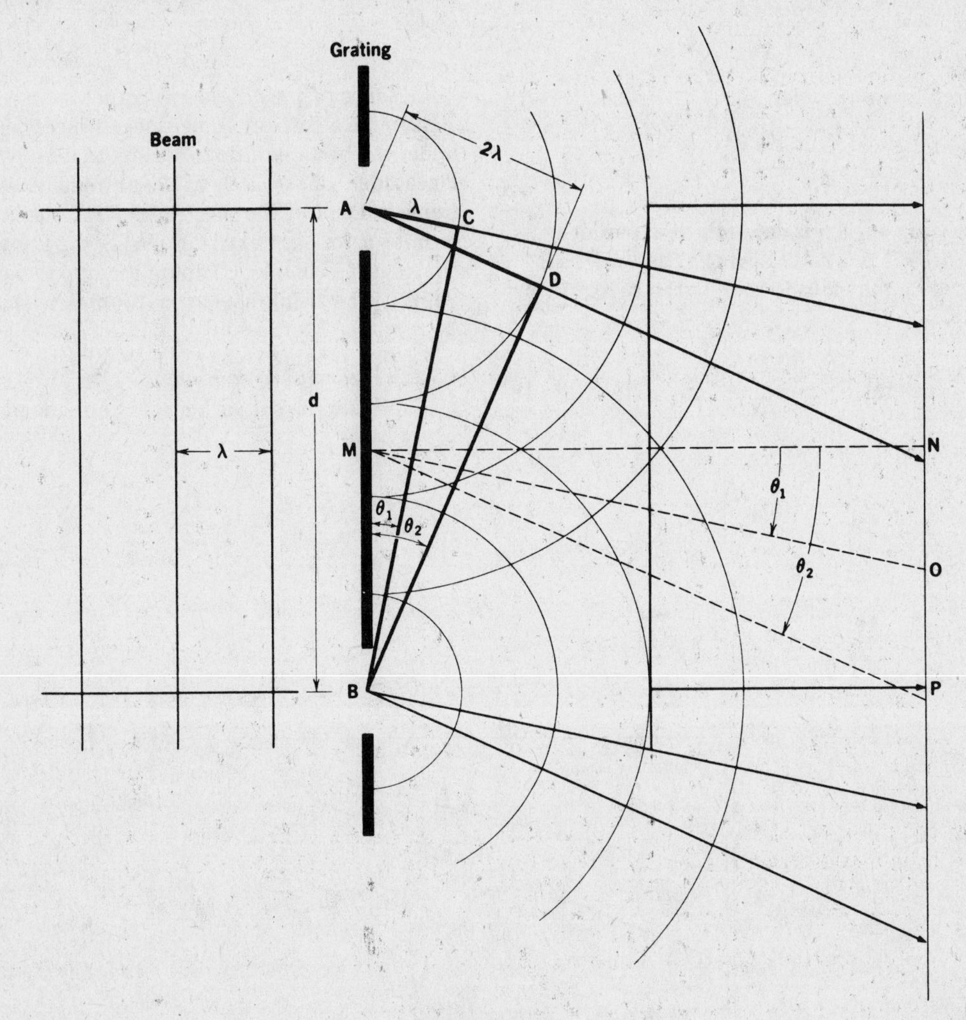

Figure 42-1

of decreasing intensity will be seen equally spaced on opposite sides of the slit opening. The first pair is known as the *first order* images, the second pair as *second order* images, etc.

In Fig. 42-1, A and B are parallel slits in a transmission grating uniformly separated by the distance d, known as the *grating constant.* A beam of monochromatic light from a distant slit incident normally on the grating gives rise to secondary wavelets simultaneously at each grating slit. A wave front of these wavelets travels along MN and, if converged, produces a direct image of the slit opening at N. Wave front CB representing a given wave from B and the first preceding wave from A travels along MO and, if converged, produces a *first order* image of the slit opening at O. A given wave from B and the second preceding wave from A produces a wave front BD which yields a *second order* image at P, etc. From the right triangle ABC, in which AC is equal to the wavelength λ of the incident light and the diffraction angle θ_1 is the angle of the diffracted wave front from the grating plane, it is evident that

$$\lambda = d \sin \theta_1 \tag{1}$$

Side AD of triangle ABD is equal to 2λ, as in the case of a second-order image,

$$\lambda = \frac{d \sin \theta_2}{2} \tag{2}$$

In the general case of n orders,

$$\lambda = \frac{d \sin \theta_n}{n} \tag{3}$$

The diffraction angle θ is determined experimentally as shown in Fig. 42-2, I_1 being a first-order slit image located on the metric scale. In the right triangle MNI_1,

$$\tan \theta_1 = \frac{NI_1}{MN} = \frac{s_1}{l} \tag{4}$$

and

$$\theta_1 = \arctan \frac{s_1}{l} \tag{5}$$

Knowing the order of image n observed, the diffraction angle θ_n, and the grating constant d, the wavelength λ can be computed from Equation 3.

PROCEDURE:

1. Grating constant

Mount the scale and slit on one end of the optical bench and illuminate the slit with white light. Mount the grating near the opposite end of the optical bench and, with the eye close to the grating observe the first-order spectra. Locate a rider on the scale at the yellowest part in each first-order spectrum. Adjust the grating about its vertical axis to position these two corresponding regions equidistant from the slit. Read the distances s and l as accurately as possible, take the wavelength λ of this yellowest portion of the continuous spectrum as 5800 Å (1 Å = 10^{-8} cm), and compute the grating constant d from Equations 5 and 3. Record the required data in Data Table 1. If your instructor can provide the known number of lines per centimeter for your grating, enter the known value of d also.

2. Range of the visible spectrum

Examine a first-order continuous spectrum critically under the best conditions of illumination attainable. Place riders on the scale at the extreme violet and extreme red end of the spectrum. Read s and I and determine θ for each extreme. Using the known grating constant if available, compute the upper and lower limits of the visible spectrum. Complete Data Table 2.

3. Wavelength of sodium

Illuminate the slit with a sodium flame placed

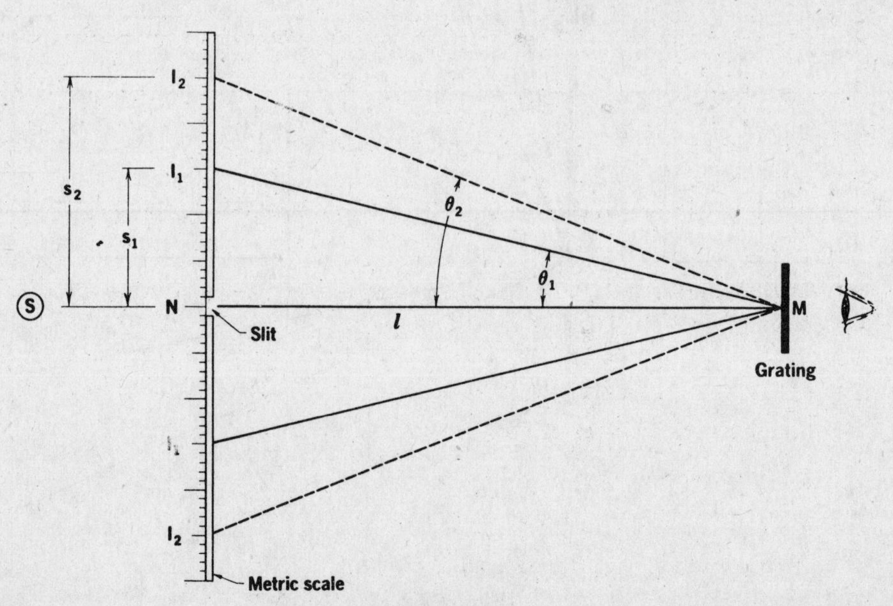

Figure 42-2

about 5 cm away and, if necessary, adjust the grating about its vertical axis as before. Carefully position a rider on a first-order image and, if possible, on a second-order image. Compute the wavelength of sodium light from each set of readings and record. Substitute the white-light source for the sodium flame.

4. Optional

Produce other bright-line spectra and measure the wavelengths of the prominent lines. Secure an unknown bright-line source from your instructor and attempt to identify it by comparing the wavelengths of the prominent lines with those of the characteristic spectra of the elements. Record all data in Table 3.

DATA *(Table 1)*

LIGHT SOURCE	λ (Å)	n	s (cm)	l (cm)	θ_n (°)	$\sin\theta_n$	d Experimental (cm)	d Actual (cm)

DATA *(Table 2)*

LIGHT SOURCE	COLOR *(extreme)*	n	s (cm)	l (cm)	θ_n (°)	$\sin\theta_n$	d (cm)	λ Experimental (Å)	λ Actual (Å)

DATA *(Table 3)*

LIGHT SOURCE	n	s (cm)	l (cm)	θ_n (°)	$\sin\theta_n$	d (cm)	λ Experimental (Å)	λ Actual (Å)

QUESTIONS: Your answers should be complete statements. Problem solutions should be set down in the space provided below.

1. Recognizing that θ cannot exceed 90°, show that a coarse-ruled grating will yield a larger number of orders than a fine-ruled grating.

2. What would have been the effect of a grating with a smaller grating constant on the measurements in this experiment?

....

....

.... 2

3. Using Fig. 42-1 as a basis, explain why the red end of the continuous spectrum was observed to be farther from the slit than the violet end when an incandescent lamp was used to illuminate the slit.

....

....

....

.... 3

4. A single slit is illuminated by monochromatic light. Explain the fact that several orders of slit images may be seen with a grating but only one with a prism.

....

....

.... 4

NAME PERIOD DATE INSTRUCTOR'S COMMENTS

43 EXPERIMENT The Polarization of Light

PURPOSE: (1) To show some ways in which light may be polarized. (2) To illustrate several uses of polarized light.

APPARATUS: 2 Polaroid disks; piece of calcite (Iceland spar); piece of ferrotype plate or metal plate enameled on one surface; 12 glass plates, 5 cm square; cellophane strips mounted between 5-cm square glass plates as shown in Fig. 43-1 or cellulose tape on a square glass plate; U-shaped piece of transparent plastic; glass from broken molded bottle.

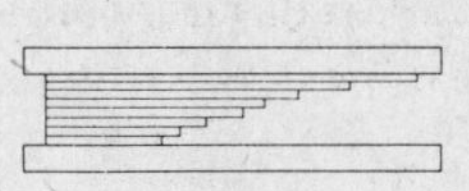

Figure 43-1

INTRODUCTION: Light waves which vibrate in only one plane are said to be polarized. There are several natural crystals, among them tourmaline and calcite, which possess the property of polarizing light which passes through them. Light may be partially polarized by reflection. Polaroid is a synthetic material which is used to polarize light. Through the use of polarized light and the means of detecting it, many scientific observations not possible with ordinary light may be made. Polarized light is used in identifying certain chemical compounds, in detecting strains in structural materials, and in determining the thickness of crystals and fibers. Polaroid is used in some sunglasses and in some types of reading lamps to reduce glare.

PROCEDURE:

1. Production and detection of polarized light with Polaroid

Look through one of the Polaroid disks at one of the walls of your laboratory. Rotate the disk. Does the intensity of light vary?

.... Do the same with the other disk. Do you get the same result? .... How does the intensity of light coming through a single disk compare with that reaching your eye directly?

....

....

....

....

The light coming through a Polaroid disk is plane-polarized. That is, the light waves passing through are all vibrating in the same plane.

Now hold both disks together, and rotate one of them while you look through them at the wall. What effect do you observe?

....

....

....

....

How much do you have to turn one disk to go from maximum to minimum brightness? .... When the Polaroid disks are placed so that a minimum of light passes through, the disks are said to be crossed. Explain why crossed polarizers transmit a minimum of light.

....

....

....

....

2. Production of polarized light by crystals

Look through a transparent piece of calcite (Iceland spar) at the period at the end of this sentence. What do you see?

....

Slowly turn the crystal. What movement do you observe?

....

....

Now look through a single rotated Polaroid disk at the images in the crystal. What happens?

....

....

....

Explain. ....

....

....

3. Production of polarized light by reflection

Place a piece of ferrotype plate, or a metal plate painted with black enamel, on your laboratory desk. Find a position where the maximum amount of light from the window is reflected from the surface of the plate as "glare." Examine this glare through a single rotated Polaroid disk. Is the glare light polarized?

.... How can you tell?

....

....

Examine the glare from a stack of 12 glass plates in a similar manner. What do you observe?

....

....

4. Elimination of glare by the use of Polaroid

In what plane is the glare reflected from a polished surface polarized?

....

....

....

Hold the Polaroid disks, oriented so that their transmitting plane is at right angles to the plane in which the glare is polarized, one over each eye. Compare the amount of glare through the Polaroid disks with the amount you see without the Polaroid disks.

....

....

What practical application does this observation have?

....

....

....

....

....

5. Use of polarized light to observe interference phenomena (Qualitative)

Cross the two Polaroid disks and place between them the glass slide containing the cellophane strips of varying thickness. Rotate the glass slide until you observe the brightest colors. How does the brightness of the color vary from the thin strips to the thick ones?

....

....

6. Use of polarized light to determine structural strains

Examine a small U-shaped piece of transparent plastic between crossed Polaroid disks. What do you observe?

....

....

Now pinch the open ends of the piece of plastic between your fingers. What happens?

....

....

How do the colors help you to detect the places where the strain is the greatest?

....

....

Examine a piece of glass broken from a molded bottle between crossed polarizers. Are there strains in the glass?

....

....

NAME PERIOD DATE INSTRUCTOR'S COMMENTS

44 EXPERIMENT Static Electricity

PURPOSE: To show certain properties of static electricity.

APPARATUS: Pith ball electroscope; ebonite or hard-rubber rods; cat's fur or wool pad; glass rod; silk pad; suspension support for hard-rubber rod; gold leaf electroscope; bare copper wire; silk thread; brass ball; demonstration capacitor.

INTRODUCTION: Whenever two dissimilar materials are rubbed together static electricity is produced, one material acquiring a positive charge and the other a negative charge. In most instances these charges are of negligible magnitude and go unnoticed. Sometimes, however, the accumulation of an electrostatic charge is quite significant, and its presence is readily detected.

In this experiment we shall produce charges, examine some of the properties of charged bodies, and verify the laws of electrostatics.

SUGGESTION: This experiment may be performed as a demonstration by the teacher or by a student. The most satisfactory results are obtained on clear, dry days.

PROCEDURE:

1. Production of negative electricity

Rub an ebonite rod with a piece of cat's fur and bring it close to a pith ball electroscope as in Fig. 44-1.

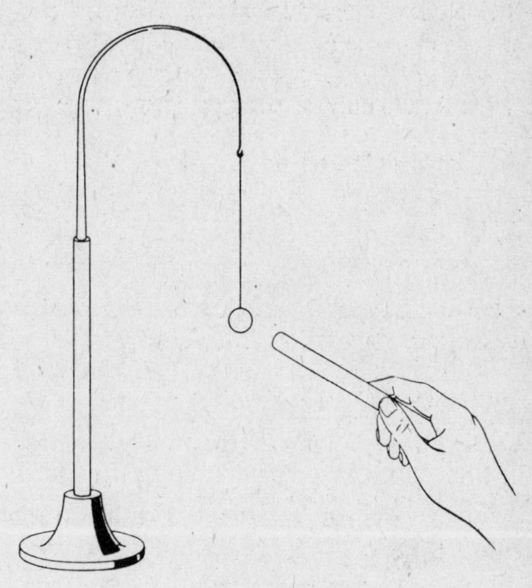

Figure 44-1

Result? ....

....

Let the pith ball momentarily touch the ebonite rod.

Result? ....

....

Explain the movement of the pith ball. ....

....

....

2. Production of positive electricity.

Rub the glass rod with silk and repeat the procedure in (1) with the pith ball.

Result? ....

....

....

Explain. ....

....

How is the pith ball charged when it is repelled by the glass rod? ....

....

If a charged ebonite rod were brought near such a charged pith ball, what should happen?

....

....

Explain. ....

Try it and see if your prediction is true.

Result? ....

....

3. First law of electrostatics

Suspend a charged ebonite rod as in Fig. 44-2. Bring a second charged ebonite rod near one end of the suspended rod.

Result? ....

....

Try the other end.

Result? ....

....

Bring a charged glass rod near one end of the suspended charged ebonite rod. What do you observe?

....

....

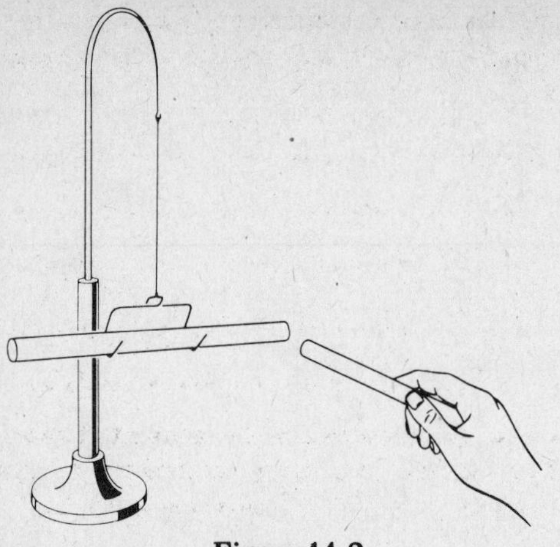

Figure 44-2

Try the other end.

Result? ....

....

Is the charge the same at both ends of the suspended ebonite rod?

....

....

Is it the same charge that you produced on the second ebonite rod?

....

....

How can you determine this?

....

....

What conclusion do you draw with regard to the behavior of electric charges toward each other?

....

....

4. Charging an electroscope

a. By conduction. Charge an ebonite rod and scrape it against the knob of an electroscope.

Result? ....

....

Explain. ....

....

What kind of charge is on the electroscope? ....

....

Explain. ....

....

The charge may be removed from the electroscope by touching the knob with the hand. Explain this action.

....

....

b. By induction. Bring a charged ebonite rod *near* the electroscope knob. What happens to the leaves of the electroscope?

....

....

With the ebonite rod still near, touch the knob of the electroscope with your finger momentarily. Explain what happens.

....

....

Remove the rod. Is the electroscope charged?

....

What kind of charge does it have?

....

Discharge the electroscope and repeat, using a glass rod charged by rubbing it with silk.

5. Conductors and insulators

Set up the apparatus as shown in Fig. 44-3. See if you can charge the electroscope either by contact or by induction using the ball B the same way you would ordinarily use knob A.

Result? ....

....

Is copper a conductor?

....

Replace the copper wire between A and B with a silk thread. Attempt to charge the electroscope using ball B as above.

Result? ....

....

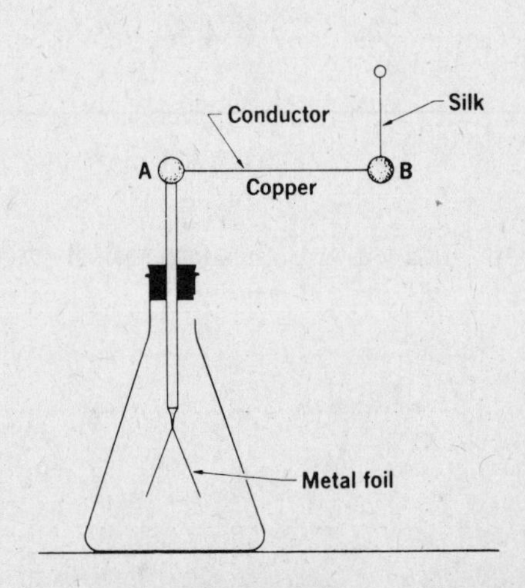

Figure 44-3

Is silk a conductor?

....

6. Action of a capacitor

Connect the knob of an electroscope to one plate of a demonstration capacitor by a short piece of copper wire. Charge this combination negatively. Bring the second plate of the capacitor near the first. What happens to the leaves of the electroscope?

....

Explain. ....

....

Remove the second capacitor plate. Explain the results.

....

....

Bring the second plate near the first again. What must you do to cause the electroscope leaves to diverge the same amount as they did with the single capacitor plate?

....

....

Try it.

Result? ....

....

How does a capacitor affect the amount of charge which must be placed on the electroscope knob to cause divergence of the electroscope leaves?

....

....

What is the function of a capacitor?

....

....

NAME PERIOD DATE INSTRUCTOR'S COMMENTS

45 EXPERIMENT Electrochemical Cells

PURPOSE: (1) To study the reactions in a voltaic cell. (2) To study the reactions in a storage cell.

APPARATUS: Simple demonstration cell, consisting of some type of battery stand; 2 zinc electrodes; carbon electrode; copper electrode; d-c voltmeter, 0-3 v; d-c ammeter, 0-3 a; enameled pan; 2 lead electrodes; 2 dry cells; rheostat; knife switch, SPST; annunciator wire, 18 ga, for connections; dilute sulfuric acid, about 1 part acid to 14 parts water; amalgamating fluid; electric bell. See Appendix A for instructions in the use of meters.

Amalgamating fluid is made by reacting 200 g of mercury with a mixture of 175 ml of concentrated nitric acid and 625 ml of concentrated hydrochloric acid (hood). After the *poisonous fumes* have disappeared, the solution can be kept in a glass-stoppered bottle and used from year to year until the mercury is exhausted. CAUTION: The solution is extremely corrosive.

INTRODUCTION: There are two general types of electrochemical cells, the primary cell and the storage (secondary) cell. The essential difference is that the reacting materials of a storage cell are renewable.

Electrochemical reactions are oxidation-reduction reactions which occur spontaneously, and thus can be used as a source of direct current. During the chemical action, electrons are removed from the cathode element, an *oxidation* process, and electrons are acquired by the anode element, a *reduction* process. In the storage cell this process is reversed to restore the electrodes, using electric energy from an outside source.

PROCEDURE:

1. Primary cell. The cathode

Place a strip of unamalgamated zinc in a tumbler one-third full of dilute sulfuric acid. CAUTION: *Dilute acid is very corrosive.* What action do you observe?

....

....

....

....

Since zinc is an element, composed of zinc only, what must have been the source of the gas bubbles?

....

....

....

The chemical formula for sulfuric acid is H_2SO_4. Zinc displaces hydrogen from the sulfuric acid. The zinc enters the solution as positively charged zinc ions, Zn^{++}. The hydrogen is released as bubbles of gas. Dip one end of a similar strip of zinc into a tumbler two-thirds full of amalgamating fluid for a few seconds. (*Caution*) Rinse with water and wipe the strip dry. What is the appearance of the amalgamated strip?

....

....

Dip it into the sulfuric acid solution and let it remain for two or three minutes. Do you observe any action?

....

Remove the amalgamated strip from the acid, rinse it, and lay it in the enameled pan. Take care not to get any acid on your clothing or on the table.

2. The anode

Insert a strip of copper in the acid. Does the acid appear to act on the copper?

....

Remove the copper strip and use in its place a carbon electrode. Is there any action of the acid on the carbon?

....

....

....

....

3. A voltaic cell on open circuit

While the carbon rod is still in the acid solution, add the amalgamated zinc electrode. They must not touch each other. Is there any apparent action on open circuit when carbon is used with amalgamated zinc?

....

....

Repeat, using carbon and unamalgamated zinc. What action appears to be taking place?

....

....

....

4. A voltaic cell on closed circuit.

Rinse the copper and amalgamated zinc strips and

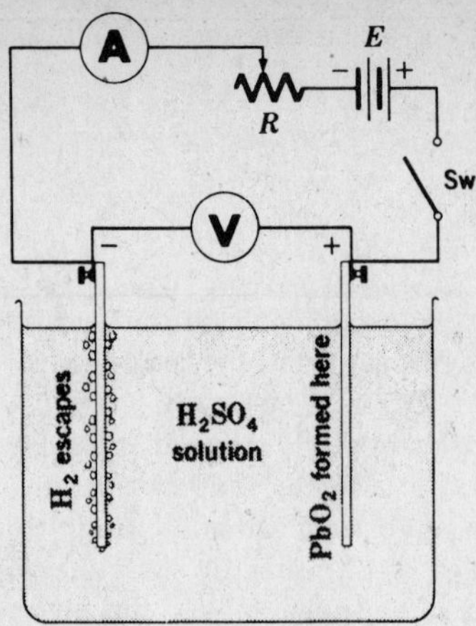

Figure 45-1

place them in a battery stand. Connect their terminals by means of short pieces of wire to a voltmeter. Read the voltmeter when the elements are first dipped into the acid solution. Record the voltage.

.... volts.

Repeat the experiment, using the carbon rod and the amalgamated zinc strip. Record the voltage.

.... volts. Where do the hydrogen bubbles appear to be liberated?

....

....

....

5. The lead storage cell

Sandpaper the lead strips until they are bright and clean. Clamp them in position in the demonstration cell and immerse them in the tumbler two-thirds full of dilute sulfuric acid. Does the voltmeter connected across the terminals show any difference in potential?

....

....

....

Next connect the lead strips in series with two dry cells, an ammeter, and a rheostat as shown in Fig. 45-1. Adjust the resistance so that about one ampere of current will flow through the circuit. Maintain this charging rate for approximately 5 minutes. Then lift the plates from the acid and examine them. Describe the changes that took place in the cell during charging. What is the difference of potential between the

electrodes now? .... volts.

Connect the cell you have just charged with an electric bell, a small motor, or a voltmeter, and permit it to discharge. Record your observations.

....

....

....

....

NAME PERIOD DATE INSTRUCTOR'S COMMENTS

46 EXPERIMENT Combinations of Cells - Internal Resistance

PURPOSE: (1) To learn how to group cells to supply proper current and potential difference for different loads. (2) To measure the internal resistance of a battery.

APPARATUS: Three No. 6 dry cells; d-c ammeter, 0-1/10 a; d-c voltmeter, 0-7.5 v; tubular rheostat; approximately 25 ohms; brass connectors, double; annunciator wire, No. 18, for connections; 2 momentary contact switches, SPST. See Appendix A for instructions in use of meters.

SUGGESTION: If the supply of dry cells is limited to no more than 3 per group, one or two groups may be asked to hook up the circuit for Part 3 and take measurements as a demonstration, after each student has determined his own circuit and drawn his circuit diagram. Sufficient ammeters will then be available for the demonstration circuit to make all current readings simultaneously. Two laboratory periods may be required to complete the entire experiment.

INTRODUCTION: A battery is composed of two or more cells connected either in series or in parallel. Cells are generally connected in series to provide a higher potential difference than that of a single cell. They are generally connected in parallel to increase the capacity for delivering continuous current to the external circuit. Only identical cells should ordinarily be connected in parallel. When the laboratory supply is sufficient, the standard No. 6 dry cell should not be required to furnish more than 0.25 a of continuous current. *To conserve the life of the laboratory dry cells, the circuit should be opened when not actually taking meter readings.*

PROCEDURE:

1. Cells in series

Connect 1 dry cell, an ammeter (0-10 a range), a voltmeter, a switch, and a rheostat in a circuit as shown in Fig. 46-1. *Observe the polarity markings on d-c meters when connecting them in the circuit.* With the rheostat resistance R_L at the maximum, momentarily close the switch and note the meter deflections. The voltmeter should read about 1.5 v and the ammeter should show little noticeable deflection. Adjust the rheostat for approximately 0.5 a of current, open the switch and change the ammeter connection to the 0-1 a range. Adjust the rheostat for precisely 0.5 a. Does the voltmeter reading change during the load adjustment? *Keep switch open except when taking readings.* The ammeter reads total current I_T in the circuit and the voltmeter reads the potential difference across the external circuit. Record both readings in the data table. Disconnect the negative lead of the voltmeter from the circuit and, *with switch closed,* momentarily touch it directly to the negative terminal of the dry cell. Record the voltage reading as V_1. (The detailed instructions concerning the circuit given above will not be repeated in the following paragraphs, but should be followed closely each time you perform the manipulation called for.)

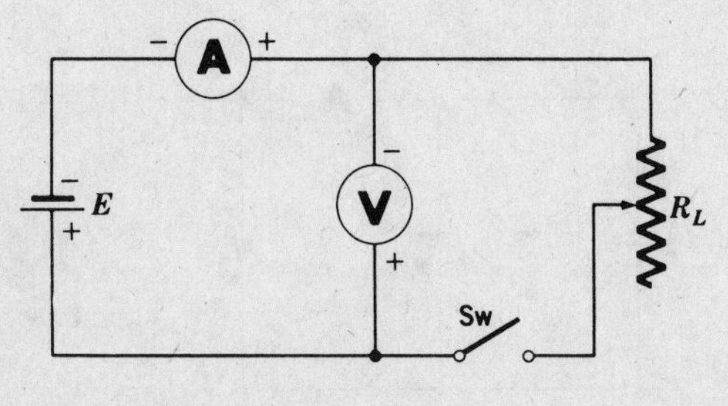

Figure 46-1

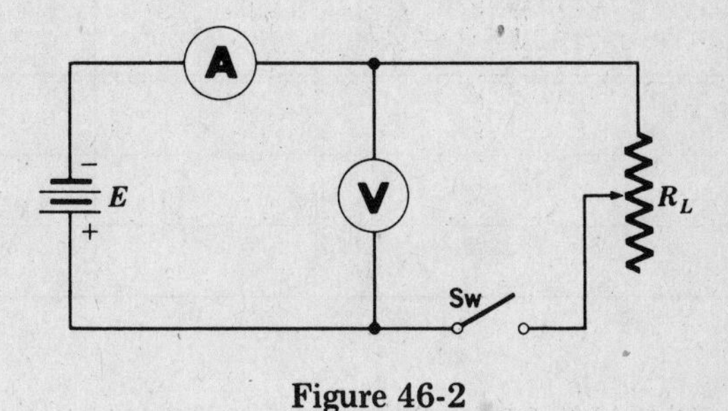

Figure 46-2

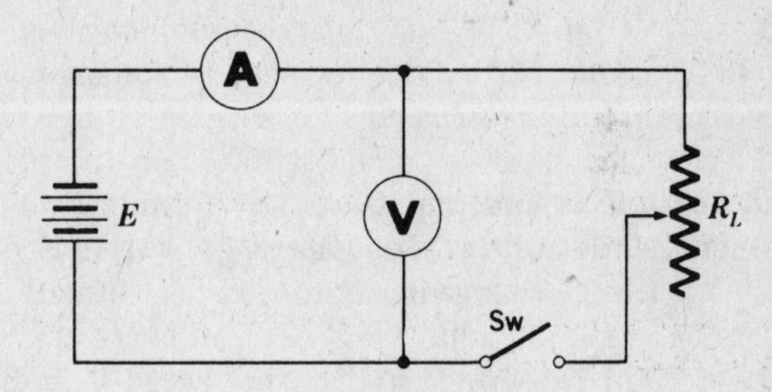

Figure 46-3

Place a second dry cell in series with the one now in circuit by connecting the negative terminal of the first to the positive terminal of the second. The negative terminal of the second cell is then connected to the external circuit. See Fig. 46-2. Close the switch and read both meters. How does the change in I_T compare with the change in V? Adjust R_L to give an I_T of 0.5 a. Does V change during this adjustment?

Note the voltage across each cell as before. Remove both voltmeter leads from the circuit this time. *Open switch.* Record I_T, V, V_1, and V_2. How does the sum of V_1 and V_2 compare with V?

Place a third dry cell in the battery in series with the other two. See Fig. 46-3. Note changes in I_T and V. Are they what you expected? Adjust R_L to give an I_T of 0.5 a. Record I_T, V, V_1, V_2, and V_3 in the data table. How do the individual cell voltages compare with the battery voltage?

DATA *Cells in Series*

NUMBER OF CELLS	I_T (a)	V (v)	V_1 (v)	V_2 (v)	V_3 (v)
1					
2					
3					

2. Cells in parallel

Remove two of the cells leaving the circuit as in Fig. 46-1. Note the changes in I_T and V and quickly adjust R_L to give an I_T of 1.0 a, then immediately open the switch. When the open-circuit voltage reading is normal, record I_T and V.

Place a second cell in parallel with the one now in the circuit by connecting the negative terminals together and the positive terminals together as in Fig. 46-4. Is there a change in either I_T or V? Should there be? Make an adjustment of R_L if necessary to give an I_T of 1.0 a, *open switch,* and record I_T and V. Remove the ammeter from the circuit replacing it with a brass connector to close the circuit and determine the current in each of the two cells of the battery by inserting the ammeter first in position *1* and then in position *2* as shown in Fig. 46-4. Record as I_1 and I_2 and return the ammeter to its original location in the circuit. How do the individual cell currents compare with the total current in the circuit?

Place a third cell in parallel in the battery as shown in Fig. 46-5. Note any change in I_T or V. Adjust I_T to 1.0 a if necessary and record readings of I_T, V, I_1, I_2, and I_3. How do the individual cell currents compare with the total current in the circuit?

3. Battery problem

Suppose an external load consisting of the rheostat R_L requires a current of 0.75 a and a potential difference across it of 3 v. Arrange a battery of No. 6 dry cells which will supply these requirements without any cell having a current in it in excess of 0.25 a. Connect the load and switch to the battery with the voltmeter to read the potential difference across the load and the ammeter to read the total current in the external circuit. Have maximum resistance in the circuit when the switch is closed. Adjust R_L for an I_T of 0.75 a. Remove the ammeter as before and use it to read the cell currents. Draw a diagram of your circuit showing an appropriate meter symbol in *each* location a meter reading was taken. Write in the current or voltage reading at each meter location.

DATA *Cells in Parallel*

NUMBER OF CELLS	V_T (v)	I_T (a)	I_1 (a)	I_2 (a)	I_3 (a)
1					
2					
3					

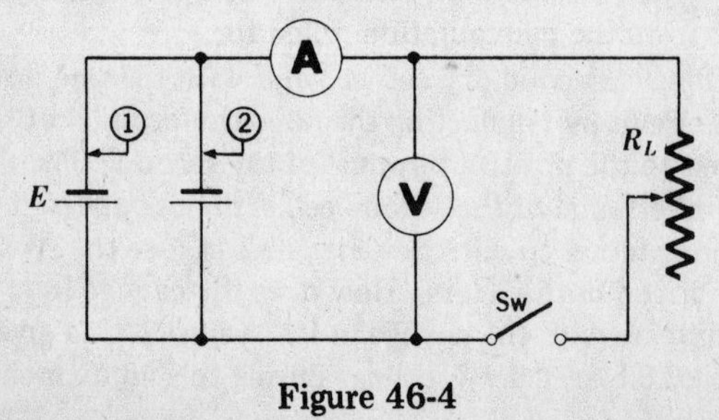

Figure 46-4

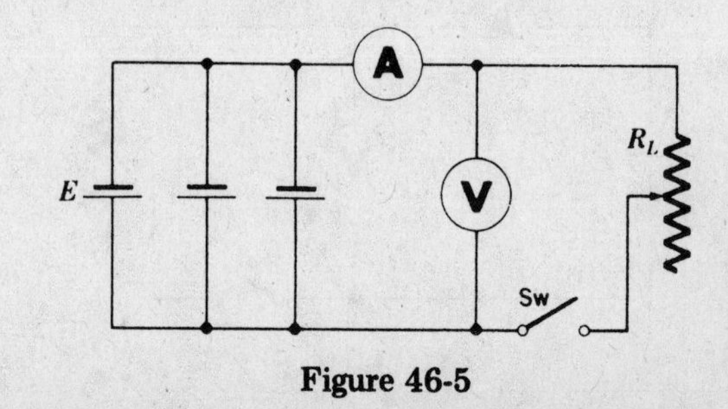

Figure 46-5

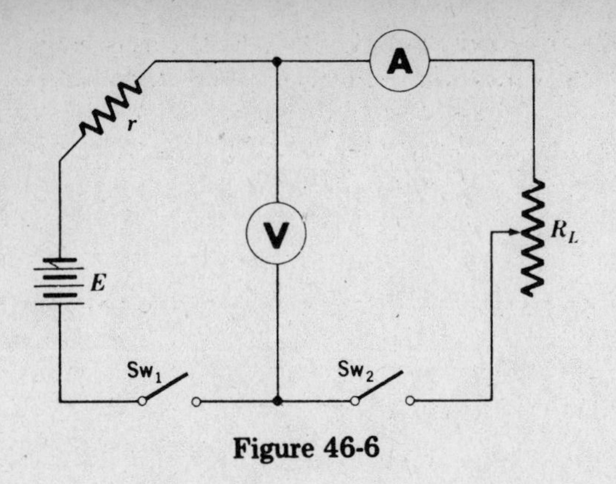

Figure 46-6

4. Internal resistance

Make up a battery of three No. 6 dry cells connected in series. Connect the voltmeter across the battery and the ammeter (0-10 a range) in series with the rheostat as shown in Fig. 46-6. The internal resistance of the battery is represented as a resistance r in series with the battery. Close sw_1 and read the voltage across the battery. The voltmeter will draw a negligible current and we will assume that it reads the open circuit emf of the battery when sw_2 is open. Close sw_2 and adjust I_T to read 0.2-0.3 a and immediately open sw_2. Open sw_1 and allow about one minute for the cells to recover.

With sw_2 open, close sw_1, and read the open-circuit emf of the battery and record as E in the data table, reading the meter to the nearest tenth of the smallest scale division. Close sw_2 and read both meters to the nearest tenth of the smallest scale division. Record as I_T and V. The meter now reads the potential difference across the battery terminals with the circuit current in the battery. How can you explain the difference in the voltmeter readings of E and V?

The circuit current in the internal resistance of the battery r develops a potential difference $I_T r$ which is in opposition to the emf of the battery E, and the closed-circuit terminal potential difference V applied to the external circuit is the difference voltage. Thus

$$V = E - I_T r$$

Or,

$$I_T r = E - V$$

Whence

$$r = \frac{E - V}{I_T}$$

Make the necessary computations to complete the data table. Assume the three cells composing the battery to be identical.

DATA

NUMBER OF CELLS IN SERIES	E (v)	I_T (a)	V (v)	$E - V$ (v)	r battery (Ω)	r cell (Ω)

QUESTION: Your answer should be a complete statement.

What difficulty would you encounter in attempting to determine the internal resistance of a battery consisting of three cells in parallel by the method used in Part 4?

....

....

....

....

....

NAME PERIOD DATE INSTRUCTOR'S COMMENTS

47 EXPERIMENT Measurement of Resistance: Voltmeter-Ammeter Method

PURPOSE: (1) To measure the resistance of conductors by the voltmeter-ammeter method. (2) To determine the effect of length, diameter, and material on the resistance of a conductor.

APPARATUS: Battery, 6 v (four No. 6 dry cells in series will serve); d-c ammeter, 0-1/10 a; d-c voltmeter, 0-7.5 v; tubular rheostat; annunciator wire, 18 ga for connections; 3 nickel-silver resistance spools, 30 ga-200 cm, 28 ga-200 cm, 30 ga-160 cm (Constantan or German-silver spools may be used); 1 copper resistance spool, 30 ga-2000 cm; brass connectors, double; momentary contact switch, SPST. See Appendix A for instructions in the use of meters.

SUGGESTION: A convenient board for permanent use may be made as follows. Select a board from 10 to 12 inches wide and 15 inches long. Near one end, at *A* and *B*, Fig. 47-1, fasten two universal binding posts about 3 inches apart. At *C* and *D* fasten two universal binding posts. Attach a momentary contact switch to the board at *sw* and connect its terminal at *A* and *C* by means of 14 ga insulated copper wire.

INTRODUCTION: Various substances offer different amounts of resistance to the passage of an electric current. In general, metals are good conductors of electricity, but even among the metals we find a wide range of conductivity. Silver and copper, for example, are much better conductors than iron, lead, nickel-silver (an alloy of copper and nickel), or nichrome (an alloy of nickel, iron, chromium, and carbon), which is used in electric heaters.

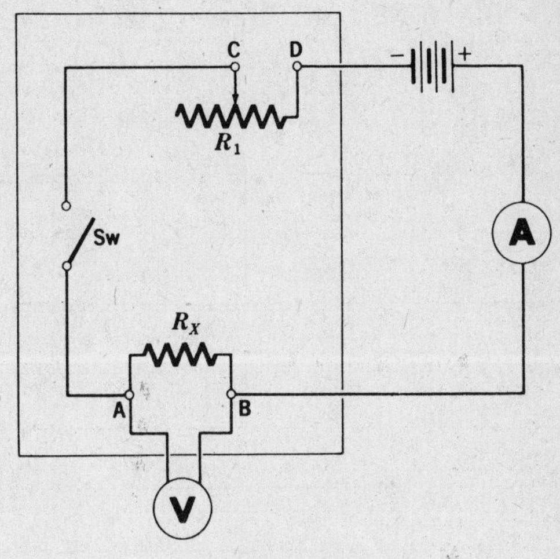

Figure 47-2

Physicists have discovered that there are four factors which determine the resistance of a conductor to the passage of electricity. These factors are (1) the temperature of the conductor; (2) the length of the conductor; (3) the cross-sectional area of the conductor; and (4) the material of which the conductor is made.

In this experiment we shall observe the effect the length, the diameter, and the material have on the resistance of a conductor. Care must be exercised to keep the resistances at room temperature to avoid the introduction of an error as a result of the increase in temperature of the resistance.

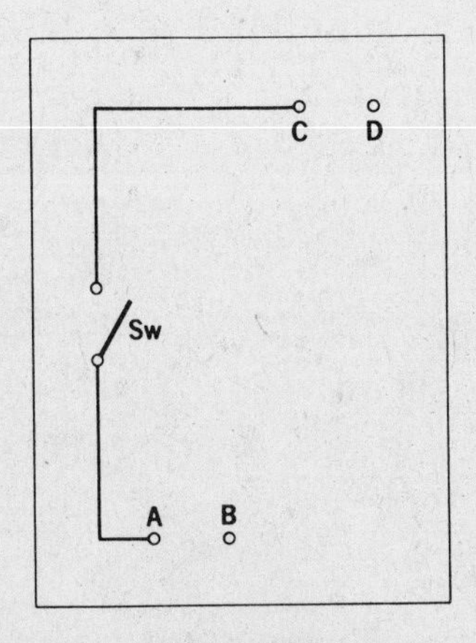

Figure 47-1

PROCEDURE: Before you connect a wire to a binding post or to a piece of apparatus, you must remove the insulation from about one-half inch of the end of the wire, and scrape the exposed metal with a knife or rub it with sandpaper until it is bright and clean. Tarnish on a metal offers considerable resistance. Connect the apparatus as shown in Fig. 47-2. The voltmeter *V* is connected in *parallel* with the spool of wire R_x whose resistance is to be measured. The ammeter *A* is connected in *series* with the spool of wire, the battery, and the rheostat R_1, which can be varied and serves merely to limit the circuit current.

With the rheostat resistance R_1 all in the circuit and the 200-cm spool of 30 ga nickel-silver wire connected across the terminals AB as R_x, gradually reduce the resistance R_1 until the ammeter reads 0.3-0.5 a. Read both the voltmeter and the ammeter, estimating to the nearest tenth of the smallest scale division. Then open the switch immediately. Can you give two reasons for adjusting R_1 for the small current?

Substitute for the coil just used each of the other coils in turn, adjust the resistance R to a suitable value, and read the voltmeter and ammeter to the precision each instrument allows.

CALCULATIONS: The voltmeter gives the potential difference across the resistance spool, while the ammeter gives the current in the resistance spool. By Ohm's Law,

$$R_x = \frac{V_x}{I}$$

calculate the resistance for each spool to the precision your measurements allow. The diameters and cross-sectional areas of wires are given in Table 21 of Appendix B. Determine the resistivity of the metal composing each spool, compare with the accepted constants (Table 20, Appendix B), and determine your error.

DATA

TRIAL	*Metal*	*Gauge number*	*Length* (cm)	*Cross-sectional area* (cm^2)	V_x (v)	I (a)	R_x (Ω)
1							
2							
3							
4							

TRIAL	*Resistivity (experimental)* (Ω cm)	*Resistivity (accepted value)* (Ω cm)	*Absolute error* (Ω cm)	*Relative error* (%)
1				
2				
3				
4				

NAME PERIOD DATE INSTRUCTOR'S COMMENTS

48 EXPERIMENT Measurement of Resistance: Wheatstone Bridge Method

PURPOSE: To learn to use the Wheatstone bridge for precision measurements of resistance.

APPARATUS: Wheatstone bridge; No. 6 dry cell or other d-c power source; galvanometer; resistance box; 3 nickel-silver resistance spools, 30 ga-200 cm, 28 ga-200 cm, 30 ga-160 cm (Constantan or German-silver spools may be used); 1 copper resistance spool, 30 ga-2000 cm; annunciator wire, 18 ga, for connections; mementary contact switch, SPST; brass connectors, double; bare copper wire, 30 ga for galvanometer shunt. See Appendix A for instructions in the use of meters.

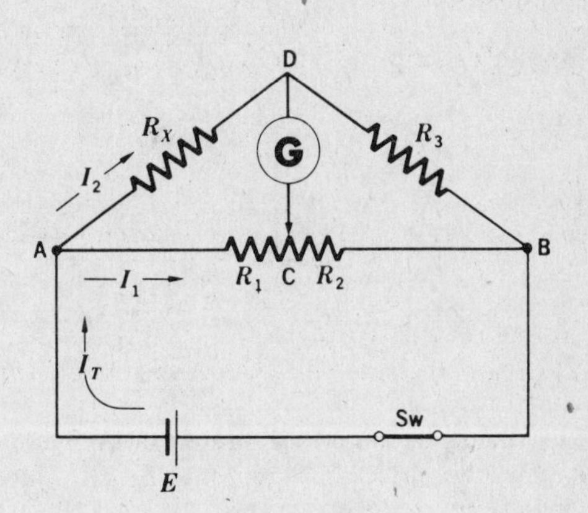

Figure 48-1

INTRODUCTION: The Wheatstone bridge provides a very precise means of measuring resistance. It is a simple *bridge* circuit consisting of a source of emf, a galvanometer, and a network of four resistors. By balancing the bridge and knowing three of the resistances, the fourth resistance can be calculated.

Referring to Fig. 48-1, an unknown resistance R_x may be balanced against known resistances R_1, R_2, and R_3 by adjusting R_1, R_2, and R_3 until a galvanometer *bridged* across the parallel branches shows zero current. When the bridge is balanced, there can be no difference of potential between points C and D. Thus

$$V_{AD} = V_{AC} \text{ and } V_{DB} = V_{CB}$$

Then $I_2 R_x = I_1 R_1$ and $I_2 R_3 = I_1 R_2$

$$\frac{R_x}{R_3} = \frac{R_1}{R_2} \text{ or } R_x = R_3 \frac{R_1}{R_2}$$

In the laboratory form of the Wheatstone bridge, Fig. 48-2, resistances R_1 and R_2 take the form of a uniform resistance wire, the position of contact C determining the lengths comprising R_1 and R_2. Since the resistance of the uniform wire is directly proportional to its length,

$$\frac{l_1}{l_2} = \frac{R_1}{R_2} \quad \text{Therefore, } R_x = \frac{l_1}{l_2}$$

The Wheatstone bridge permits a more precise measurement of resistance than the voltmeter-ammeter method. It requires, however, more careful work to insure the precision of which it is capable. By using the same resistance spools measured previously by the voltmeter-ammeter method, a comparison of the relative precision of these two methods of measuring resistance can be made. Resistivity constants will be determined in metric units.

PROCEDURE: Set up the apparatus as shown in Fig. 48-2, making sure that all contact points are clean and bright. Use as short pieces of wire as possible for all connectors. One terminal of the galvanometer G is connected to the binding post at D; the other terminal of the galvanometer is connected to the slide contact C by a wire long enough to permit the contact key to touch any point of the wire AC. The resistance box R_3 is connected to the binding posts on the brass strips at one end of the Wheatstone bridge, and the unknown resistance R_x is connected to the binding posts at the other end of the bridge. A dry cell of 1.5 volts is connected in series with a momentary contact switch to the brass strips at the ends of the bridge which serve as terminals of the wire AB.

For the first trial, use a 200-cm spool of 30 ga nickel-silver wire as R_x whose resistance is to be measured. Place a low resistance shunt across the galvanometer by winding several turns of 30 ga bare copper wire around the terminals. Why? Unplug some resistance from R_3, momentarily close the switch, and press the contact key near the 50-cm mark. If the galvanometer shows no deflection for several positions of the contact key, remove some of the shunting wire from the terminals. Allow the switch to remain closed for short periods only.

When the galvanometer shows some deflection, try moving the contact key to the right. Does the deflection increase or decrease? If the deflection increased, move the contact key to the left. Is there a region on the wire where the deflection reverses as the contact key is moved through it? Try to locate one, changing the resistance of R_3 if necessary.

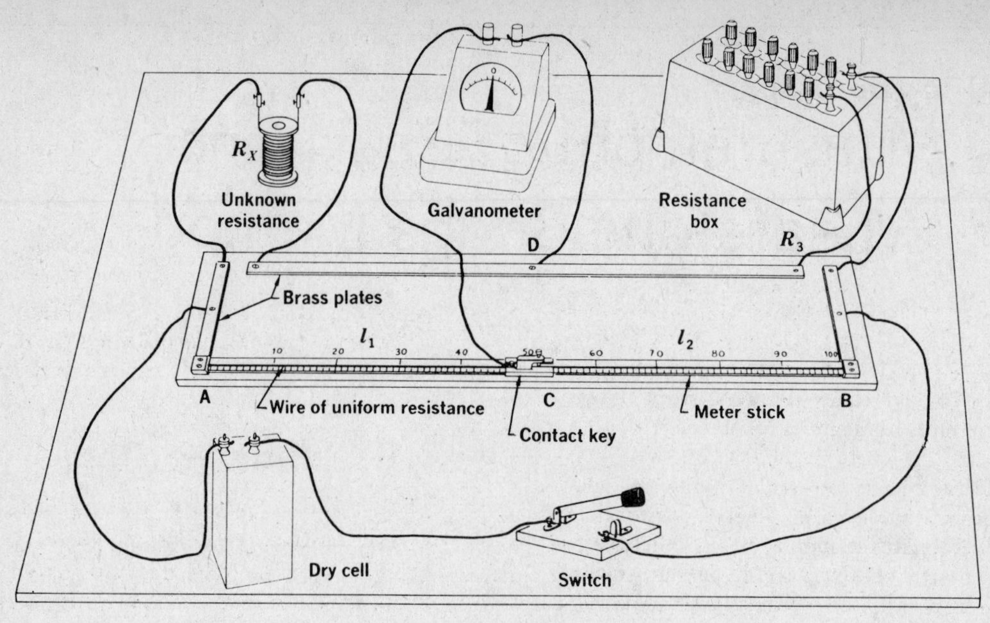

Figure 48-2

If the balance region is located near either end of the slide wire, considerable error may be introduced. If l_1 is much shorter than l_2, then R_x must be smaller than R_3 since $R_x/R_3 = l_1/l_2$. Use this relation as a basis for determining what adjustment to make in R_3 to move the balance region between the 40-cm and 60-cm marks on the meter stick.

As the bridge is brought nearer the final balance, remove shunting turns from the galvanometer to increase its sensitivity. Finally locate the final balance point, with all shunting turns removed, by holding the contact key down and sliding it back and forth to locate precisely the position of zero deflection on the galvanometer. Record the necessary data in the data table.

Replace the resistance spool just measured with the 200-cm spool of 28 ga nickel-silver wire. Balance the bridge as before and record the required data. In a similar manner measure the resistance of the 160-cm spool of 30 ga nickel-silver wire and finally that of the 2000-cm spool of 30 ga copper wire. Record the necessary data.

CALCULATIONS: Calculate the resistance of each spool to the precision your measurements will allow. Consult Table 21, Appendix B, for cross-sectional areas of the wires used and determine the resistivity of the metal composing each spool. Compare with accepted constants in Table 20, Appendix B, and express your error in each case.

DATA

TRIAL	*Metal*	*Gauge number*	*Length* (cm)	*Cross-sectional area* (cm^2)	l_1 (cm)	l_2 (cm)	R_3 (Ω)	R_x (Ω)
1								
2								
3								
4								

TRIAL	*Resistivity (experimental)* (Ω cm)	*Resistivity (accepted value)* (Ω cm)	*Absolute error* (Ω cm)	*Relative error* (%)
1				
2				
3				
4				

NAME PERIOD DATE INSTRUCTOR'S COMMENTS

49 EXPERIMENT Effect of Temperature on Resistance

PURPOSE: (1) To determine the temperature coefficient of resistance of a metallic conductor. (2) To study the effects of a change in temperature on the resistance of metallic and nonmetallic conductors.

APPARATUS: Wheatstone bridge; dry cell or other d-c power source; galvanometer; resistance box; momentary contact switch, SPST; annunciator wire, 18 ga, for connections; bare copper wire, 30 ga, for galvanometer shunt; temperature coil, copper, approximately 3 ohms; thermometer; magnifier; calorimeter; crushed ice; boiler or large beaker; burner; tripod; carbon lamp, 32 candle; tungsten lamp, 40 watts; lamp socket, standard base; fuse block, double, with 6-a fuses; split-line extension cord and plug; a-c milliammeter, 0-500 ma; a-c voltmeter, 0-150 v; cross-section paper.

INTRODUCTION: The resistance of a conductor depends on its composition, its length, its cross-sectional area, and its temperature. In the case of pure metals and most metallic alloys, resistance increases rapidly with a rise in temperature. Many nonmetals, on the other hand, show a decrease in resistance with a rise in temperature. Special alloys may have resistances which are practically independent of temperature.

The resistance of a conductor composed of a given material and at a constant temperature is equal to the product of a dimensional constant, called resistivity ρ (rho), and the ratio of the length l to the cross-sectional area A.

$$R = \rho \frac{l}{A} \quad \text{whence } \rho = \frac{RA}{l}$$

ρ has the dimension Ω cm.

The change in resistivity of a conductor which occurs over a moderate temperature range defines a useful quantity characteristic of the substance: the **temperature coefficient of resistivity,** α *(alpha); the ratio of the change in resistivity due to a change of temperature of 1° C to its resistivity at 0° C.*

$$\alpha = \frac{\Delta\rho / \Delta T}{\rho_o} \tag{1}$$

The value of α depends on the temperature on which ρ_o is based, conventionally 0° C. The change in resistivity $\Delta\rho$ is, of course, expressed in terms of the resistivity ρ_T at temperature T and the resistivity ρ_o at 0° C. $\Delta\rho = \rho_T - \rho_o$. Similarly, $\Delta T = T - 0° = T$. Substituting these values for $\Delta\rho$ and ΔT, Equation 1 becomes

$$\alpha = \frac{\rho_T - \rho_o}{\rho_o T} \tag{2}$$

Since it is the resistivity of a substance that changes with temperature, the resistance R of a given conductor changes accordingly. Thus R may be substituted for ρ in Equation 2.

$$\alpha = \frac{R_T - R_o}{R_o T} \tag{3}$$

When R_T is the measured resistance in ohms at temperature T, R_o the resistance in ohms at 0° C, and T the final temperature in °C, α is the *temperature coefficient of resistance* whose dimension depends only upon the unit of T.

PROCEDURE:

1. Temperature coefficient of resistance

Set up the Wheatstone bridge apparatus as in Experiment 48, connect the temperature coil as R_x, and balance the bridge when the coil temperature has adjusted to that of the ice water. How can you tell when the coil temperature is constant? Record the bridge measurements and the temperature as precisely as possible. Use the magnifier to help you read the thermometer to the nearest 0.1°.

Measure the resistance of the coil while it is submerged in boiling water and record the temperature and the bridge measurements as before.

Construct a graph of resistance as a function of temperature on rectangular coordinate paper using the vertical axis for resistance and the horizontal axis for temperature. Locate 0° at the origin and extend the temperature scale to 100° C. Provide a suitable resistance scale on the vertical axis to yield resistance readings slightly above and below the two values found for R_x (observe that the origin *cannot* serve to locate zero resistance). Plot the two known points and join them with a straight line; for a pure metal the curve is known to be essentially linear.

By extrapolation (extending the curve beyond known data), read off the resistance at 0° C and at 100° C and record as R_o and R_T. The final temperature T is recorded as 100° C. Compute the temperature coefficient of resistance and compare it with the accepted value for copper.

DATA

MATERIAL	Temp. (°C)	Bridge measurements l_1 (cm)	l_2 (cm)	R_3 (Ω)	R_x (Ω)	R_o (Ω)	R_T (Ω)	T (°C)	α experimental ($C^{\circ -1}$)	α accepted ($C^{\circ -1}$)	Relative error (%)
Copper											

2. *Resistance of cold lamp filaments*
Measure the resistance of a carbon-filament lamp with the Wheatstone bridge using two or three dry cells in series if necessary for satisfactory bridge performance. Record the required data. Repeat the measurement with a 40-watt tungsten lamp.

3. *Resistance of hot lamp filaments.*
Arrange the circuit as shown in Fig. 49-1 (**Caution:** *Completely deenergize the circuit by removing the plug from the supply receptacle before manipulating the circuit components in any way*) and measure the resistance of the hot filaments of the carbon and tungsten filament lamps by the voltmeter-ammeter method. Record the required data in the table below.

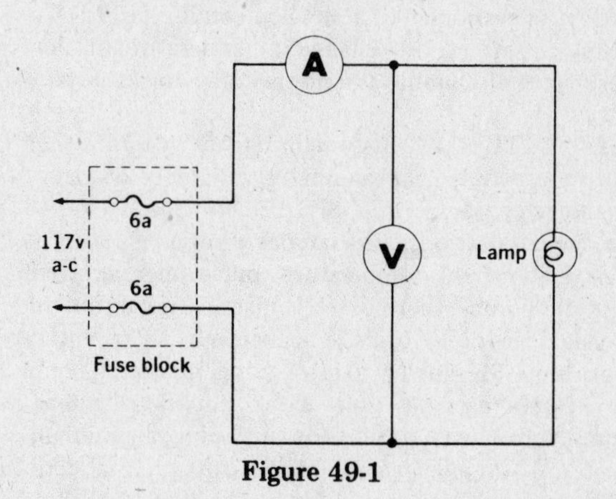

Figure 49-1

DATA

MATERIAL	Cold filament l_1 (cm)	l_2 (cm)	R_3 (Ω)	R_x (Ω)	Hot filament potential difference (v)	current (a)	R (Ω)
Carbon							
Tungsten							

QUESTIONS: Your answers should be complete statements. Problem solutions should be set down in the space provided below.

1. Considering the change in resistance with temperature of the carbon and tungsten filaments, how do their temperature coefficients of resistance compare?

....

....

.... 1

2. The resistance of a certain tungsten wire at 0° C is 79.00 Ω and the temperature coefficient of resistance of tungsten is 0.0045/C°. What is the resistance of the wire at 100.0° C?

2

3. How is the fact that all pure metals have an a of the order of $0.004/C^\circ$ related to the phenomenon of *superconductivity* of pure metals?

....

....

....

....

....

....

....

....

....

....

.... 3

4. **What temperature coefficient of resistance would be desirable for a wire used to make the resistance coils of a resistance box?**

....

....

....

.... 4

NAME PERIOD DATE INSTRUCTOR'S COMMENTS

50 EXPERIMENT Resistances in Series and Parallel

PURPOSE: (1) To measure the resistance of resistance elements joined in series. (2) To measure the equivalent resistance of resistance elements joined in parallel.

APPARATUS: The same as for Experiment 47 or 48, depending on the method used.

SUGGESTION: The instructor may wish to substitute resistance spools other than those used in the previous experiments. If equipment is limited, half the group may use the Wheatstone bridge method and the other half the voltmeter-ammeter method and a further comparison of the relative precision of the two methods in resistance measurements may be made.

INTRODUCTION: When resistance coils are connected in series, the circuit current I_T is in each coil. Each succeeding coil adds its resistance to that of the others in series, and their combined resistance equals the sum of all the separate resistances. If R_1, R_2, R_3, etc., are individual resistances joined in series, the total resistance R_T is found from the equation $R_T = R_1 + R_2 + R_3 \ldots$

When two or more resistance coils are joined in parallel, more paths are provided for the electric current and consequently the equivalent resistance equals the reciprocal of the equivalent resistance. If R_1, R_2, R_3, etc., are individual resistances joined in parallel, the equivalent resistance R_{eq} is found from the equation $1/R_{eq} = 1/R_1 + 1/R_2 + 1/R_3 \ldots$

Provisions are made in this experiment for determining the resistance of the series and parallel combinations using the resistances of the separate spools as calculated from the resistivity table. Knowing the kind of metal, length, gauge number, and resistivity constant for each spool, the resistance of the spool may be calculated using the following expression:

$$R = \rho \frac{l}{A}$$

Where ρ is the resistivity constant in ohm centimeters, l is the length of wire in centimeters, and A is its cross-sectional area in square centimeters; R is the resistance of the spool in ohms.

PROCEDURE:

1. Wheatstone bridge method

Set up the Wheatstone bridge as it was used in Experiment 48. Insert the series combination of 200 cm of 30 ga nickel-silver and 200 cm of 28 ga nickel-silver as R_x and determine their combined resistance.

Join the same spools of wire in parallel and measure their equivalent resistance. Continue the experiment by measuring the combined resistance of the 160-cm spool of 30 ga nickel-silver wire joined in series with the 2000-cm spool of 30 ga copper wire. Join these spools of wire in parallel and measure the equivalent resistance.

Record all pertinent data and solve for R_x. Also record R_x in the Data Summary table of Part 3.

2. Voltmeter-ammeter method. (Alternate method)

As an alternate experiment, the resistances may be found by the voltmeter-ammeter method in the same manner as in Experiment 47. Record all pertinent

DATA *Table 1 (Wheatstone Bridge Method)*

METAL	*Length* (cm)	*Gauge*	TYPE OF CIRCUIT	l_1 (cm)	l_2 (cm)	R_3 (Ω)	R_x (Ω)
1.							
2.							
1.							
2.							
3.							
4.							
3.							
4.							

data and solve for R_x in each case. Also record R_x in the Data Summary table of Part 3.

3. Enter the data from either Part 1 or Part 2 in the Data Summary table. Record the resistivity constant for each spool using the resistivity tables in Appendix B. Calculate the resistance of each spool using the resistivity constant and the wire dimensions. Calculate the value of R_x for each series and parallel combination and compare with the corresponding experimental values of R_x expressing the error in each instance as a relative error.

DATA *Table 2 (Voltmeter-ammeter method)*

METAL	*Length* (cm)	*Gauge*	TYPE OF CIRCUIT	V_x (v)	*I* (*a*)	R_x (Ω)
1.						
2.						
1.						
2.						
3.						
4.						
3.						
4.						

DATA SUMMARY

SPOOL NUMBER	TYPE OF CIRCUIT	*Resistivity* (Ω cm)	*Separate resistances (calculated)* (Ω)	R_x *(calculated)* (Ω)	R_x *(experimental)* (Ω)	*Relative error* (%)
1	series					
2						
1	parallel					
2						
3	series					
4						
3	parallel					
4						

NAME PERIOD DATE INSTRUCTOR'S COMMENTS

51 EXPERIMENT Simple Networks

PURPOSE: To study the relationships between potential difference, current strength, and resistance throughout a simple network.

APPARATUS: 3 d-c ammeters, 0-1 a; 4 d-c voltmeters, 0-3 v; 3 resistance boxes; 4 dry cells or other d-c power source; momentary contact switch; annunciator wire, 18 ga, for connections. See Appendix A for instructions in the use of meters.

SUGGESTION: Since the experiment calls for three ammeters, four voltmeters, and three resistance boxes, the instructor may wish to assign it as a demonstration experiment.

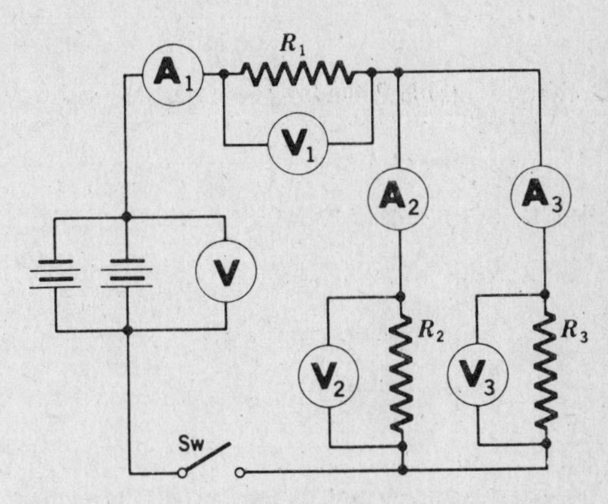

Figure 51-1

INTRODUCTION: No matter how simple or how complicated an electric network is, Ohm's law expresses the relationships between resistance, potential difference, and current in each part of the circuit and over the entire circuit.

PROCEDURE: 1. Set up the apparatus as shown in Fig. 51-1. The ammeters, A_1, A_2, and A_3, connected in series with each resistance, show the magnitude of current in each resistance. The voltmeters, V_1, V_2, and V_3, connected in parallel with each resistance, show the potential difference across each resistance, while voltmeter V shows the potential difference across the external circuit.

2. Set resistance boxes R_2 and R_3 at zero, but set R_1 at 10.0 ohms. Close the switch, read each of the meters, and record the values in the data table. You should be able to read to the nearest hundredth of an ampere and to the nearest hundredth of a volt. Also record the values of the three resistances. As a check on your setup, the reading of ammeter A_2 and ammeter A_3 should be the same. The reading of voltmeter V_2 and voltmeter V_3 should be zero. The reading of voltmeter V_1 should be the same as voltmeter V. If these values are not obtained, open the switch and check the wiring and the resistance boxes for loose connections or loose plugs.

3. Set the resistances in the three boxes as follows: R_1, 10.0 ohms; R_2, 10.0 ohms; R_3, 10.0 ohms. Close the switch, and make all readings. Note that all three resistances have the same value.

4. Set the resistances in the three boxes as follows. R_1, 10.0 ohms; R_2, 20.0 ohms; R_3, 10.0 ohms. Close the switch and make all readings. In this trial one of the resistances in parallel has twice the resistance of the other.

5. Set the resistances in the three boxes with R_1, 10.0 ohms; R_2, 5.0 ohms; R_3, 10.0 ohms. Close the switch and make all readings.

6. Set the resistances as R_1, 10.0 ohms; R_2, 2.5 ohms; R_3, 10.0 ohms. Close the switch and make all readings. In this case one of the resistances in parallel has four times the resistance of the other.

7. In each case calculate the equivalent resistance of R_2 and R_3 joined in parallel. Also calculate the total resistance of the entire circuit.

DATA

TRIAL	V (v)	I_1 (a)	V_1 (v)	R_1 (Ω)	I_2 (a)	V_2 (v)	R_2 (Ω)	I_3 (a)	V_3 (v)	R_3 (Ω)	R_{eq} (Ω)	R_T (Ω)
1.												
2.												
3.												
4.												
5.												

QUESTIONS: Your answers should be complete statements.

What conclusions can you draw about:

1. the total current in the circuit and the current in the parallel branches?

....

.... 1

2. the potential difference across each parallel branch?

....

.... 2

3. the potential difference across the various parts of a circuit and across the entire circuit?

....

....

.... 3

4. resistance and current in two parallel branches?

....

.... 4

5. the equivalent resistance of two parallel branches, the potential drop across the branches, and the total current in the branches?

....

....

.... 5

6. the potential difference across a circuit, the total resistance of the circuit, and the current in the circuit?

....

....

.... 6

NAME PERIOD DATE INSTRUCTOR'S COMMENTS

52 EXPERIMENT Electric Equivalent of Heat

PURPOSE: To determine the number of calories per joule of electric energy.

APPARATUS: Stop watch; d-c voltmeter, 0-7.5/15 v; d-c ammeter, 0-10 a; tubular rheostat; knife switch, SPST; electric calorimeter with a 2.5 to 5.0 ohm heating coil; thermometer; magnifier; platform balance; set of masses; 6-volt or 12-volt storage battery, or other low voltage d-c source capable of delivering approximately 3 a of continuous current; annunciator wire, 18 ga, for connections. See Appendix A for instructions in the proper use of meters.

INTRODUCTION: Electric energy is converted into heat energy by the resistance of a conductor. In this experiment, we will use an electrically heated coil to warm a known mass of water a measured number of Celsius degrees. By observing the time required to heat the water, the current in the heating coil, and the potential difference across its terminals, we can calculate the number of calories per joule of electric energy, the electric equivalent of heat. Since 1 joule of electric energy = 1 watt sec and 1 watt = 1 volt ampere, it is apparent that the electric energy converted to heat energy in the resistance coil is

$$W = VIt$$

All work done by a current in a resistance appears as heat. Thus, the electric energy expended in the resistance is directly proportional to the heat energy that appears.

$$W = JQ$$

Where W is electric energy in joules and Q is the quantity of heat produced in calories, J is the proportionality constant having the dimensions joules/calorie which we shall call the *electric equivalent of heat.* Solving this equation for Q, we have a general expression for Joule's law.

$$Q = W/J \qquad \text{(Joule's law)}$$

PROCEDURE: Set up the apparatus as shown in Fig. 52-1. A 6-v or 12-v storage battery may be used to furnish the electric energy. Adjust the rheostat for a current of about 2 amperes and immediately open the switch.

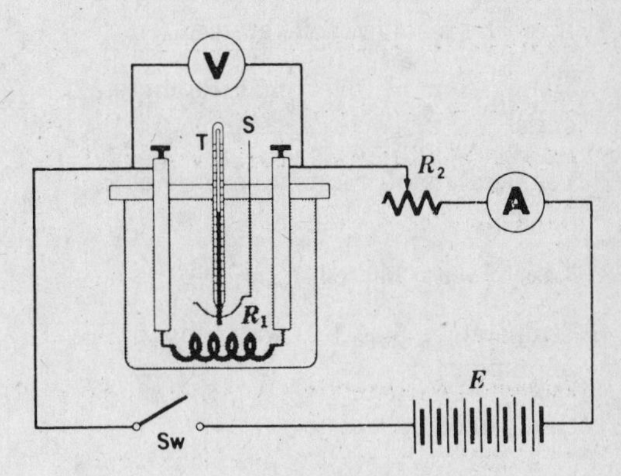

Figure 52-1

The heating coil usually has a resistance of 2.5 or 5.0 ohms, the latter value requiring a 12-volt source for most satisfactory operation. If the heating coil you are to use has different current and voltage requirements, your instructor will provide the information.

Determine the mass of the calorimeter cup. Then fill the calorimeter cup about three-fourths full of water that is about 10 degrees below room temperature. Find the mass of the calorimeter cup and water, then seat the cup in its outer jacket. Place the cover on the calorimeter, stir the water thoroughly, and determine its temperature accurately. Use the magnifier to help you read the thermometer to the nearest tenth of a degree. Observe the time to the nearest second or start the stopwatch just as you close the switch. Immediately read the voltmeter and ammeter as accurately as possible. Keep the current as nearly constant as possible by slight adjustments of the rheostat.

Take both voltmeter and ammeter readings at one-minute intervals and record their final average values in the data table. Stir the water occasionally. Continue the heating until the water is about 10 Celsius degrees above room temperature. Open the switch and record the time to the nearest 0.1 second. Stir the water thoroughly and read the thermometer accurately to the nearest 0.1 degree when it reaches its highest temperature, and immediately lift the heating unit out of the water.

Take data from a second trial if time permits. Record all pertinent data and make the necessary computations.

DATA

	TRIAL 1	TRIAL 2
1. Mass of calorimeter cup	....................g	....................g
2. Specific heat of calorimeter	....................cal/g C°	....................cal/g C°
3. Mass of calorimeter cup and water	....................g	....................g
4. Voltmeter readings, final average	v	v
5. Ammeter readings, final average	....................a	....................a
6. Temperature of water and calorimeter, initial	.... °C	.... °C
7. Temperature of water and calorimeter, final	.... °C	.... °C
8. Mass of water heated	....................g	....................g
9. Temperature change of water and calorimeter cup	....................C°	....................C°
10. Heat gained by water	....................cal	....................cal
11. Heat gained by calorimeter cup	....................cal	....................cal
12. Total heat gained, Q	....................cal	....................cal
13. Total time of heat, t	....................sec	....................sec
14. Electric energy input, W	....................j	....................j
15. Electric equivalent of heat, J (experimental)	....................j/cal	....................j/cal
16. Electric equivalent of heat, J (accepted value)	....................j/cal	....................j/cal
17. Absolute error	....................j/cal	....................j/cal
18. Relative error	....................%	....................%

CALCULATION: Derive an equation from which the *heat* developed in the resistance of an electric circuit can be calculated directly from the *resistance* of the circuit, the *current* in the circuit, and the *time* the current is maintained.

NAME PERIOD DATE INSTRUCTOR'S COMMENTS

53 EXPERIMENT Electrochemical Equivalent of Copper

PURPOSE: To determine experimentally the electrochemical equivalent of copper.

APPARATUS: Copper cathode, 7.5 × 10 cm (see suggestion below); 6-v storage battery; knife switch, DPST; d-c ammeter, 0-3 a; tubular rheostat, 25 ohm; triple-beam balance, 0.01 g sensitivity; stopwatch or watch with sweep second hand; demonstration cell; copper wire, bare, 14 ga; annunciator wire, 18 ga, for connections; fine sandpaper; copper plating solution (see suggestion below); 1 beaker each of distilled water, alcohol, and ether for use by the class as dipping baths.

SUGGESTION: The copper plating solution should be made up in advance by dissolving 125 g of $CuSO_4 \cdot 5H_2O$ in 900 ml of distilled water. When dissolved, add *slowly* (while stirring) 35 ml concentrated H_2SO_4 and 50 ml of ethyl alcohol.

The copper cathode should be cut from copper sheet so as to be shaped into a cylinder. A strip should be left extending beyond the top of the cylinder to mate with the electrode clamp of the cell. The detail of a cathode blank most suitable for the particular electrode clamps to be used can best be determined by the instructor.

INTRODUCTION: The electrodeposition of copper is an electrolytic process, an oxidation-reduction chemical reaction. In this experiment, a careful determination will be made of the mass of copper deposited by the passage of a known quantity of electric charge through the electroplating cell.

Michael Faraday first observed that the mass of an element deposited during electrolysis is directly proportional to the quantity of charge that passes and to the chemical equivalent (atomic weight/valence) of the ion. A charge of 96,500 coulombs (1 faraday) deposits the chemical equivalent of any element in grams. Such quantities of elements are said to be *electrochemically equivalent.* The electrochemical equivalent (z) of an element is the mass of the element, in grams, deposited by 1 coulomb of electricity.

$$z = \frac{m}{Q} \tag{1}$$

Where m is the mass in grams of an element deposited and Q is the quantity of electricity in coulombs passed through the cell, z is the electrochemical equivalent of the element in g/c.

Since the quantity of electricity Q is determined by the product of the current in the circuit in amperes and the time in seconds,

$$z = \frac{m}{It} \tag{2}$$

PROCEDURE: Clean the copper cathode with steel wool or fine sandpaper until it has a smooth polished surface. Shape it into a cylinder suitable for clamping into the cell to be used. (Handle between the folds of a clean towel to avoid finger contact. Why?) Determine the mass of the cathode to the nearest 0.01 g and record in the data table. Clean a section of heavy copper wire and form a spiral anode of suitable length by winding it around a lead pencil. Then carefully as-

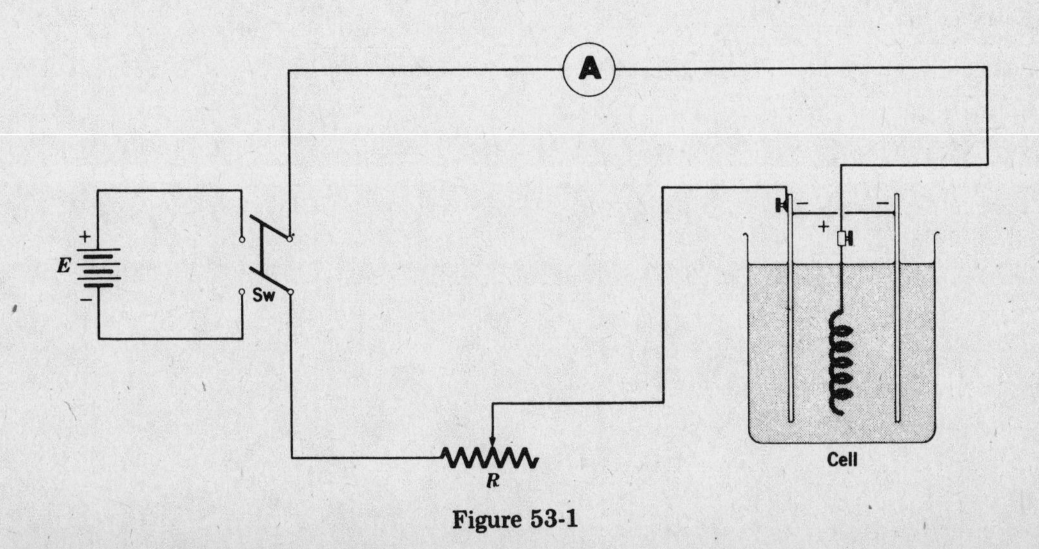

Figure 53-1

semble the cell (do not add the electrolyte until ready to put the cell in operation).

Arrange the circuit as shown in Fig. 53-1 being certain that the cylinder is connected as the cathode and the spiral as the anode. Estimate as accurately as you can the cathode area to be submerged in electrolyte and compute an appropriate current magnitude to be used on the basis of about 0.02 a per cm^2 of cathode area. Carefully mark this reading on the glass face of the ammeter to aid in monitoring the current during the timed operation of the cell.

With the rheostat resistance all in the circuit, add the electrolytic solution to the predetermined level, *close the switch just long enough to quickly adjust the rheostat to give approximately the current reading indexed.* Open the switch immediately and prepare to start a timed operation of 20 to 30 minutes duration.

Start the timed operation by closing the switch and quickly adjust the rheostat for the precise current reading. Record the initial time (if stopwatch is not the timing device) and monitor the circuit to maintain the current reading indexed on the ammeter.

At the end of the timed operation open the circuit, record the time, remove the cathode from the cell, and carefully dip it in distilled water, in alcohol, and finally in ether. When dry, determine the mass of the cathode to the nearest 0.01 g and record.

CALCULATIONS: From the data collected, compute the experimental value of the electrochemical equivalent of copper. Enter the accepted value from Table 19, Appendix B, and compute your relative error.

DATA

Item		Unit
1. Time, initial	....	
2. Time, final	....	
3. Elapsed time	....	sec
4. Average current	....	a
5. Quantity of electricity passed	....	c
6. Mass of cathode, initial	....	g
7. Mass of cathode, final	....	g
8. Mass of copper deposited	....	g
9. z for copper (experimental)	....	g/c
10. z for copper (theoretical)	....	g/c
11. Absolute error	....	g/c
12. Relative error	....	%

NAME PERIOD DATE INSTRUCTOR'S COMMENTS

54 EXPERIMENT Magnetic Field about a Conductor

PURPOSE: To study the effect of the magnetic field which is set up in vertical and horizontal conductors by currents in them.

APPARATUS: Galvanoscope, with binding posts for a single turn, a few turns, and many turns of wire (Fig. 54-1); annunciator wire, 18 ga, for connections; dry cell; d-c ammeter, 0-1 a; rheostat, approximately 25 ohms; momentary contact switch, SPST; 1 large compass; Ampère's law stand; 4 small compasses. (A galvanoscope can be improvised by wrapping a single turn, a few turns, and many turns of insulated wire into a coil large enough to accommodate the large compass.)

SUGGESTION: For best results, all apparatus should be isolated from local magnetic fields, such as those created by d-c power supplies. For that reason, dry cells are recommended.

INTRODUCTION: Hans Christian Oersted (1777-1851), a Danish physicist, was the first to show that a current in a conductor produces a magnetic field about the conductor, proving that there is a relation between electricity and magnetism. A magnetic needle placed near a conductor carrying an electric current is deflected. In this experiment we shall study the direction and amount of such deflection.

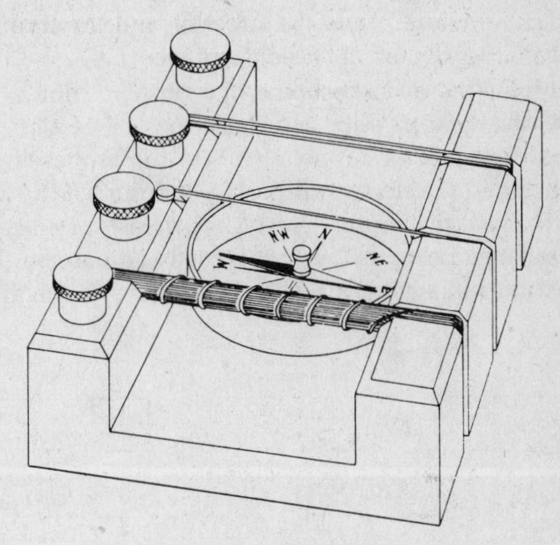

Figure 54-1

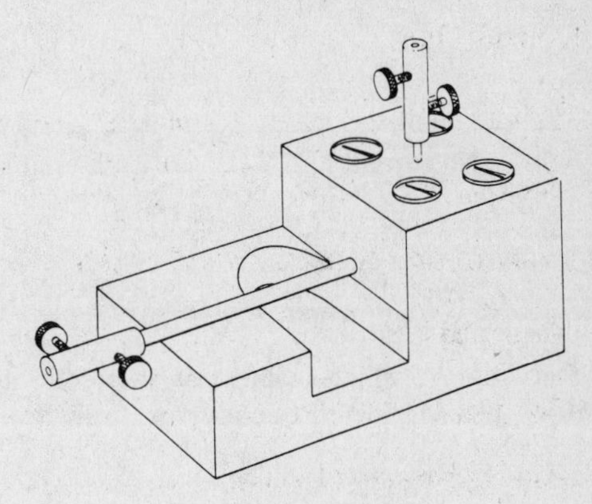

Figure 54-2

PROCEDURE:

1. Horizontal conductors

Place the large compass inside the galvanoscope coil of many turns. Turn the galvanoscope until the turns of wire lie in the North-and-South plane as indicated by the compass needle. Next turn the compass around until its N-pole is directly above the zero index of the compass card. Connect a dry cell in series with a rheostat, ammeter, and switch to the binding posts serving the coil of many turns in such a manner that electron flow will be from *south to north* through the loop segments *above* the needle. *With all the resistance of the rheostat in the circuit*, close

DATA *Horizontal conductor*

COIL	*Current* (a)	*Electron flow*	*Deflection direction*	*Deflection* (°)
Many turns		S to N		
Many turns		N to S		
Few turns		S to N		
Few turns		N to S		
Single turn		S to N		
Single turn		N to S		

the switch and adjust the current to a magnitude which just produces the maximum deflection of the compass needle. Record the current, the direction in which the N-pole of the compass needle is deflected, and the number of degrees of deflection. *By rheostat adjustments, maintain this same magnitude of current for all galvanoscope tests in Part 1.*

Reverse the direction of the electron flow and adjust the rheostat if necessary to supply the same current as before. Note the direction and the amount of the deflection of the needle and record.

Keeping the galvanoscope in the same position, move the compass until it is inside the coil of a few turns. Connect the supply circuit to the proper binding posts so the electron flow will be from south to north above the needle. Adjust the rheostat to give the same current previously recorded and observe the direction and amount of deflection. Reverse the direction of electron flow, and again observe and record the readings.

Repeat the experiment, using the coil of one turn, with the electrons first flowing from south to north above the needle, and then from north to south.

2. Vertical conductor

Arrange the 4 small compasses on the Ampère's law stand as shown in Fig. 54-2. Connect the terminals to the supply circuit so that the electrons flow upward. Adjust the rheostat to supply the *least* current that will yield conclusive deflections of the compass needles. Note the position taken by the needles. Reverse the direction of the electron flow. Observe the directions indicated by the N-poles of the compass needles, and make sketches in the labeled squares to show them for each case. Do not leave the switch closed any longer than is required to make your observations.

Vertical conductor

No current	Electrons flowing downward	Electrons flowing upward

QUESTIONS: Your answers should be complete statements.

1. State the rule which enables you to predict the direction in which the N-pole of a compass will be deflected if it is placed beneath a conductor through which electrons are flowing.

....

....

.... 1

2. The flow of electrons through a conductor is from south to north. In which direction will the N-pole of a compass needle placed over the conductor be deflected?

....

.... 2

3. State the rule that enables you to predict the direction in which the magnitude lines of flux circle a vertical conductor through which electrons are flowing.

....

....

.... 3

4. A compass is placed to the east of a vertical conductor. In which direction must electrons flow through the conductor to cause the N-pole of the compass to point south?

....

.... 4

NAME PERIOD DATE INSTRUCTOR'S COMMENTS

55 EXPERIMENT Galvanometer Constants

PURPOSE: To determine the resistance, current sensitivity, and voltage sensitivity of a galvanometer.

APPARATUS: Wheatstone bridge, slide-wire form; 4 resistance boxes, 1 of the order of 100 ohms, 2 of 1000 ohms, 1 of 10,000 ohms; galvanometer; tubular rheostat, approximately 200 ohms for meter resistance of the order of 100 ohms, approximately 500 ohms for meter resistance of the order of 40 ohms; knife switch, SPST; contact key; annunciator wire, 18 ga, for connections; 2 new No. 6 dry cells; d-c voltmeter, 0.3 v.

SUGGESTION: Part 1 should be performed by each laboratory group. However, Part 2 may be performed as a demonstration due to the number and variety of resistance boxes required.

INTRODUCTION: There is a fall of potential between two points on a conductor carrying a current, due to the resistance of the conductor. If a circuit is connected across two such points, a current proportional to the IR drop between these points will be established in the circuit, the difference of potential being used as a source of emf.

A galvanometer connected in the slide circuit between two points on a length of resistance wire as shown in Fig. 55-1 enables us to observe the relation between the current in the slide circuit and the distance between the contact points on the resistance wire. We shall assume that the deflection of the galvanometer is directly proportional to the magnitude of current in it and, since its resistance remains constant, to the voltage applied across the slide circuit.

The sensitivity of a galvanometer may be expressed in several ways. Collectively, these expressions are known as the galvanometer constants. The most commonly used constants are *current sensitivity* and *voltage sensitivity*. Current sensitivity k is defined as the current required to produce a deflection of one scale division and is expressed in *microamperes per scale division*. Voltage sensitivity is defined as the potential difference across the terminals which produces a deflection of one scale division. Voltage sensitivity may be derived from the current sensitivity and resistance of the galvanometer according to Ohm's law.

In the circuit of Fig. 55-2, we shall assume that the internal resistance of new dry cells is negligible and that R_3 is very large compared to R_2 so that the current in the galvanometer is negligible compared to that in R_1 and R_2. The battery current in R_1 and R_2 is

$$I = \frac{E}{R_1 + R_2}$$

The potential difference across R_2 is

$$V_2 = IR_2.$$

Substituting the value of I:

$$V_2 = \frac{ER_2}{R_1 + R_2}.$$

The current in the galvanometer I_G is

$$I_G = \frac{V_2}{R_3 + R_G},$$

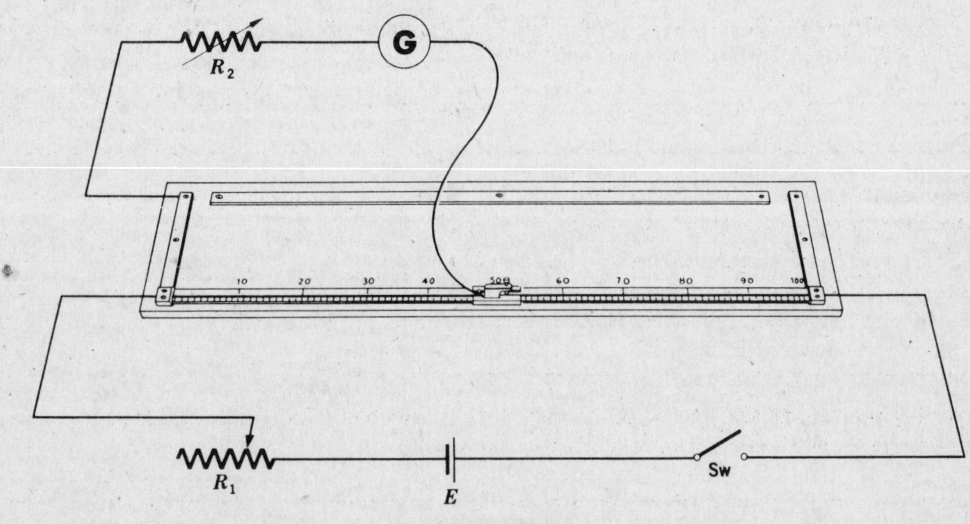

Figure 55-1

DATA *Table 1 (Galvanometer resistance)*

TRIAL	*Original slider position* (cm)	*Original deflection* (relative)	*Final slider position* (cm)	*Final deflection* (relative)	R_G (Ω)
1					
2					
3					
Average					

where R_G is the galvanometer resistance. If the current I_G causes a pointer deflection s, then the current sensitivity k is

$$k = \frac{I_G}{s}.$$

Substituting:

$$k = \frac{1}{s} \times \frac{ER_2}{(R_1 + R_2)(R_3 + R_G)}.$$

PROCEDURE:

1. Galvanometer resistance

Connect the circuit as shown in Fig. 55-1. The battery should consist of at least two dry cells connected in *parallel.* The resistance box, 111 ohms, is connected as R_2.

Set R_2 at zero resistance and adjust the rheostat R_1 so that all resistance is in the circuit. Close the switch and readjust R_1 for a convenient galvanometer deflection less than midscale when the slider contacts the resistance wire at midpoint, the 50-cm mark. Record the deflection and the initial slide position. *Keep the knife switch open at all times except when actually taking a reading.*

Move the slider to the 100-cm mark. The length of resistance wire across which the galvanometer is connected is now doubled. Thus, the voltage drop is doubled, and the potential difference placed across the galvanometer circuit, the current in the galvanometer, and the galvanometer deflection are doubled. Record the deflection and the final slide position. Adjust the resistance of R_2 to return the galvanometer deflection back to its initial reading. According to Ohm's law, if V is doubled, R must be doubled in order to maintain I constant. The original resistance in the side circuit was that of the galvanometer itself. Thus, the value of R_2 must equal the galvanometer resistance R_G.

Repeat the entire procedure several times, selecting new initial slider positions, adjusting R_1 as necessary, and then double the slider setting, adjusting R_2 to return the galvanometer to the initial deflection. Record three trials which are in close agreement and average the resistance of R_2 to give an average value of galvanometer resistance R_G.

2. Current sensitivity

Connect the circuit as shown in Fig. 55-2. R_1 and R_2 are resistance boxes of the order of 1000 ohms each; R_3 is a resistance box of the order of 10,000 ohms. The battery should consist of at least two new dry cells in *parallel.* The contact key K is closed only to take occasional readings of E to be sure there is no variation during the experiment. *Keep the switch Sw open at all times when not taking galvanometer readings.* Set R_1 at 300 ohms and leave constant for all trials. For the initial trial, set R_2 at 500 ohms and R_3 at 8000 ohms. Momentarily close the knife switch observing whether the galvanometer deflection is a convenient readable value. If not, make other initial settings of resistances, but keep $R_1 + R_2$ fairly large and R_3 at least 2 or 3 thousand ohms larger than R_2. Record R_1, R_2, R_3, and the galvanometer deflection s.

Make at least four trials with different combinations of R_1, R_2, and R_3 and record the pertinent data required.

CALCULATIONS: Using the expression for the current sensitivity k derived in the Introduction, solve for k for each trial. The galvanometer resistance R_G is the average value determined for your instrument in Part 1.

3. Voltage sensitivity

Using the average value of the galvanometer resistance R_G and the average value of the current sensitivity k, determine the voltage sensitivity of your galvanometer.

Assuming enough significant digits in your experimental results, calculate to the nearest 0.1 ohm the amount of resistance required in series with your galvanometer in order for a potential difference of 1 volt to produce a deflection of one scale division.

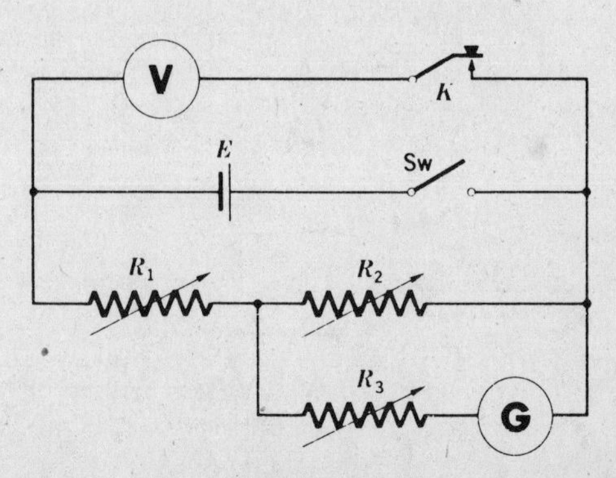

Figure 55-2

DATA *Table 2 (Current sensitivity)*

TRIAL	R_1 (Ω)	R_2 (Ω)	R_3 (Ω)	s (relative)	k (μa/div)
1					
2					
3					
4					
5					
6					
Average					

NAME PERIOD DATE INSTRUCTOR'S COMMENTS

56 EXPERIMENT Electromagnetic Induction

PURPOSE: (1) To study some of the phenomena of electromagnetic induction. (2) To determine the factors which influence the magnitude and direction of induced currents. (3) To recognize the principle of the electric generator.

APPARATUS: Pair of bar magnets, or a large horseshoe magnet; compact coil of approximately 100 turns wound on brass or plastic spool through which the bar magnet will pass; insulated copper wire, 22 ga and 28 ga; test tubes, one 200 mm × 25 mm, one 150 mm × 18 mm for making electromagnets for Part 2 (student-type primary and secondary coil set with iron core may be used for Part 2); twine; galvanometer, zero center; contact key, or push button; dry cell; tubular rheostat, about 10 ohms; iron rod suitable as a core.

SUGGESTIONS: Suitable coils of wire for Parts 1 and 3 may be made up and tied with twine to make them permanent. The coils for Part 2 may be made as follows. Wrap 80 turns of 28 ga insulated copper wire in a single layer around a large test tube, 200 mm × 25 mm. Forty turns of 22 ga wire are wrapped in a single layer around a regular size tube, 150 mm × 18 mm. The free ends of the coils may be secured by rubber bands. The wrapped smaller test tube will slip inside the larger one for the experiment.

INTRODUCTION: After Oersted discovered that electricity and magnetism were related, two men, Michael Faraday in England and Joseph Henry in the United States, began investigating the use of magnets in the generation of electricity. Working independently, both found at nearly the same time that an emf can be induced across a conductor moving through a magnetic field in such a manner that there is relative motion between the conductor and the magnetic flux. Faraday is usually given credit for the pioneer work that led to the invention of the electric generator, while Henry pioneered in developing the electromagnet and the principles of self-induction.

PROCEDURE:

1. Induction with a permanent magnet

Connect the ends of a compact coil to a sensitive galvanometer, and then thrust the coil over the N-pole of a magnet, as shown in Fig. 56-1.

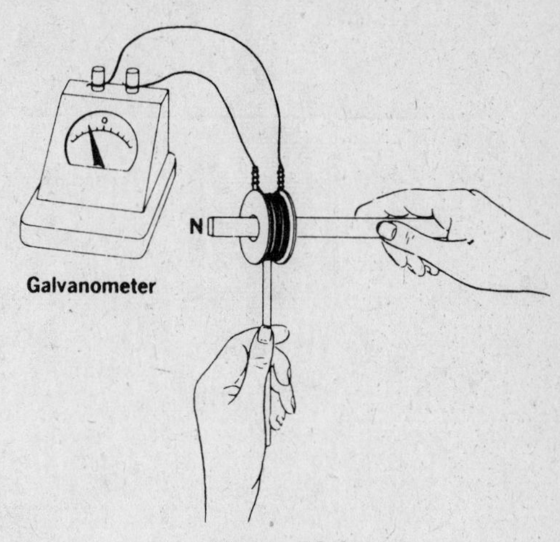

Figure 56-1

Result? ....

....

Remove the coil quickly. Repeat both motions more slowly.

Result? ....

....

Repeat the experiment, but thrust the coil down over the S-pole of the magnet, and then remove it quickly.

Holding the coil stationary, thrust first one pole of the magnet into the coil, remove it quickly, and then try the other pole. Repeat both motions more slowly.

Result? ....

....

2. Induction with an electromagnet

Attach the terminals of a large coil of about 80 turns of fine insulated wire, about 28 ga, to a galvanometer, as shown in Fig. 56-2. Connect the terminals of a small coil of about 40 turns of insulated 22 ga copper wire in series with a contact key, a dry cell, and a rheostat. Slip this coil inside the large coil, adjust the rheostat as necessary to provide an appropriate deflection, and observe the galvanometer: 1. when the circuit is closed; 2. when the circuit remains closed for a few seconds; 3. when the circuit is interrupted suddenly; 4. when the magnitude of the current is varied; 5. when the direction of electron flow is reversed by changing the battery connections; 6. when an iron rod is placed inside the small coil and the circuit is opened and closed. Record your observations.

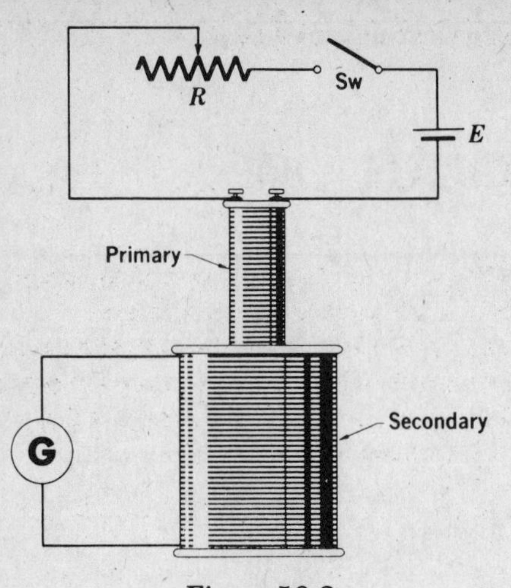

Figure 56-2

3. The generator principle

Wind about 20 turns of 28 ga insulated copper wire into a coil like that shown in Fig. 56-3. The oblong coil should be of such a size that it can be rotated between the poles of the magnet that you are using. Tie the separate turns together with twine to make a compact coil, and leave at least 18 inches at each end of the loop so it can be easily attached to the galvanometer. Hold the end of the coil between the thumb and finger so that the coil is between the poles of a horseshoe magnet or a V-magnet formed by two bar magnets. Give the coil a sudden twist so that its upper half A will move about 60° past the N-pole of the magnet. Note the direction in which the galvanometer needle is deflected. Continue to turn the coil until that part of the loop B is nearly adjacent to the N-pole. Note the deflection of the needle. See whether you can find a position for the loop in which the turning of its coils will not cause any deflection of the needle. Rotate the loop in the opposite direction and observe the results. Record all observations.

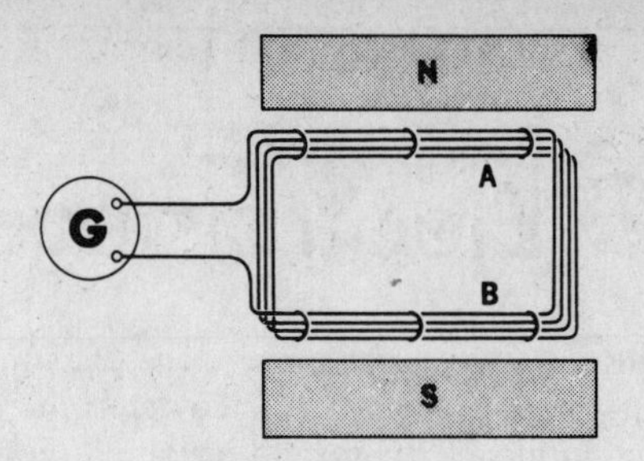

Figure 56-3

NAME PERIOD DATE INSTRUCTOR'S COMMENTS

57 EXPERIMENT The Electric Motor

PURPOSE: To study the construction and operation of an electric motor.

APPARATUS: Demonstration motor (St. Louis type is satisfactory); dry cell or storage cell; compass; 2 dual range d-c ammeters, 0-3/30 a; annunciator wire, 18 ga, for connections.

INTRODUCTION: A d-c motor consists of three parts: (1) the field magnet, which sets up a stationary magnetic field; (2) the armature, which is free to turn on its axis when attracted and repelled by the stationary magnetic field of the field magnet; and (3) the commutator, which acts as a current reverser, and the brushes, by means of which the current enters the armature. When the commutator changes the direction of the current, the poles of the armature also change polarity. In order to keep the armature turning, the current must change direction in the armature at such a time that the poles of the armature will be continuously repelled and attracted by the poles of the field magnet. That means that the N-pole of the armature is always repelled by the N-pole of the field magnet and attracted to the S-pole of the field magnet. It means, too, that at the same time, the S-pole of the armature is repelled by the S-pole and attracted by the N-pole of the field magnet. The polarity of the armature is reversed at just the right time to keep the rotation continuous.

PROCEDURE:

1. The commutator and the armature

Remove the bar magnets or electromagnet from the demonstration motor, and connect the armature to the cell. Use a small compass to test the polarity of the armature in different positions as you turn it slowly through a complete revolution. At what position of the armature does its polarity change? What causes this change in polarity?

Observation: ....

....

In the diagram Fig. 57-1 indicate the polarity of the armature at 30° intervals of its rotation.

Reverse the dry cell connections to the armature. Test the polarity of the armature now throughout a complete revolution. How does the polarity compare with that found first?

Observation: ....

....

In the diagram Fig. 57-2 indicate the polarity of the armature at 30° intervals of its rotation. Restore the armature and dry cell connections to their original arrangement.

2. Magnetic field furnished by permanent magnets

Place the permanent magnets in their holders with the N-pole of one magnet to the left of the armature and the S-pole of the other magnet to the right of the armature. Give the armature a gentle push to start it rotating. Mark in Fig. 57-1 the polarity of the field magnets, and the direction of rotation of the armature.

Reverse the connections between the dry cell and the armature. Mark in Fig. 57-2 the polarity of the field magnets, and the direction of rotation of the armature.

Reverse the polarity of the field magnets, making the pole to the left of the armature an S-pole and that to the right of the armature an N-pole.

Observation: ....

....

Reverse the armature connections once again.

Observation: ....

....

What keeps the armature rotating when the poles of the armature and the field are all in a straight line?

....

....

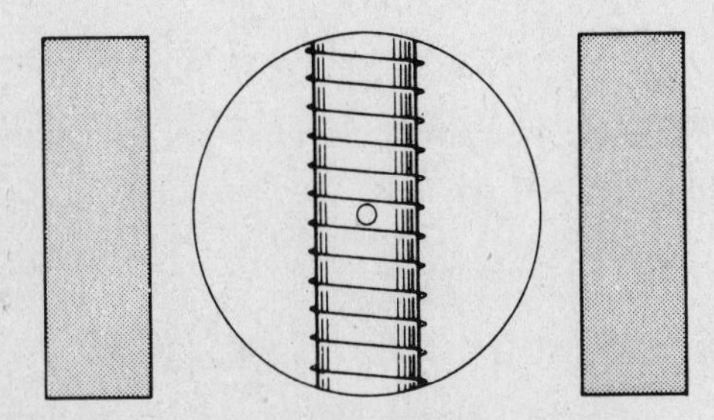

Figure 57-1

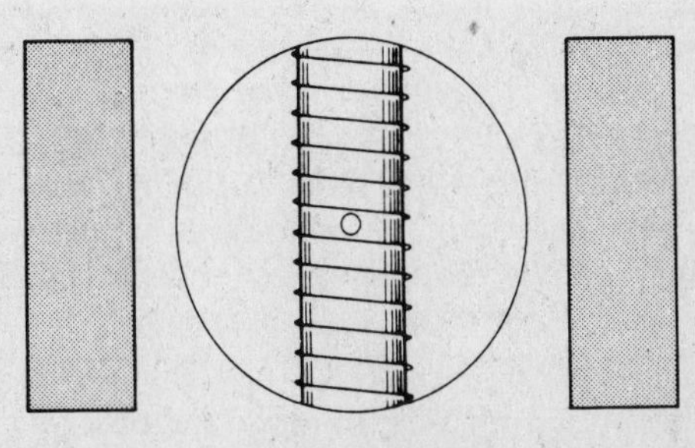

Figure 57-2

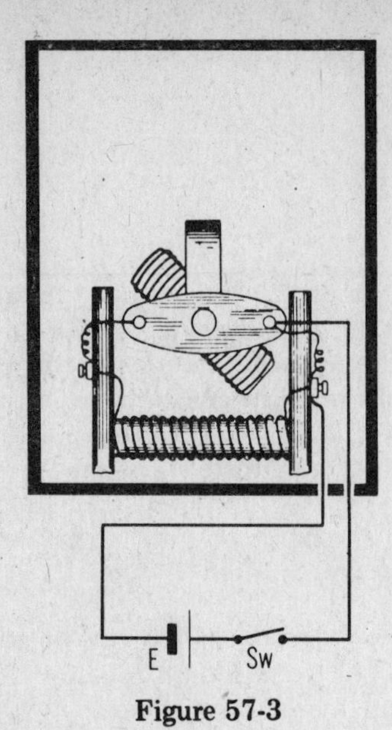

Figure 57-3

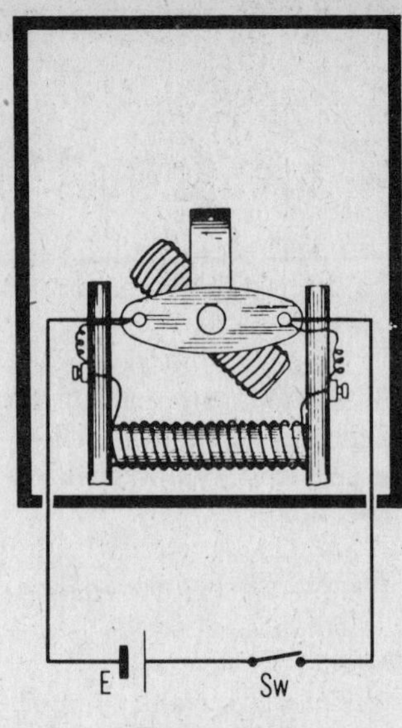

Figure 57-4

3. The series-wound motor

Replace the bar magnets with the electromagnet, and connect the cell in such a manner that the armature and field are in series, as in Fig. 57-3. Test the polarity of the field magnet and the polarity of the armature as the armature makes a complete revolution.

Observation: ....

....

Permit the motor to run freely and note the direction of rotation of the armature. Reverse the connections to the dry cell. What effect does this have on the direction of rotation of the motor?

....

....

Reverse the connections to the field magnet only. What effect does this have on the direction of rotation of the motor?

....

....

....

In what other way can such a motor be reversed?

....

....

Connect an ammeter in series (highrange first) with the dry cell and motor. Prevent the armature from rotating, and read the ammeter. Armature stationary

....

Now permit the armature to rotate and observe the ammeter readings as the armature comes up to speed.

Ammeter reading at full speed ....

....

Explain. ....

....

4. The shunt-wound motor

Connect the cell with the motor so that the field magnet and the armature are connected in parallel. See Fig. 57-4. Test the polarity of the field magnet and the polarity of the armature as the armature makes a complete revolution.

Observation: ....

....

....

Insert an ammeter in each branch of this circuit, so that it is possible to measure the current in the field coil and in the armature separately. Hold the armature to prevent it from rotating, and read the ammeters.

Field coil ....

Armature ....

Now let the motor come up to speed. Read the ammeters again.

Field coil ....

Armature ....

Explain these observations. ....

....

....

....

NAME PERIOD DATE INSTRUCTOR'S COMMENTS

58 EXPERIMENT Capacitance in a-c Circuits

PURPOSE: (1) To observe the effects of capacitance in a-c circuits. (2) To determine the reactance of capacitors.

APPARATUS: Knife switch, DPST; 4 dry cells or other d-c power source; d-c milliammeter, low range (a galvanometer shunted with fine wire will serve); d-c voltmeter, 0-7.5 v; 6-v lamp, No. 40 miniature screw base, 150 ma; lamp base, miniature screw; 2 test leads with alligator clips; capacitors (25 v), 1-25 μf, 2-50 μf; 6- volt filament transformer, or other 6-vac source; a-c milliammeter, 0-200 ma; a-c voltmeter, 0-7.5 v; annunciator wire, 18 ga, for connections.

INTRODUCTION: A *capacitor* is a combination of conducting plates separated by a dielectric and is used in electric circuits to store electric charge. The ratio of the charge on either plate to the potential difference between the plates is called *capacitance* and is a constant for any fixed capacitor, being expressed ordinarily in microfarads. The difference in the effect of capacitance in a-c and d-c circuits will be observed in this experiment.

A good quality capacitor offers essentially pure reactance to an alternating current and thus the capacitive reactance can be determined directly from the current in the capacitor circuit and the potential difference across the capacitor. This voltage may be assumed to lag the circuit current by 90°.

PROCEDURE:

1. Capacitors in d-c circuits

Make up a 6-volt battery with standard dry cells to serve as a d-c source. Arrange a No. 40 pilot lamp, a 25 μf capacitor, and a d-c milliammeter in series and connect to the battery through a DPST knife switch as shown in Fig. 58-1. Attach short test leads with alligator clips on the free ends to a d-c voltmeter. While observing the milliammeter, close the switch. Was there any initial movement of the milliammeter

pointer? .... Describe it. ....

....

....

Does the lamp light up? ....

Giving attention to the polarity of the voltmeter terminals, clip the voltmeter across the lamp and then across the capacitor. Result?

....

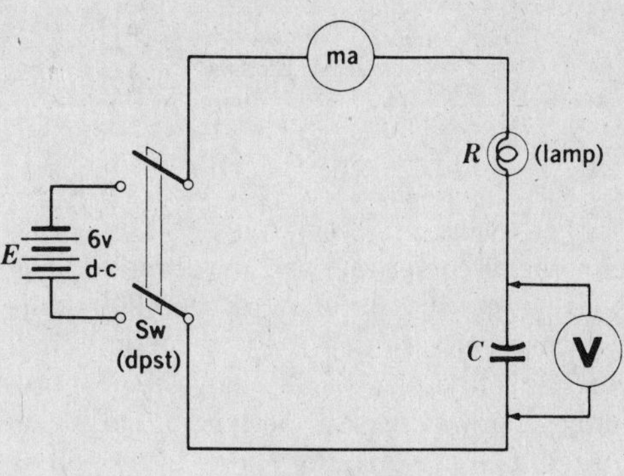

Figure 58-1

....

Explain the two voltmeter readings and the milliammeter reading.

....

....

....

....

....

With the voltmeter clipped across the capacitor, open the switch. Explain the action of the voltmeter.

....

....

What is the discharge path of the capacitor when the switch is open?

....

Charge and discharge the capacitor several times observing closely the rate at which the meter pointer falls to zero. Can you detect any change in rate?

....

If a 20,000 ohm per volt d-c voltmeter is available, place it across the capacitor and observe the discharge rate.

Replace the 25 μf capacitor with a 50 μf capacitor and repeat the measurements and observations. Place a second 50 μf capacitor in *parallel* with the one in the circuit. What is the total capacitance now in the circuit?

....

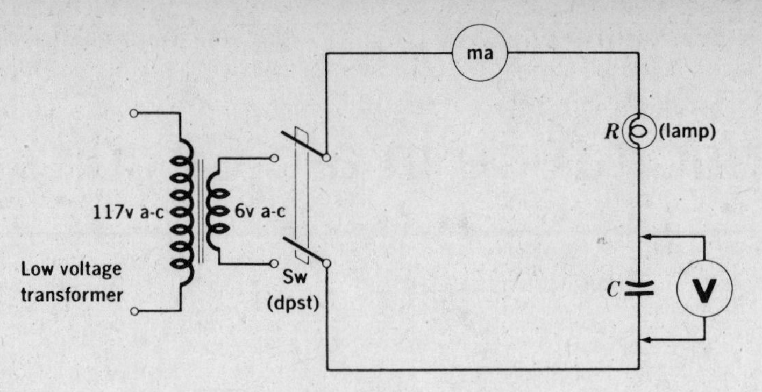

Figure 58-2

Observe the discharge rate on the voltmeter. (This may not be very successful unless a 20,000 ohm per volt voltmeter can be used.) As the meter pointer approaches zero during the discharge cycle, observe carefully. Can you suggest the type of curve that would result from a graph of V_C as a function of time during the discharge of C?

....

....

2. Capacitors in a-c circuits

Replace the d-c milliammeter and the d-c voltmeter with a-c instruments and replace the battery circuit with a 6-vac source. The circuit should be arranged as shown in Fig. 58-2. The lamp will serve both as a resistance and visual indicator in the circuit. The resistance of the filament will change, however, with the filament temperature.

With the two 50 μf capacitors in parallel, measure the circuit current I, the potential difference V applied across the series combination of R and C, the voltage V_R across the lamp, and the voltage V_C across the capacitor. Read each value to the degree of precision your instruments will allow and record in the data table along with the total capacitance used. How do the voltages across the lamp and the capacitor compare with the applied voltage?

....

....

Replace one of the 50 μf capacitors with a 25 μf capacitor, connecting it in parallel with the remaining 50 μf capacitor, repeat all measurements and record. Next, record all measurements with 50 μf capacitance in the circuit, then with 25 μf in the circuit.

What would be the total capacitance in the circuit if you connected a 50 μf and a 25 μf capacitor in series?

....

From the data you have already collected, predict the magnitude of circuit current and the potential drop across the series capacitors. Try this series combination in the circuit. Result?

....

....

After measuring V_C, predict how the voltage V_C is divided between the two capacitors in series. Measure the potential drop across each. Do the measured voltages agree with your prediction? ....

Explain. ....

....

....

....

Do not record these data in the data table.

CALCULATIONS: 1. From the circuit current I and the potential difference across the lamp V_R, calculate the resistance of the lamp filament R for each set of observations and record the values of R in the data table.

2. Similarly, calculate the values of the capacitive reactance X_C and record in the data table. A good quality capacitor acts as an essentially pure reactance and thus the potential difference V_C measured across

DATA

C (μf)	I (ma)	V (v)	V_R (v)	V_C (v)	R (Ω)	X_C (Ω)	Z (Ω)	ϕ (°)

it may be considered to be a reactance voltage.

3. Capacitive reactance produces a lagging voltage and is plotted in a negative direction producing a negative phase angle. Knowing capacitive reactance and the series resistance, the phase angle ϕ is an angle whose tangent is $-X_C/R$.

$\phi = \arctan \dfrac{-X_C}{R}$. Determine the phase angle ϕ in each case and record.

4. The magnitude of the impedance Z of each circuit may now be found by any one of several methods since R, X_C, and ϕ are known. Record these magnitudes in the data table.

5. Using a straight edge and a protractor, construct voltage vector diagrams for each set of data. Choose a scale which will allow the four diagrams to be arranged on one page. Identify each by the capacitance value used and show the impedance magnitude and phase angle as found graphically.

QUESTIONS: Your answers should be complete statements.

1. Compare the performance of a capacitor in d-c and a-c circuits.

....

....

....

....

....

.... 1

2. How does capacitive reactance vary with capacitance and with frequency?

....

....

....

....

.... 2

3. How do you explain the voltages measured across each capacitor when the 25 μf and 50 μf capacitors were connected in series?

....

....

....

....

.... 3

4. Explain the fact that you found the lamp filament to have a different resistance for each different value of capacitance placed in the series circuit.

....

....

....

....

.... 4

NAME PERIOD DATE INSTRUCTOR'S COMMENTS

59 EXPERIMENT Inductance in a-c Circuits

PURPOSE: (1) To observe the effects of inductance in a-c circuits. (2) To determine the reactance of inductors.

APPARATUS: Storage battery, 6 volt; knife switch, DPDT; primary and secondary coil set with iron core, student type; lamp, 6 volt carbon filament; lamp base with socket; d-c ammeter, 0.3 a; d-c milliammeter, 0-100 ma; d-c voltmeter, 0-7.5 v; 2 test leads with alligator clips; 6-volt filament transformer, or other 6-vac source; tubular rheostat, about 10 ohms; a-c ammeter, 0-3 a; a-c milliammeter, 0-50 ma; a-c voltmeter, 0-7.5 v; annunciator wire, 18 ga.

INTRODUCTION: An inductor is simply a coil of wire in its usual form. Because the wire has resistance, it offers resistive opposition to current in both d-c and a-c circuits. In a-c circuits there is the additional effect of inductance, an inertia-like property opposing any change in current. The effects of inductance will be observed in this experiment.

The inductance of a coil may be varied by varying the amount of iron in the core; by withdrawing the iron core the inductance is decreased and by inserting the iron core it is increased. Because an inductor offers both resistance and inductive reactance to an alternating current, the current does not lag the voltage by the full 90°. If we know the d-c resistance of the coil and can measure the voltage across it and the current in the inductor circuit, the impedance of the circuit can be found. Knowing both the impedance magnitude and the phase angle, the inductive reactance and the inductance of the coil may be computed.

PROCEDURE:

1. Inductor in a d-c circuit

Arrange the circuit as shown in Fig. 59-1 being careful to observe the proper polarity in connecting the d-c ammeter and voltmeter. *Do not connect the transformer to the a-c line at this time.* Use the primary coil of a small induction coil set as the inductor L. Connect test leads to the voltmeter and clip it across the coil.

Throw the switch to the d-c source and observe the lamp while inserting and withdrawing the iron core of the coil.

Result? ....

....

Can you determine the resistance of the coil with this circuit?

....

....

If the voltage reading across the coil is too low to read directly, determine it indirectly by using the more accurate midscale region of the meter.

2. Inductor in an a-c circuit

Before connecting the circuit to the a-c source, replace the d-c meters with a-c instruments, using an a-c ammeter, 0-3 a range. Throw the switch to the a-c source and adjust the rheostat to provide the same voltage as the battery. Observe the lamp intensity while inserting and withdrawing the iron core.

Result? ....

....

Read the circuit current and the potential difference across the coil as the inductance is varied. How do you explain the meter readings?

....

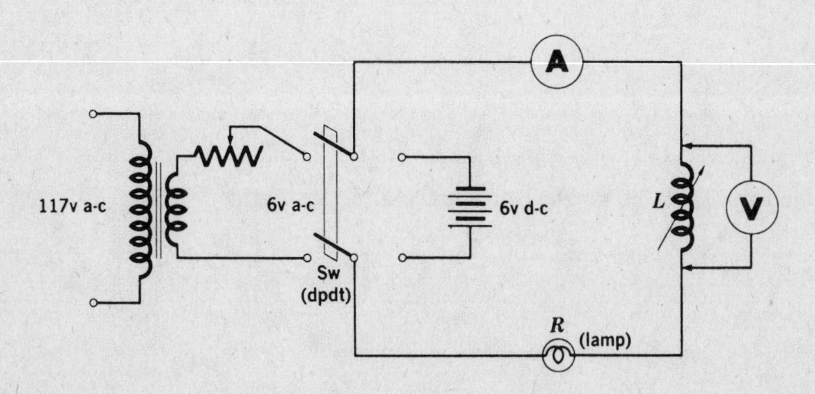

Figure 59-1

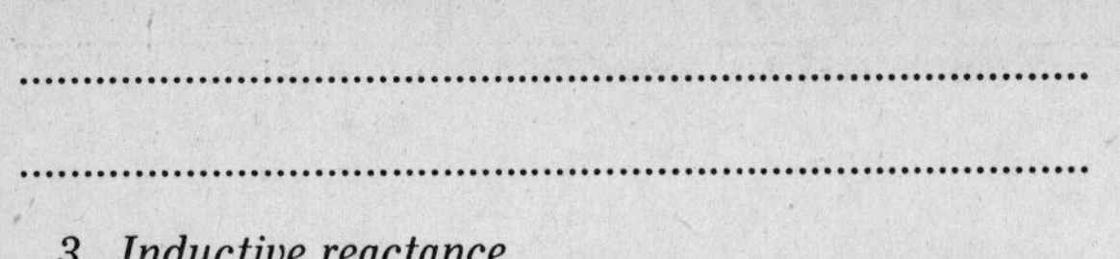

3. Inductive reactance

Replace the primary coil with the larger secondary coil and remove the lamp from the circuit entirely. Determine the resistance R_L of the coil by replacing the a-c meters with a d-c voltmeter and a d-c milliammeter, 0-100 ma. De-energize the a-c source completely and switch to the d-c source. Read the meters as accurately as possible, compute the resistance of the inductor, and record as R_L in the data table.

Replace the d-c milliammeter with an a-c milliammeter, 0-50 ma, and the d-c voltmeter with the a-c voltmeter. Place the iron core in the secondary coil and switch to the a-c source. Adjust the rheostat for a convenient current reading and record I and V_L, reading as accurately as possible.

CALCULATIONS: 1. Knowing the circuit current I and the potential difference V_L across the circuit, calculate the circuit impedance and record.

2. The power factor in an a-c circuit is defined as the ratio of the circuit resistance to the circuit impedance, this ratio being the cosine of the phase angle ϕ

$$\text{pf} = \frac{R}{Z} = \cos \phi.$$

Therefore, the phase angle is an angle whose tangent equals the power factor

$$\phi = \arccos \frac{R}{Z}.$$

Determine the phase angle ϕ and record in the data table.

3. Using a separate sheet of unlined paper, a straight edge, and a protractor, construct the impedance vector diagram. Find the magnitude of X_L graphically and record. Attach this sheet to your laboratory report.

4. Inductive reactance is directly proportional to both the frequency and the inductance and is expressed as $X_L = 2\pi fL$.

Assume the frequency of your a-c source to be 60 hz and calculate the inductance L of your coil. Record this value in the data table.

DATA

I (ma)	V_L (v)	R_L (Ω)	Z (Ω)	ϕ (°)	X_L (Ω)	L (h)

QUESTIONS: Your answers should be complete statements.

1. Explain the difference in the effect of an inductor in d-c and a-c circuits.

....

....

....

....

.... 1

2. What would be the effect on the current in the a-c circuit of Part 3 if the frequency of the source were doubled?

....

....

....

.... 2

3. How could the impedance diagram be converted to a voltage diagram?

....

....

.... 3

NAME PERIOD DATE INSTRUCTOR'S COMMENTS

60 EXPERIMENT Series Resonance

PURPOSE: To observe the condition of electric resonance in a series circuit containing inductance, capacitance, and resistance.

APPARATUS: Capacitor decade box, 0.01 μf to 1.1 μf in 0.01 μf steps; 3 inductors, filter chokes ranging from 8 to 30 h; lamp, 117 v, clear glass 7 to 25 watts; lamp base with socket; plugs and split line for 117-v receptacle; a-c milliammeter, 0-50/300 ma (determined by lamp used); a-c voltmeter, 0-600 v, with insulated test leads; a-c voltmeter, 0-150 v; annunciator wire, 18 ga, for connections.

SUGGESTION: Because of the equipment requirements and the danger of electric shock from exposed connections to the 117-vac service, it is recommended that the instructor perform this experiment as a demonstration.

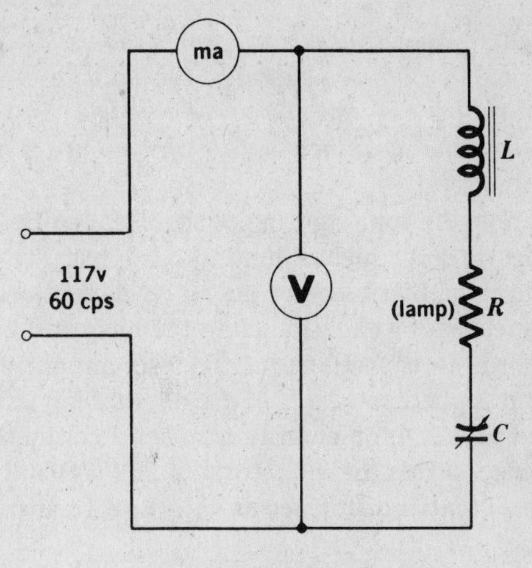

Figure 60-2

INTRODUCTION: A practical inductor offers both resistance and inductive reactance to an alternating current. We may consider the resistance and reactance to act in series. Thus, the voltages appearing across an inductance and a capacitance in series in an a-c circuit are of opposite polarity. If X_L is larger than X_C, the load is inductive and the potential difference across the circuit leads the current. However, if X_C is the larger reactance, the load is capacitive and the circuit voltage lags the current. In general, the impedance Z of the circuit has a magnitude and a phase angle ϕ which may be expressed as follows:

$$Z = \sqrt{R^2 + (X_L - X_C)^2}\,,$$

$$\phi = \arctan\,(X_L - X_C)/R.$$

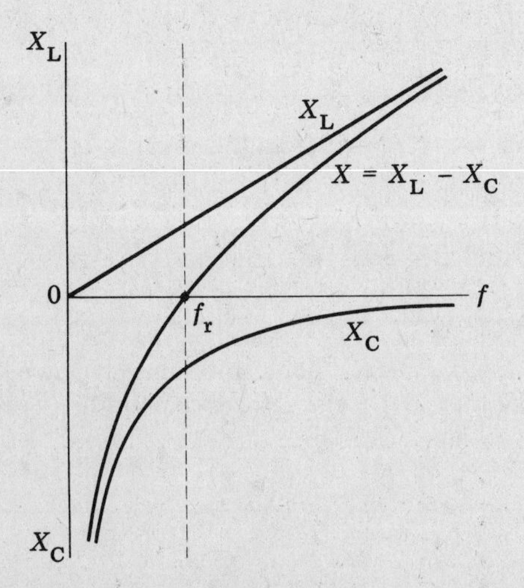

Figure 60-1

Figure 60-1 shows that over a range of low frequencies the load is capacitive and over a range of high frequencies the load of an L, R, C series circuit is inductive. At some intermediate frequency the inductive and capacitive reactances are equal and the reactance of the circuit ($X_L - X_C$) equals zero, the condition of *series resonance.* The impedance is equal to the resistance, the circuit current is maximum, and the voltage across the circuit is in phase with the circuit current.

For a particular combination of L and C there is one resonant frequency f_r. At this frequency

$$X_L = X_C,\ 2\pi f_r \mathrm{L} = \frac{1}{2\pi f_r C},\ f_r^{\,2} = \frac{1}{4\pi^2 LC},\ f_r = \frac{1}{2\pi\sqrt{\mathrm{LC}}}$$

Where L is in henrys and C is in farads, f_r is expressed in hertz.

In this experiment we shall supply a signal voltage at a fixed frequency, 60 hz, to a series circuit consisting of a fixed inductor, an indicator lamp, and a variable capacitor. See Fig. 60-2. We will select the capacitance which allows the maximum (resonant) current in the circuit and, by solving for L, we will be able to determine the inductance of the coil.

PROCEDURE: Mark each inductor for identification purposes. Filter chokes commonly range from about 7 henrys to nearly 30 henrys. Assuming a frequency of 60 hz, compute the limiting values of capacitance necessary to produce resonance over this range of inductance. By using a capacitance decade box of suitable range in which the smallest capacitance is 0.01 μf, you are able to find the desired value of capacitance to the nearest 0.01 μf.

Connect one inductor, the indicator lamp, the capacitance decade box, and a milliammeter of suitable range (if available) in series. Connect the voltmeter across the entire load. When the circuit components are arranged for maximum accessibility and all exposed connections are secure, plug the circuit into the 117-vac receptacle.

CAUTION: *Dangerous electric shock is possible from the exposed circuit connection at this voltage. A standard rubber covered plug should be used and removed from the receptacle before any change is made in the circuit.*

Determine the rough setting of the decade box which provides the highest luminous intensity of the lamp and then, by observing the meter deflection, find the resonating capacitance to the nearest 0.01 μf. Record this capacitance C, the resonant current I, and the applied voltage V in the data table. Using insulated probes, momentarily connect a voltmeter, 600-v range, across the inductor (*CAUTION*) and record the potential difference as V_L. Record also V_R and V_C.

Unplug the circuit before making any circuit changes. What function does the voltmeter connected across the circuit serve when the plug is removed from the receptacle? First, short out the inductor and then the capacitor and observe the effect of the other in the circuit.

Results? ....

....

....

....

In a similar manner test at least two other inductors, recording the necessary data in the table.

CALCULATIONS: 1. The impedance of a series L, R, and C circuit is minimum at resonance and is equal to R_T, which is composed of the resistance of the lamp filament and the inherent resistance of the inductor turns. Compute the impedance of each circuit and record. From the data recorded, determine the approximate resistance of each inductor.

2. Assuming the frequency of the applied voltage to be 60 hz, compute the inductance of each inductor used.

DATA

INDUCTOR USED	C (f)	I (ma)	V (v)	V_L (v)	V_R (v)	V_C (v)	$Z = R_T$ (Ω)	R_L (Ω)	L (h)

QUESTIONS: Your answers should be complete statements.

1. How do you explain the voltages recorded across different components of the resonant circuit?

....

....

....

....

....

.... 1

2. What would be the effect on the resonant current if the lamp were removed from the circuit?

....

....

.... 2

3. Since the capacitance in the circuit is not continuously variable, the precise point of resonance may not be obtainable using the procedure of this experiment. Assuming that the precise resonance point had been obtained, should V_L and V_C be expected to read precisely the same? Explain.

....

....

.... 3

NAME PERIOD DATE INSTRUCTOR'S COMMENTS

61 EXPERIMENT Thermionic Emission

PURPOSE: To determine the relation between the electric emission of a vacuum tube and the cathode temperature as controlled by the filament voltage.

APPARATUS: Diode, type 6H6, or triode, type 6J5; octal socket, mounted; d-c milliammeter, 0-50 ma; d-c voltmeter, 0-7.5 v; rheostat, 50 ohms, 25 watt; knife switch, SPST; contact key; "A" battery, 12-v storage battery or 6 dry cells; "B" battery, 22.5-v and 45-v terminals (Burgess type 5308, or equivalent); annunciator wire, 18 ga, for connections; cross-section paper.

INTRODUCTION: The conductivity of metals is the result of the movements of free electrons within the material. Free electrons move about inside the conductor with a speed that increases with temperature. Attractive forces at the surface of the material normally restrain the free electrons and keep them within the material.

If an electron is to escape from the surface of a conductor, it must do the work necessary to overcome the surface forces, and the energy must come from the electron's kinetic energy resulting from its motion. When an electron's kinetic energy exceeds the work it must perform to overcome the surface forces, it can escape into the space beyond the surface. At high temperatures, where the average kinetic energy of the free electrons is large, an appreciable number will be able to escape from the surface of the material. Thus the number of electrons escaping from the surface of a material is related to the nature of the material (surface forces) and its absolute temperature. The escaping of electrons from the surface of a hot body is called *thermionic emission.*

Electron emission in practically all receiving-type vacuum tubes is derived from cathodes coated with a mixture of barium and strontium oxides over which is formed a surface layer of metallic barium and strontium. Such emitters may be heated to their operating temperature either indirectly by radiation from an incandescent tungsten filament or directly by the conduction of filament current.

An electron-emitting cathode in a vacuum tube is surrounded by an electron cloud which constitutes a negative space charge. If the cathode is surrounded by a plate or anode, as in a diode, electrons from the space charge are attracted to the plate when a positive plate potential is applied. This movement of electrons constitutes a plate current and is conveniently measured by a milliammeter in the plate-cathode circuit. In this experiment we shall hold the plate voltage constant and measure the change in plate current as the temperature of the cathode filament is varied, this variation in temperature being accomplished by varying the potential difference across the filament.

PROCEDURE: Set up a data table on separate paper using the heading shown at the end of the procedure. Include this data sheet with your report of this experiment.

One section of a 6H6 twin diode or a 6J5 may be used. The triode may be operated as a diode by connecting the grid directly to the plate. The socket connection diagrams (bottom views) for both tube types are given in Fig. 61-1. If another tube type is to be used, you should have modifying instructions from your instructor before proceeding.

Connect the vacuum tube in the test circuit as shown in Fig. 61-2 being careful to follow the proper socket diagram for pin connections. Return the plate to the 22.5-v terminal of the "B" battery and record this voltage as your initial value of E_P. When the circuit connections are completed, place all the resistance of the rheostat R in the filament circuit, leave the switch SW_1 open, and request the instructor to inspect your circuit before proceeding further.

Close the contact key SW_2 when milliammeter readings are required. Does the milliammeter show any current in the plate-cathode circuit while the cathode-heater circuit is open? The contact key SW_2 is placed

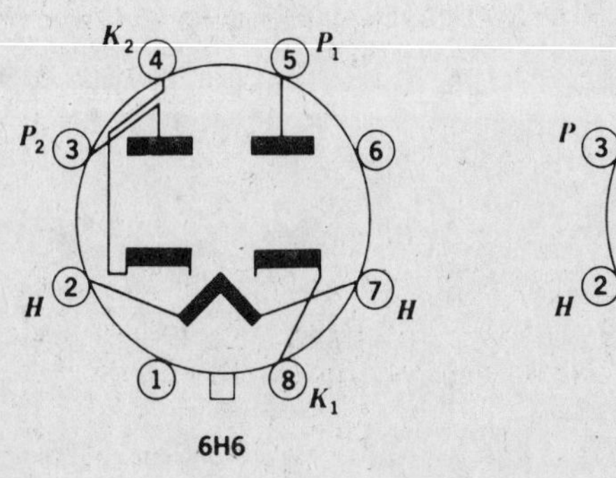

Figure 61-1

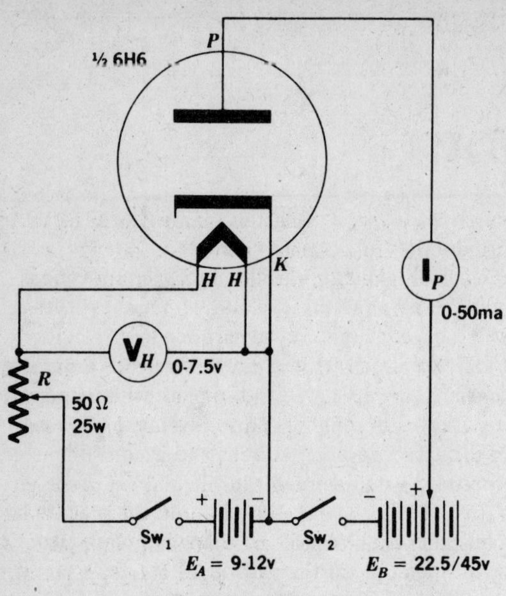

Figure 61-2

in the plate circuit to protect the tube from excessive plate current when the heater voltage is increased beyond the rated value of approximately 6 volts. When taking plate current readings above this heater voltage, close the plate circuit only long enough to get a steady milliammeter reading. Close the heater switch and read the initial voltage across the filament. Did you expect it to be zero with maximum R in the circuit? Explain the voltmeter reading.

Adjust the reheostat for the nearest even half volt across the filament and record this voltage as your initial heater voltage reading E_{H} in the proper E_{P} column of the data table. Record also the milliammeter reading for the plate current I_{P}. Increase the heater voltage in half-volt increments allowing sufficient time between adjustments to permit the filament to reach a constant temperature. When the first indication of plate current appears (SW_2 must be closed), be sure that the milliammeter pointer is steady before recording data. Record data for filament voltages to 7.5 v, opening the heater switch after the final reading.

Return the plate to the 45-v terminal on the "B" battery and repeat the procedure, recording E_{H} and I_{P} as before.

DATA

TUBE TYPE			
E_{P} = ________ v		E_{P} ________ v	
E_{H} (v)	I_{P} (ma)	E_{H} (v)	I_{P} (ma)

DATA REDUCTION: Prepare a graph of plate current as a function of heater voltage for each value of plate voltage used. Plot E_{H} as abscissas and I_{P} as ordinates and construct smooth curves, both on the same sheet. Attach this graph, properly labeled, to your regular report.

QUESTION: Your answer should be a complete statement.

Explain why your two curves had separated (assumed different slopes) by the time the design voltage of the heater, approximately 6 v, was reached.

NAME PERIOD DATE INSTRUCTOR'S COMMENTS

62 EXPERIMENT Diode Characteristics

PURPOSE: To determine the change in plate current of a diode with increases in plate voltage.

APPARATUS: Diode, type 6H6, or triode, type 6J5, with grid connected to the plate; octal socket, mounted; d-c milliammeter, 0-15 ma; d-c voltmeter, 0-7.5 v; d-c voltmeter, 0-150 v, 20,000 ohms/volt or VTVM; "A" battery, 12-v storage battery or 6 dry cells in series; "B" battery, 135 v (3 Burgess type 5308, or equivalent); rheostat, 50 ohms, 25 watts; potentiometer, 4000 ohms, 10 watts; knife switch, SPST; contact key; 4 resistors, 5 watts, 10,000 ohms, 20,000 ohms, 40,000 ohms, and 80,000 ohms; annunciator wire, 18 ga; cross-section paper.

INTRODUCTION: When the plate of a diode is made positive with respect to the cathode, space-charge electrons are attracted to the plate and the tube *conducts.* Emission from an oxide-coated cathode is quite abundant and, at normal plate voltages, electrons are supplied to the space charge by the emitting cathode as rapidly as they are removed by the plate. When a full space charge is present, the plate current depends upon the plate voltage. The modern diode would probably be damaged by the excessive plate voltage before this voltage could be raised enough to cause the total electron emission to be in transit to the plate.

In this experiment the heater temperature of a diode will be maintained at the normal level and the positive voltage applied to the plate will be varied. By measuring the plate current through a range of plate voltages you will be able to plot a characteristic curve of diode plate current as a function of plate voltage. Different values of plate-load resistance can be introduced and a family of characteristic curves can be plotted.

PROCEDURE: Arrange a data table using a heading similar to the one below. Provide for 14 lines of data. Include this data sheet with your report of this experiment.

Arrange the diode circuit as shown in Fig. 62-1 using one section of a 6H6 twin diode or a 6J5 triode connected as a diode. Refer to the tube-base diagrams of Experiment 61 (Fig. 61-1) for proper pin connections. The potentiometer R_2 is connected across the plate supply battery through the contact key SW_2 to conserve the "B" battery. Plate voltage adjustments and plate current readings are made with the contact key closed. At other times, this key should be left open.

Adjust R_2 for an initial plate voltage of zero. Close the knife switch SW_1, adjust the rheostat R_1 for a heater voltage of 6 v, and allow the tube to warm up to normal operating temperature. Take milliammeter readings at plate voltage intervals of 5 v from 0 v to 20 v. Record E_P and I_P in the data table in the appropriate column. Open the heater circuit between series of readings.

Place a resistance R_1 of 10,000 ohms in the plate circuit by inserting the resistor between the diode plate and the milliammeter. Take plate current readings at plate voltage intervals of 10 v from 0 v to 120 v and record in the data table. Repeat the readings for an R_L of 20,000 ohms, of 40,000 ohms, and of 80,000 ohms.

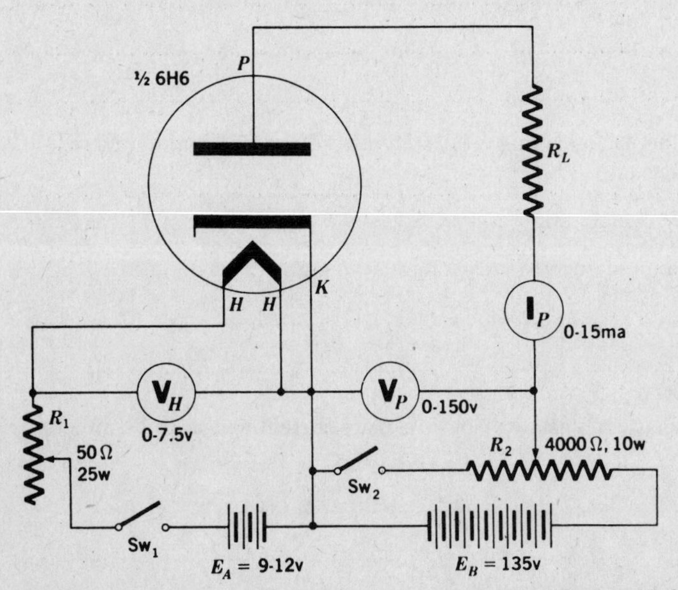

Figure 62-1

DATA

TUBE TYPE:						HEATER VOLTAGE:			v
$R_L = 0\ \Omega$		$R_L = 10{,}000\ \Omega$		$R_L = 20{,}000\ \Omega$		$R_L = 40{,}000\ \Omega$		$R_L = 80{,}000\ \Omega$	
E_P (v)	I_P (ma)	E_P (v)	I_P (ma)	E_P (v)	I_P (ma)	E_P (v)	I_P (ma)	E_P (v)	I_P (ma)

DATA REDUCTION: Prepare a family of characteristic diode curves of plate current as a function of plate voltage on a sheet of rectangular coordinate paper, plotting I_P as ordinates and E_P as abscissas for each value of R_L. Attach the graph, properly labeled, to your regular report.

QUESTIONS: Your answers should be complete statements.

1. What is the source of the energy that is delivered to the diode (energy input) during its operation in a circuit?

....

.... 1

2. How is this energy first expended in the tube and what is the energy transformation?

....

....

.... 2

3. What is the evidence that the plate is heated, other than the proximity of the plate to the cathode heater, and what is the energy transformation?

....

....

.... 3

4. What could cause the plate to be warmed to a red heat during the tube operation?

....

....

....

.... 4

5. Why was the plate voltage limited to a low value when making plate-current readings with a minimum of resistance in the plate circuit (no load resistance added)?

....

....

....

.... 5

6. From the diode characteristic curves, under what circumstance does the plate current appear to be directly proportional to the plate voltage?

....

....

.... 6

NAME PERIOD DATE INSTRUCTOR'S COMMENTS

63 EXPERIMENT Triode Characteristics

PURPOSE: (1) To study the action of the control grid of a triode. (2) To determine the important characteristics of a triode.

APPARATUS: Triode, type 6J5; octal socket, mounted; d-c milliammeter, 0-15 ma; d-c voltmeter, 0-7.5 v; d-c voltmeter, 0-15 v; d-c voltmeter, 0-150 v, 20,000 ohms/volt or VTVM; "A" battery, 12-v storage battery or 6 dry cells in series; "B" battery, 135 v (3 Burgess type 5308, or equivalent); "C" battery, 9-v; rheostat, 50 ohms, 25 watts; potentiometer, 4000 ohms, 10 watts; potentiometer, 1000 ohms, 4 watts; 2 knife switches, SPST; contact key; annunciator wire, 18 ga, for connections; cross-section paper.

SUGGESTION: The procedure for this experiment should be studied carefully as an advance assignment and each student should prepare appropriate data tables before undertaking the experiment. Two data tables are necessary, one for Parts 1 and 2 and another for Part 3. The two tables should be ruled in smooth form on separate sheets of unlined paper and made a part of the experiment report. The data table for Experiment 62 may be used as a guide.

INTRODUCTION: The triode is a three-electrode vacuum tube consisting of a cathode, a control grid, and a plate. The control grid is an open spiral of fine wire interposed between the cathode and plate which can have an important controlling effect on electrons in transit from cathode to plate while offering a negligible physical barrier to these electrons. The grid is normally operated at a negative potential with respect to the cathode and thus attracts no electrons (produces no grid current), but its electrostatic field effectively controls the number of electrons that pass between its loops. If the grid is made sufficiently negative with respect to the cathode, all electrons are repelled toward the cathode, the plate current is zero, and the tube is said to be *cut off.* The smallest negative grid voltage, which causes the tube to cease to conduct is known as the *cut-off bias.*

The most important characteristics of triodes are the relationships between:

1. plate current and plate voltage with constant grid voltage, and

2. plate current and grid voltage with constant plate voltage. We shall determine these relationships in this experiment and plot their characteristic curves.

PROCEDURE:

1. Grid control

Arrange the circuit as shown in Fig. 63-1. Observe that each battery circuit has a switch included. Use them to conserve battery power when a series of readings is not being taken. Refer to the tube-base diagrams in Experiment 61 (Fig. 61-1) for the pin connections for the 6J5 tube. Adjust the heater voltage to 6 v and allow the tube to warm to normal operating temperature.

The grid bias battery E_C is connected to supply negative grid voltages (with respect to the cathode) across the potentiometer R_1. With the plate voltage E_P set at 20 v, vary the grid bias value of E_G through its full range of negative values and observe the varia-

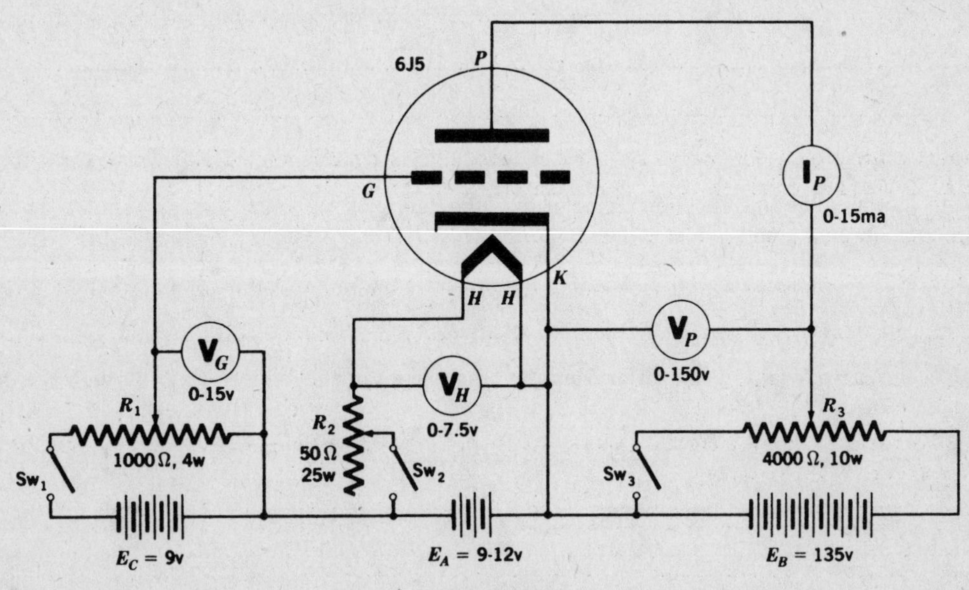

Figure 63-1

tion in plate current. Carefully determine the minimum negative grid voltage, at E_P = 20 v, which causes the tube to *cease* to conduct. What is this voltage called? Record the values of E_G and I_P in the E = 20 v column of your data table for Parts 1 and 2.

Advance the plate voltage in 20-v steps and determine for each step the minimum negative grid voltage that cuts off the plate current. Record both E_G and I_P for each value of E_P as before.

2. I_P vs. E_G characteristic

Set the plate voltage at 20 v and the grid voltage at the cut-off voltage; then reduce the grid voltage to the next even half-volt position and record E_G and I_P. Reduce the grid voltage to zero in half-volt steps, recording E_G and I_P for each step.

Increase the plate voltage in 20-v steps and repeat the measurements for each step. All grid voltages recorded up to this point are *negative*.

In order to make the grid positive with respect to the cathode, the bias battery connections to the potentiometer R_1 must be reversed. (The voltmeter V_G connections must be reversed also unless a zero-center meter is being used.) Arrange the grid circuit for positive grid voltages, set the grid bias as 0 v, and advance the voltage in half-volt steps to +2 v, taking readings for each setting of E_P and record as before.

3. I_P vs. E_P characteristic

In this procedure, the plate current will be measured over the range of plate voltages with constant values of grid voltage. Set the grid voltage at +2 v and read the plate current for plate voltage advanced in 20-v steps from 0 to 120 v. Record in the data table you have prepared for Part 3.

Reduce the grid bias to 0 v and repeat the measurements, recording as before.

Change the grid bias battery back to the original arrangement for negative grid voltages and repeat the series of readings, increasing the grid voltage in a negative sense in 2-v steps until the cut-off bias for an E_P = 120 v is reached. Record the required data in your table.

4. Amplification factor

Set the grid bias at -1.0 v, the plate voltage at 60 v, and read the plate current as precisely as possible. Increase the plate voltage to 80 v and then vary the grid voltage as necessary to return the plate current precisely to the original value. From these data compute the amplification factor of the tube. (Observe the proper sign convention in your computation.)

DATA REDUCTION:

1. I_P vs. E_G characteristic

Plot the data tabulated in Part 2 on a sheet of rectangular coordinate paper with E_G as abscissas and I_P as ordinates. Construct a family of smooth curves labeling each for the constant value of E_P used.

Plot the data collected in Part 4 on this graph labeling ΔE_P and ΔE_G. Show your computation for the amplification factor in smooth form on an unused portion of the graph paper.

2. I_P vs. E_P characteristic

Plot the data tabulated in Part 3 on a second sheet of rectangular coordinate paper with E_P as abscissas and I_P as ordinates. Label each curve for the constant value of E_G used. Label both graphs properly and include them with your report.

QUESTIONS: Your answers should be complete statements.

1. Does your family of I_P vs. E_G curves support your measured value of *mu?* How can you determine this?

....

....

....

....

.... 1

2. Can you determine the amplification factor of the tube directly from your I_P vs. E_G family? Show one such construction on your graph. What is the result?

....

.... 2

3. How does the curve for E_G = +2 v differ from the other curves of the family? Can you suggest a possible explanation for this?

....

....

....

.... 3

NAME PERIOD DATE INSTRUCTOR'S COMMENTS

64 EXPERIMENT The Size of a Molecule

PURPOSE: (1) To determine the order of magnitude of the size of an oleic acid molecule. (2) To determine the number of molecules in one mole of a molecular substance.

APPARATUS: Large dish, such as ripple tank used in Experiment 27; medicine dropper; metric ruler; oleic acid in alcohol solution, 1:500, lycopodium powder or fine talcum powder; concentrated hydrochloric acid.

INTRODUCTION: Molecules are much too small to observe directly. The electron microscope makes it possible to photograph only the largest molecules. However, because some molecules will form layers which are one molecule thick on water, it is possible with ordinary laboratory equipment to determine the order of magnitude of the size of such molecules.

A mole of a molecular substance is the quantity of the substance equal to its molecular weight expressed in grams. The number of molecules in a mole of a molecular substance is an important physical constant known as Avogadro's number. Its value is 6.0219×10^{23} molecules per mole. In this experiment we shall also attempt to verify Avogadro's number.

SUGGESTION: Before the laboratory period, the instructor should prepare a 1:500 solution of oleic acid in denatured alcohol.

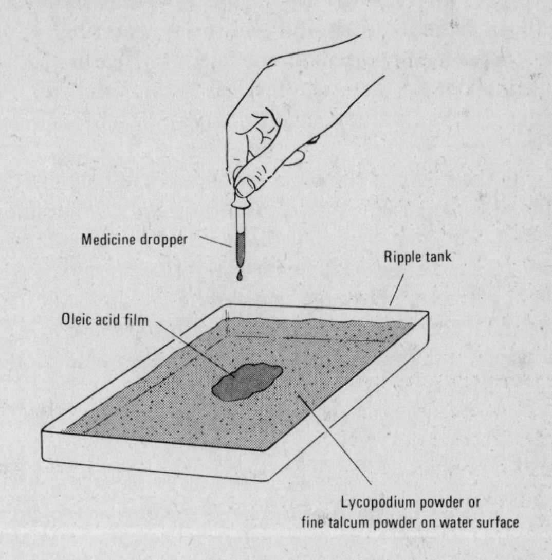

Figure 64-1

PROCEDURE: Pour water to a depth of about one centimeter into a ripple tank and dust the water surface very lightly with lycopodium powder or talcum powder. Using a medicine dropper, place one drop of oleic acid solution on the water surface as in Fig. 64-1. The oleic acid pushes the powder outward so the area of the oleic acid layer is visible. Measure the diameter of the layer and calculate its area.

Experimentally determine the number of drops of oleic acid solution per cubic centimeter, and calculate

DATA

TRIAL	*Diameter of layer* (cm)	*Area of layer* (cm^2)	*Number of drops of oleic acid solution* (per cm^3)	*Volume of one drop of oleic acid solution* (cm^3)
in water				
in acid solution				

Volume of oleic acid in one drop of solution (cm^3)	*Thickness of layer* (cm)	*Volume of an oleic acid molecule* (cm^3)	*Volume of one mole of oleic acid* (cm^3)	*Avogadro's Number* (experimental)	(accepted)	*Relative error* (%)

the volume of one drop of solution. Using the dilution ratio of the solution, 1:500, compute the volume of oleic acid in the drop of solution used. This is the volume of the surface layer. Using this volume and the area of the layer, calculate its thickness.

Repeat the experiment, but this time add 20 ml of concentrated hydrochloric acid to the water before dusting it with lycopodium or talcum. Measure the diameter of the oleic acid layer as before, and calculate its area, volume, and thickness. Record all data.

The diameter of the oleic acid layer will be considerably larger in the acid solution than it is in plain water. This indicates that the oleic acid molecule is elongated. In water, it floats upright because of a slight attraction between one end of the molecule and water. Hydrochloric acid destroys this attraction, however, and in the acid solution the oleic acid molecule floats on its side. Hence a good approximation of the shape of the molecule is a cylinder.

Use the thickness of the oleic acid layer on water as the height of the cylindrical molecule and the thickness on the acid solution as the diameter. Calculate the volume of a single oleic acid molecule, using the equation

$$V = \frac{1}{4} \pi d^2 h$$

The mass density of oleic acid is 0.895 g/cm^3 and one mole has a mass of 282 g. Calculate the volume occupied by one mole of oleic acid. From this value and the volume of an oleic acid molecule, calculate the number of molecules in a mole of oleic acid. Compare your results with the accepted value for Avogadro's number and calculate your relative error. Record all data and show your calculations in your report.

QUESTIONS: Your answers should be complete statements.

1. What is the ratio of the height of the oleic acid molecule to its diameter?

....

....

.... 1

2. How could you make sure that the oleic acid, and not the alcohol, is pushing the powder inside?

....

....

.... 2

3. What factors determine the precision of this experiment?

....

....

.... 3

4. What properties of oleic acid make it desirable for this experiment?

....

....

.... 4

5. How does the surface tension of water help in this experiment? How does it hinder the experiment?

....

....

....

.... 5

NAME PERIOD DATE INSTRUCTOR'S COMMENTS

65 EXPERIMENT Radioactivity

PURPOSE: (1) To study the effect of radioactive materials on an electroscope and on a Geiger counter. (2) To study the absorptive effect of air, cardboard, aluminum, and lead on beta radiation.

APPARATUS: Simple electroscope; cat's fur; hard-rubber rod; stopwatch, or watch with second hand; Geiger tube and associated counting apparatus; radioactive sample; wrist watch with illuminated dial; uranium metal, or uranium compound; radioactive mineral sample, or radioactive isotope, if available; 20 cardboard sheets, 10 cm square, 1.0 mm thick; 15 aluminum sheets, and 15 lead sheets of the same size.

INTRODUCTION: The nuclei of the atoms of radioactive materials break down spontaneously with the emission of particles and rays, consisting of alpha particles, beta particles, and gamma rays. Radiation can be detected by an electroscope and by a Geiger tube. While alpha particles ionize gas molecules in the air and discharge an electroscope, their penetrating power is not great enough to affect a Geiger tube. Beta particles and gamma rays are detected by the tube.

In this experiment, the radioactivity of several materials will be determined. The penetrating power of beta particles given off from radioactive materials and the effect of cosmic rays on a Geiger tube will also be observed.

PROCEDURE:

1. *Effect of radioactive materials on an electroscope*

Charge an electroscope by induction using the cat's fur and hard-rubber rod. Observe the rate of discharge of the electroscope by determining how much the leaves collapse during a fifteen-minute period. Now recharge the electroscope and place a radioactive sample beneath the leaves of the electroscope. Observe the rate of discharge of the electroscope during a fifteen-minute period. (Other parts of the experiment may be performed in the intervals between these observations.) Record your observations.

Observation: ....

....

....

2. *Effect of radioactive materials on a Geiger tube*

a. Background count. Set the Geiger counter apparatus in operation. Note the frequency of clicks when there are no radioactive materials near the tube. In accurate work this background count must be subtracted from all other measurements.

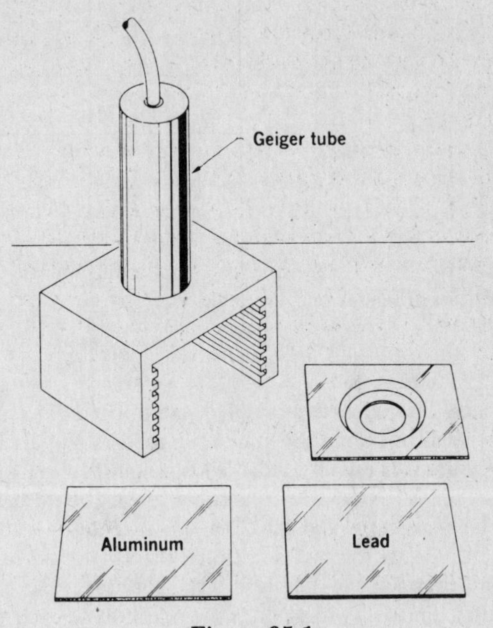

Figure 65-1

Observation: ....

....

....

....

b. Intensity of radiation. Place the radioactive sample at such a distance from the Geiger tube that you obtain the maximum count reading which the apparatus can measure. Record this reading. Reading: .... Remove the radioactive sample, and place the illuminated dial of a wrist watch the same distance from the tube. Record the result. Reading: .... If some uranium metal or uranium compound is available, place it the same distance from the tube. Again record the count. Reading: ....
If some radioactive ore or a radioactive isotope is available, test its radioactivity in a similar fashion.

Reading: ....

3. *Absorptive effect of air on beta particles*

Place a radioactive sample 5.0 cm from the tube. Determine the count. Now place the sample 10.0 cm from the tube. Again determine the count. Continue to move the sample away from the tube in 5.0 cm intervals up to a distance of 60.0 cm. Make a count reading for each position and record the measurements in the data table below. Do not forget to subtract the background count in each case if it is significant.

DATA

Distance	Count	Distance	Count	Distance	Count

GRAPH: Plot a curve of these data, using the distances as abscissas and the counts as ordinates.

4. *Absorptive effect of other materials on beta radiation*

Place the radioactive sample close to the counter tube so as to obtain the maximum reading of which the counter is capable. Record this count in the data table below. Now place one sheet of cardboard between the sample and the tube. Make a count reading. Continue to place additional sheets of cardboard between the sample and the counter tube, making a count reading after the addition of each sheet. Use a total of 20 sheets of cardboard.

Repeat the experiment, but use sheets of aluminum instead of cardboard. Make readings when 1, 2, 3, 4, 5, 7, 10, 12, and 15 sheets of aluminum are placed between the counter and the radioactive sample.

Repeat the experiment, but use sheets of lead this time. Make readings when 1, 2, 3, 4, 5, 7, 10, 12, and 15 sheets of lead are used.

Plot curves of these data on the same sheet of coordinate paper, using the number of sheets of various materials as abscissas and the counts as ordinates. Use different colored pencils to draw the graph lines so that the curves may be identified. Which is the most effective material for absorbing beta radiation?

....

....

....

....

....

....

DATA

CARDBOARD				ALUMINUM		LEAD	
No. sheets	*Count*	*No. sheets*	*Count*	*No. sheets*	*Count*	*No. sheets*	*Count*
0		11		0		0	
1		12		1		1	
2		13		2		2	
3		14		3		3	
4		15		4		4	
5		16		5		5	
6		17		7		7	
7		18		10		10	
8		19		12		12	
9		20		15		15	
10							

QUESTIONS: Your answers should be complete statements.

1. What is the source of background radiation?

....

....

.... 1

2. How far can beta particles travel through the air without being absorbed? What type of radiation from the radioactive sample still reaches the tube?

....

....

....

.... 2

3. What is the most effective material for absorbing beta radiation?

....

....

.... 3

4. What must be the relative thickness of cardboard, aluminum, and lead to absorb beta radiation effectively?

....

....

....

.... 4

NAME PERIOD DATE INSTRUCTOR'S COMMENTS

66 EXPERIMENT Half-Life

PURPOSE: To measure the half-life of a radioactive element.

APPARATUS: Geiger tube and associated counting apparatus; radioactive isotope; sample holder; clock or watch with second hand; graph paper.

INTRODUCTION: The half-life of a radioactive element is the length of time during which half a given number of atoms of the element will decay. In this experiment, the activity of a radioisotope will be measured with a Geiger counter over a period of several days. The resulting data will be plotted on a graph and the half-life determined by inspection. Half-life does not depend on the number of atoms in the sample. Consequently, the size of the radioactive sample is not important. It is important, however, to obtain a sample with a half-life in the order of magnitude of a few days (rather than a few seconds or several years).

PROCEDURE: Familiarize yourself with the operation of the Geiger tube and its associated apparatus. Set the apparatus in operation and note the frequency of clicks per minute when there is no radioactive material near the tube. This is the background count. Record it on your data table.

Place the radioactive sample in a holder and position it near the Geiger tube. Note the exact time and record it. Mark the position of the tube and sample so that they can be duplicated for future readings. Measure the count for exactly one minute and record it.

At the same time the next day, make another count for exactly one minute and record it. Repeat this procedure for about one week.

DATA

Date and time	*Count* (min^{-1})	*Background count* (min^{-1})	*Corrected count.* (min^{-1})

GRAPH: Plot a graph of your data, using elapsed time as abscissas and the corrected count/minute as ordinates. Detailed instructions for the preparation of graphs are given in the Introduction. Read the half-life from the graph. If necessary, extend the graph to obtain this reading. Compare it with the accepted value of the half-life furnished by your instructor.

QUESTIONS: Your answers should be complete statements.

1. Why is it important to maintain the same relative position of Geiger tube and sample in this experiment?

....

....

.... 1

2. Are the clicks of the counting apparatus uniform during the one-minute counts? Explain. Does the rate vary in different directions?

....

....

.... 2

3. Do you think temperature affects the results? Elaborate.

....

....

....

.... 3

4. Does the sensitivity of the counting apparatus affect the results? Explain.

....

....

....

.... 4

NAME PERIOD DATE INSTRUCTOR'S COMMENTS

67 EXPERIMENT Simulated Nuclear Collisions

PURPOSE: (1) To simulate the relationship between the kinetic energy of a nuclear particle and its stopping distance in a detection device. (2) To show that energy momentum are conserved in nuclear collisions.

APPARATUS: Ring stand, tall, or table support; masonite or plywood, about 30 cm × 40 cm; manila folder; nickels; paper, legal size; masking tape.

INTRODUCTION: In a nuclear collision, all or part of the kinetic energy of the incident particle is imparted to a target particle. In particle detection devices, such as the cloud or bubble chamber, these energies are proportional to the distances which the particles travel in the device before coming to rest. If the masses of the particles are known, their kinetic energies as well as their momenta can then be calculated from these distances.

In this experiment, nickels will be used to simulate incident and target nuclear particles with equal masses. A sheet of paper will act as the detection device for the measurement of distances. Kinetic energy will be imparted to the incident nickel by sliding it down an incline. The collisions, both in one and two dimensions, will then be analyzed in much the same way as were the motions in Experiments 11 and 12.

PROCEDURE:

1. Energy measurements

Set up the nuclear collision simulator as shown in Fig. 67-1. The slope of the inclined path should be about 70°. The masonite or plywood backing is used to make the inclined path as straight and smooth as possible. The curvature at the bottom of the incline should be sharp enough to serve as a good reference for distance measurements, but not so sharp that the nickel will jump when it passes over it. A little practice is required in order to adjust this part of the simulator properly.

Release a nickel near the top of the incline and note how far it slides on the "detector" paper. Mark the initial and final positions of the nickel. Repeat this procedure with several nickels until you have two nickels that slide the same distance when released from the same height. *(Be sure to use the same sides of the nickels throughout the experiment. You will also find that worn nickels slide better than new ones.)*

When you have selected a pair of nickels for your final trials, make a series of measurements of the relationship between release height (measured along the incline from the center of curvature at the bottom) and stopping distance on the "detection" paper (also measured from the reference line). Make three determinations for each trial and record the average in the data table.

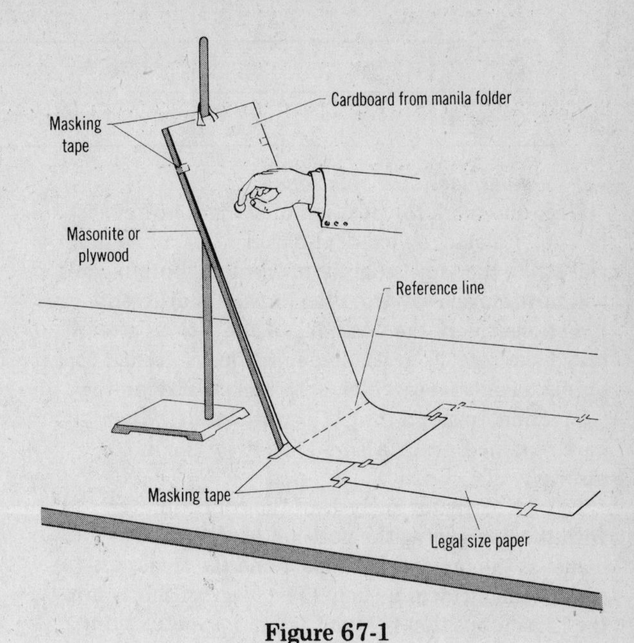

Figure 67-1

DATA

TRIAL	*Release height* (cm)	*Stopping distance* (cm)
1		
2		
3		

2. One-dimensional collisions

Place one of the selected nickels on the "detection" paper a few cm from the reference line. Mark its position by drawing a line around it. Release the other nickel from a previously measured height and directly in line with the target nickel. Practice this collision until you can make the incident nickel stop upon impact and the target nickel move away in the same direction that the incident nickel had before impact.

After the target nickel has come to rest, mark its position and measure its distance from its starting point. Make two more determinations and record the average measurements under Trial 1 in the data table. Conduct two more trials, using different release heights and different positions for the target nickel. Record the results.

Add the second and third distances for each trial and compare the sum with the stopping distance of the nickel in Part 1 that was released from the same height. Record the difference in the data table as the deviation.

DATA

TRIAL	*Release height* (cm)	*Distance of target from reference* (cm)	*Stopping distance of target* (cm)	*Deviation* (cm)
1				
2				
3				

3. Two-dimensional collisions

Place and mark the position of a target nickel as before. Release the incident nickel along a line that will strike the target slightly off-center. Both nickels will now move away from the impact in different directions. Mark the positions of the nickels after they have come to rest. Some practice is needed to obtain a collision in which both nickels stay on the "detection" paper during the entire event. When you have obtained a satisfactory trial, draw the lines shown in Fig. 67-2.

To make a vector analysis of the collision, start with the line joining the position of the incident nickel at the moment of impact and the final position of the incident nickel. (The diagram shows how the impact position is found.) Draw a vector along this line equal in magnitude to the square root of this distance. Draw a second vector along the line that the incident nickel would have taken if it had not collided with a target and equal in magnitude to the square root of the distance it would have traveled from the point of impact. This distance can be obtained from the first data table. Now connect these two vectors with a third vector. This vector represents the momentum of the target nickel.

Square the magnitude of the final vector. Starting from the center of the position of the target nickel, draw a line with this magnitude and parallel to the third vector. The end of this line represents the calculated position of the center of the target nickel after collision. Compare this line with that drawn to the actual position of the target nickel. Record the magnitude and angular differences on your diagram.

GRAPH: Plot a graph of the data in the first table, using stopping distances as abscissas and release heights as ordinates.

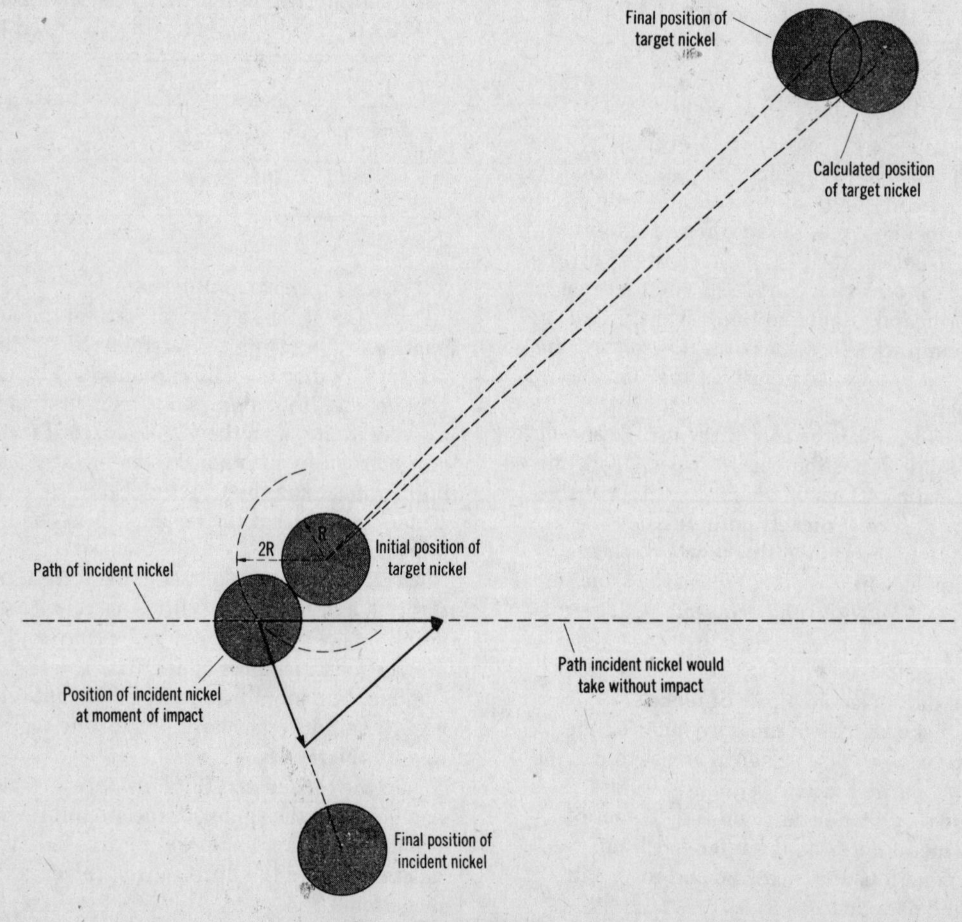

Figure 67-2

QUESTIONS: Your answers should be complete statements.

1. What is the relationship between the kinetic energies of the incident nickels in this experiment and their stopping distances?

....

.... 1

2. In the one-dimensional collisions, how is the stopping distance of the target nickel related to the kinetic energy of the incident nickel?

....

....

....

.... 2

3. Why does the incident nickel stop at the moment of impact in a one-dimensional collision?

....

....

....

....

.... 3

4. Why is the square root of the distance used in making the vector analysis of the two-dimensional collision?

....

....

....

....

....

.... 4

Appendix

Appendix A —Use of Electric Instruments

A. Introduction.

Meters used for electric measurements are delicate instruments and must be handled with great care. The greater the precision of the instrument the more fragile it is and therefore, the more easily it may be damaged by abuse or improper use. Ordinary commercial grade meters will provide readings of sufficient accuracy for the majority of experiments. Highgrade laboratory instruments may sometimes be used where a higher order of precision is essential.

The greatest care must be used to avoid an excessive current in a meter. You must keep in mind the fact that the heating effect in an electric circuit increases as the *square* of the current. It is a good idea to fuse the circuit, in which a meter is to be employed, with a fuse having a capacity less than that of the instrument itself. A knife switch which can be opened quickly may help to protect measuring instruments.

Meters for use in a-c circuits and d-c circuits are constructed differently. One type of instrument must not be placed in the other type of circuit. Instruments sometimes have more than one range over which readings may be made. In general, more accurate results are obtained by selecting a range which enables the reading to be made in the *middle* portion of the calibrated scale.

B. The voltmeter.

A meter designed to give readings in volts is used to measure the difference in potential between two points in an electric circuit. Thus, a *voltmeter is always connected in parallel with the part of a circuit across which one wishes to measure the potential difference.* If the potential difference across a dry cell is to be measured, a d-c voltmeter having a range of from 0 to 3 volts will give more accurate readings than one of higher range. For use on a commercial lighting circuit, an a-c voltmeter should have a range of from 0 to 150 volts. Perhaps you are given a three-range instrument, 0 to 3, 0 to 15, and 0 to 150, and you have no idea of the potential difference across the circuit on which the voltmeter is to be used. You begin first with the highest range, and then adjust to a range which provides approximately mid-scale readings.

Since a voltmeter is connected in parallel with a circuit, it acts to *load* the circuit. The range of the meter together with its built-in resistance determines its sensitivity in *ohms-per-volt.* If the meter has a 0 to 3-volt range and is rated at 1000 ohms-per-volt, it has a loading effect of 3000 ohms when placed across a circuit. In practice, the loading effect should be held as low as possible (the meter resistance should be high compared to the circuit resistance). The meter resistance should be at least 10 times the resistance of the circuit across which it is connected in order to avoid too much change in the circuit constants. The higher the ohms-per-volt rating of the meter, the lower will be its loading effect on a circuit across which it is placed.

C. The ammeter.

The commercial type of ammeter may consist of a coil pivoted between the poles of a permanent magnet. Since its resistance is very low, a shunt is connected across its terminals to prevent its being "burnt out." The instrument may have different shunts to give different ranges.

An ammeter is connected in series in a circuit. Great care must be taken before connecting an ammeter in a circuit to see that the current in the circuit does not exceed the range of the instrument. A rheostat may be connected in series with the instrument to reduce the current strength. The resistance can then be cut out gradually when it is determined that the meter can handle the circuit current.

If an ammeter has different shunts, always connect it so that you use the highest range first. For example, if the shunts give the instrument a range of 0 to 5 amperes, and of 0 to 25 amperes, start with the higher range first. If the current in the circuit is less than five amperes, then use the 0 to 5-ampere range to secure greater accuracy in reading. *Never connect an ammeter in parallel with any other instrument.*

D. Galvanometers.

A galvanometer may be used to detect the presence of an electric current, to determine its direction, or to find its relative magnitude. For Wheatstone-bridge work and for testing induced currents, galvanometers are desirable. Galvanometers having the D'Arsonval movement, with the coil mounted on jeweled bearings instead of being suspended by means of a wire or ribbon, can be obtained at moderate cost. They are very satisfactory for student use. Since a galvanometer is not protected by a high resistance in series, or by a shunt across its terminals, it may be used safely *only with very small currents.* Your instructor may direct that a resistance coil or a shunt be used with the instrument.

E. Rheostats and resistance coils.

A rheostat is used in a circuit to increase the resistance and thus reduce the current strength, or to protect certain instruments. It may be graduated for use as a measuring instrument. It is made of wire having a high resistance which does not change appreciably with changes in temperature.

While a rheostat, or a set of resistance coils is affected only slightly by a change in temperature, great care must be taken not to heat the coils unduly. Apparatus manufacturers usually state the maximum current capacity of each resistance coil, or box of coils. The terms rheostat and resistance box (or coils) are sometimes used interchangeably, but in this manual the term *resistance box* or *resistance coils* is used to designate a measuring instrument, and the term *rheostat* to designate any resistance added to reduce the current strength. Resistance boxes are designated for use with *small* currents such as those encountered in Wheatstone-bridge circuits and should not be used as a current-limiting resistance in general circuit applications where a rheostat would be appropriate.

Appendix B –Tables

Contents

Table 1 PREFIXES OF THE METRIC SYSTEM

Factor	*Prefix*	*Symbol*
10^{12}	tera	T
10^9	giga	G
10^6	mega	M
10^3	kilo	k
10^2	hecto	h
10	deka	da
10^{-1}	deci	d
10^{-2}	centi	c
10^{-3}	milli	m
10^{-6}	micro	μ
10^{-9}	nano	n
10^{-12}	pico	p
10^{-15}	femto	f
10^{-18}	atto	a

Table 2 GREEK ALPHABET

Greek letter	*Greek name*	*English equivalent*	*Greek letter*	*Greek name*	*English equivalent*
Α α	alpha	ä	Ν ν	nu	n
Β β	beta	b	Ξ ξ	xi	ks
Γ γ	gamma	g	Ο ο	omicron	o
Δ δ	delta	d	Π π	pi	p
Ε ϵ	epsilon	e	Ρ ρ	rho	r
Ζ ζ	zeta	z	Σ σ	sigma	s
Η η	eta	ā	Τ τ	tau	t
Θ θ	theta	th	Υ υ	upsilon	ü, ōō
Ι ι	iota	ē	Φ ϕ	phi	f
Κ κ	kappa	k	Χ χ	chi	h
Λ λ	lambda	l	Ψ ψ	psi	ps
Μ μ	mu	m	Ω ω	omega	ō

Table 3 SELECTED PHYSICAL QUANTITIES AND MEASUREMENT UNITS

Physical quantity	*Quantity symbol*	*Measurement unit*	*Unit symbol*	*Unit dimensions*
		Fundamental Units		
length	l	meter	m	m
mass	m	kilogram	kg	kg
time	t	second	sec	sec
electric charge	Q	coulomb	c	c
temperature	T	degree Kelvin	°K	°K
luminous intensity	I	candle	cd	cd
		Derived Units		
acceleration	a	meter per second per second	m/sec^2	m/sec^2
area	A	square meter	m^2	m^2
capacitance	C	farad	f	c^2 sec^2/kg m^2
density	D	kilogram per cubic meter	kg/m^3	kg/m^3
electric current	I	ampere	a	c/sec
electric field intensity	$\mathscr{E}$	newton per coulomb	n/c	kg m/c sec^2
electric resistance	R	ohm	Ω	kg m^2/c^{2}sec
emf	E	volt	v	kg m^2/c sec^2
energy	E	joule	j	kg m^2/sec^2
force	F	newton	n	kg m/sec^2
frequency	f	hertz	hz	sec^{-1}
heat	Q	joule	j	kg m^2/sec^2
illumination	E	lumen per square meter	lm/m^2	cd sr/m^2
inductance	L	henry	h	kg m^2/c^2

Table 3 SELECTED PHYSICAL QUANTITIES AND MEASUREMENT UNITS (cont'd)

Physical quantity	*Quantity symbol*	*Measurement unit*	*Unit symbol*	*Unit dimensions*
luminous flux	Φ	lumen	lm	cd sr
magnetic flux	Φ	weber	wb	kg m^2/c sec
magnetic flux density	B	weber per square meter	wb/m^2	kg/c sec
potential difference	V	volt	v	kg m^2/c sec^2
power	P	watt	w	kg m^2/sec^3
pressure	p	newton per square meter	n/m^2	kg/m sec^2
velocity	v	meter per second	m/sec	m/sec
volume	V	cubic meter	m^3	m^3
work	W	joule	j	kg m^2/sec^2

Table 4 PHYSICAL CONSTANTS

Quantity	*Symbol*	*Value*
atmospheric pressure, normal	*atm*	1.01325×10^5 n/m^2
atomic mass unit, unified	u	1.66043×10^{-27} kg
avogadro number	N_A	6.02252×10^{23}/mole
charge to mass ratio for electron	e/m_e	1.758796×10^{11} c/kg
electron rest mass	m_e	9.1091×10^{-31} kg
		5.48597×10^{-4} u
electron volt	ev	1.60210×10^{-19} j
electrostatic constant	k	8.987×10^9 n m^2/c^2
elementary charge	e	1.60210×10^{-19} c
faraday	f	9.648682×10^0 c/mole
gas constant, universal	R	6.236×10^4 mm cm^3/mole °K
		8.2057×10^{-2} l atm/mole °K
		8.3143×10^0 j/mole °K
gravitational acceleration, standard	g	9.80665×10^0 m/sec^2
mechanical equivalent of heat	J	4.1868×10^0 j/cal
neutron rest mass	m_n	1.67482×10^{-27} kg
		1.0086654×10^0 u
Planck's constant	h	6.6256×10^{-34} j sec
proton rest mass	m_p	1.67252×10^{-27} kg
		1.00727663×10^0 u
speed of light in a vacuum	c	2.997925×10^8 m/sec
speed of sound in air at S.T.P.	v	3.3145×10^2 m/sec
universal gravitational constant	G	6.670×10^{-11} n m^2/kg^2
volume of ideal gas, standard	V_0	2.24136×10^1 l/mole

Table 5 CONVERSION FACTORS

LENGTH

$1\ m = 10^{-3}\ km = 10^2\ cm = 10^3\ mm = 10^6\ \mu m = 10^9\ nm = 10^{10}\ Å$

$1\ \mu = 1\ \mu m = 10^{-6}\ m = 10^{-4}\ cm = 10^{-3}\ mm = 10^3\ m\mu = 10^3\ nm = 10^4\ Å$

$1\ m\mu = 1\ nm = 10^{-9}\ m = 10^{-7}\ cm = 10^{-6}\ mm = 10^1\ Å$

$1\ Å = 10^{-10}\ m = 10^{-8}\ cm = 10^{-4}\ \mu = 10^{-1}\ m\mu = 10^{-1}\ nm$

AREA

$1\ m^2 = 10^{-6}\ km^2 = 10^4\ cm^2 = 10^6\ mm^2$

VOLUME

$1\ m^3 = 10^{-9}\ km^3 = 10^3\ l = 10^6\ cm^3$

$1\ l = 10^3\ ml = 10^3\ cm^3 = 10^{-3}\ m^3$

ANGULAR

$1° = 1.74 \times 10^{-2}$ radian $= 2.78 \times 10^{-3}$ revolution

1 radian $= 57.3° = 1.59 \times 10^{-1}$ revolution

1 revolution $= 360° = 6.28$ radians

MASS

$1\ kg = 10^3\ g = 10^6\ mg = 6.02 \times 10^{26}\ u$

$1\ g = 10^{-3}\ kg = 10^3\ mg = 6.02 \times 10^{23}\ u$

$1\ u = 1.66 \times 10^{-24}\ g = 1.66 \times 10^{-21}\ mg = 1.66 \times 10^{-27}\ kg$

TIME

$1\ hr = 60\ min = 3.6 \times 10^3\ sec$

$1\ min = 60\ sec = 1.67 \times 10^{-2}\ hr$

$1\ sec = 1.67 \times 10^{-2}\ min = 2.78 \times 10^{-4}\ hr$

VELOCITY

$1\ km/hr = 10^3\ m/hr = 16.7\ m/min = 2.78 \times 10^{-1}\ m/sec$

$1\ m/min = 10^2\ cm/min = 1.67 \times 10^{-2}\ m/sec = 1.67\ cm/sec$

$1\ m/sec = 10^{-3}\ km/sec = 3.6\ km/hr = 10^2\ cm/sec$

ACCELERATION

$1\ cm/sec^2 = 10^{-2}\ m/sec^2 = 10^{-5}\ km/sec^2$

$1\ m/sec^2 = 10^2\ cm/sec^2 = 10^{-3}\ km/sec^2$

$1\ km/hr/sec = 10^3\ m/hr/sec = 2.78 \times 10^{-1}\ m/sec^2 = 2.78 \times 10^1\ cm/sec^2 = 2.78 \times 10^2\ mm/sec^2$

FORCE

$1\ n = 10^5$ dynes

1 dyne $= 10^{-5}\ n$

PRESSURE

$1\ atm = 760.00\ mm\ Hg = 1.013 \times 10^5\ n/m^2 = 1.013 \times 10^6\ dynes/cm^2$

$1\ n/m^2 = 10\ dynes/cm^2 = 9.87 \times 10^{-6}\ atm$

ENERGY

$1\ j = 10^7\ ergs = 2.39 \times 10^{-1}\ cal = 2.39 \times 10^{-4}\ kcal = 2.78 \times 10^{-7}\ kw\ hr = 1.49 \times 10^{18}\ ev$

$1\ cal = 10^{-3}\ kcal = 4.19\ j = 1.16 \times 10^{-6}\ kw\ hr$

$1\ kcal = 10^3\ cal = 4.19 \times 10^3\ j = 1.16 \times 10^{-3}\ kw\ hr$

$1\ ev = 10^{-6}\ Mev = 1.60 \times 10^{-12}\ erg = 1.60 \times 10^{-19}\ j$

$1\ kw\ hr = 10^3\ w\ hr = 3.6 \times 10^3\ kw\ sec = 3.6 \times 10^6\ w\ sec = 8.6 \times 10^5\ cal$

$1\ w\ sec = 2.78 \times 10^{-4}\ w\ hr = 2.78 \times 10^{-7}\ kw\ hr$

MASS-ENERGY

$1\ j = 1.11 \times 10^{-17}\ kg = 1.11 \times 10^{-14}\ g = 6.69 \times 10^9\ u$

$1\ ev = 1.07 \times 10^{-9}\ u = 1.78 \times 10^{-33}\ g$

$1\ u = 1.49 \times 10^{-3}\ erg = 1.49 \times 10^{-10}\ j = 931\ Mev = 9.31 \times 10^8\ ev$

$1\ kg = 9.00 \times 10^{16}\ j = 9.00 \times 10^{23}\ ergs$

Table 6 NATURAL TRIGONOMETRIC FUNCTIONS

Angle	*Sine*	*Cosine*	*Tan-gent*	*Angle*	*Sine*	*Cosine*	*Tan-gent*
0.0	0.000	1.000	0.000				
0.5	0.009	1.000	0.009	**23.0**	0.391	0.921	0.424
1.0	0.017	1.000	0.017	**23.5**	0.399	0.917	0.435
1.5	0.026	1.000	0.026	**24.0**	0.407	0.914	0.445
2.0	0.035	0.999	0.035	**24.5**	0.415	0.910	0.456
2.5	0.044	0.999	0.044	**25.0**	0.423	0.906	0.466
3.0	0.052	0.999	0.052	**25.5**	0.431	0.903	0.477
3.5	0.061	0.998	0.061	**26.0**	0.438	0.899	0.488
4.0	0.070	0.998	0.070	**26.5**	0.446	0.895	0.499
4.5	0.078	0.997	0.079	**27.0**	0.454	0.891	0.510
5.0	0.087	0.996	0.087	**27.5**	0.462	0.887	0.521
5.5	0.096	0.995	0.096	**28.0**	0.470	0.883	0.532
6.0	0.104	0.995	0.105	**28.5**	0.477	0.879	0.543
6.5	0.113	0.994	0.114	**29.0**	0.485	0.875	0.554
7.0	0.122	0.992	0.123	**29.5**	0.492	0.870	0.566
7.5	0.131	0.991	0.132	**30.0**	0.500	0.866	0.577
8.0	0.139	0.990	0.141	**30.5**	0.508	0.862	0.589
8.5	0.148	0.989	0.149	**31.0**	0.515	0.857	0.601
9.0	0.156	0.988	0.158	**31.5**	0.522	0.853	0.613
9.5	0.165	0.986	0.167	**32.0**	0.530	0.848	0.625
10.0	0.174	0.985	0.176	**32.5**	0.537	0.843	0.637
10.5	0.182	0.983	0.185	**33.0**	0.545	0.839	0.649
11.0	0.191	0.982	0.194	**33.5**	0.552	0.834	0.662
11.5	0.199	0.980	0.204	**34.0**	0.559	0.829	0.674
12.0	0.208	0.978	0.213	**34.5**	0.566	0.824	0.687
12.5	0.216	0.976	0.222	**35.0**	0.574	0.819	0.700
13.0	0.225	0.974	0.231	**35.5**	0.581	0.814	0.713
13.5	0.233	0.972	0.240	**36.0**	0.588	0.809	0.726
14.0	0.242	0.970	0.249	**36.5**	0.595	0.804	0.740
14.5	0.250	0.968	0.259	**37.0**	0.602	0.799	0.754
15.0	0.259	0.966	0.268	**37.5**	0.609	0.793	0.767
15.5	0.267	0.964	0.277	**38.0**	0.616	0.788	0.781
16.0	0.276	0.961	0.287	**38.5**	0.622	0.783	0.795
16.5	0.284	0.959	0.296	**39.0**	0.629	0.777	0.810
17.0	0.292	0.956	0.306	**39.5**	0.636	0.772	0.824
17.5	0.301	0.954	0.315	**40.0**	0.643	0.766	0.839
18.0	0.309	0.951	0.325	**40.5**	0.649	0.760	0.854
18.5	0.317	0.948	0.335	**41.0**	0.656	0.755	0.869
19.0	0.326	0.946	0.344	**41.5**	0.663	0.749	0.885
19.5	0.334	0.943	0.354	**42.0**	0.669	0.743	0.900
20.0	0.342	0.940	0.364	**42.5**	0.676	0.737	0.916
20.5	0.350	0.937	0.374	**43.0**	0.682	0.731	0.932
21.0	0.358	0.934	0.384	**43.5**	0.688	0.725	0.949
21.5	0.366	0.930	0.394	**44.0**	0.695	0.719	0.966
22.0	0.375	0.927	0.404	**44.5**	0.701	0.713	0.983
22.5	0.383	0.924	0.414	**45.0**	0.707	0.707	1.000

Table 6 NATURAL TRIGONOMETRIC FUNCTIONS (cont'd)

Angle	*Sine*	*Cosine*	*Tan-gent*	*Angle*	*Sine*	*Cosine*	*Tan-gent*
45.5	0.713	0.701	1.018	**68.0**	0.927	0.375	2.475
46.0	0.719	0.695	1.036	**68.5**	0.930	0.366	2.539
46.5	0.725	0.688	1.054	**69.0**	0.934	0.358	2.605
47.0	0.731	0.682	1.072	**69.5**	0.937	0.350	2.675
47.5	0.737	0.676	1.091	**70.0**	0.940	0.342	2.747
48.0	0.743	0.669	1.111	**70.5**	0.943	0.334	2.824
48.5	0.749	0.663	1.130	**71.0**	0.946	0.326	2.904
49.0	0.755	0.656	1.150	**71.5**	0.948	0.317	2.983
49.5	0.760	0.649	1.171	**72.0**	0.951	0.309	3.078
50.0	0.766	0.643	1.192	**72.5**	0.954	0.301	3.172
50.5	0.772	0.636	1.213	**73.0**	0.956	0.292	3.271
51.0	0.777	0.629	1.235	**73.5**	0.959	0.284	3.376
51.5	0.783	0.622	1.257	**74.0**	0.961	0.276	3.487
52.0	0.788	0.616	1.280	**74.5**	0.964	0.267	3.606
52.5	0.793	0.609	1.303	**75.0**	0.966	0.259	3.732
53.0	0.799	0.602	1.327	**75.5**	0.968	0.250	3.867
53.5	0.804	0.595	1.351	**76.0**	0.970	0.242	4.011
54.0	0.809	0.588	1.376	**76.5**	0.972	0.233	4.165
54.5	0.814	0.581	1.402	**77.0**	0.974	0.225	4.331
55.0	0.819	0.574	1.428	**77.5**	0.976	0.216	4.511
55.5	0.824	0.566	1.455	**78.0**	0.978	0.208	4.705
56.0	0.829	0.559	1.483	**78.5**	0.980	0.199	4.915
56.5	0.834	0.552	1.511	**79.0**	0.982	0.191	5.145
57.0	0.839	0.545	1.540	**79.5**	0.983	0.182	5.396
57.5	0.843	0.537	1.570	**80.0**	0.985	0.174	5.671
58.0	0.848	0.530	1.600	**80.5**	0.986	0.165	5.976
58.5	0.853	0.522	1.632	**81.0**	0.988	0.156	6.314
59.0	0.857	0.515	1.664	**81.5**	0.989	0.148	6.691
59.5	0.862	0.508	1.698	**82.0**	0.990	0.139	7.115
60.0	0.866	0.500	1.732	**82.5**	0.991	0.131	7.596
60.5	0.870	0.492	1.767	**83.0**	0.992	0.122	8.144
61.0	0.875	0.485	1.804	**83.5**	0.994	0.113	8.777
61.5	0.879	0.477	1.842	**84.0**	0.994	0.104	9.514
62.0	0.883	0.470	1.881	**84.5**	0.995	0.093	10.38
62.5	0.887	0.462	1.921	**85.0**	0.996	0.087	11.43
63.0	0.891	0.454	1.963	**85.5**	0.997	0.078	12.71
63.5	0.895	0.446	2.006	**86.0**	0.998	0.070	14.30
64.0	0.899	0.438	2.050	**86.5**	0.998	0.061	16.35
64.5	0.903	0.431	2.097	**87.0**	0.999	0.052	19.08
65.0	0.906	0.423	2.145	**87.5**	0.999	0.044	22.90
65.5	0.910	0.415	2.194	**88.0**	0.999	0.035	28.64
66.0	0.914	0.407	2.246	**88.5**	1.000	0.026	38.19
66.5	0.917	0.399	2.300	**89.0**	1.000	0.017	57.29
67.0	0.921	0.391	2.356	**89.5**	1.000	0.009	114.1
67.5	0.924	0.383	2.414	**90.0**	1.000	0.000	. . .

Table 7 TENSILE STRENGTH OF METALS

Metal	*Tensile strength* (n/m^2)
aluminum wire	2.4×10^8
copper wire, hard drawn	4.8×10^8
iron wire, annealed	3.8×10^8
iron wire, hard drawn	6.9×10^8
lead, cast or drawn	2.1×10^8
plantinum wire	3.5×10^8
silver wire	2.9×10^8
steel (minimum)	2.8×10^8
steel wire (maximum)	32×10^8

Table 8 ELASTIC MODULUS

Metal	*Elastic modulus* (n/m^2)
aluminum, 99.3%, rolled	6.96×10^{10}
brass	9.02×10^{10}
copper, wire, hard drawn	11.6×10^{10}
gold, pure, hard drawn	7.85×10^{10}
iron, cast	9.1×10^{10}
iron, wrought	19.3×10^{10}
lead, rolled	1.57×10^{10}
platinum, pure, drawn	16.7×10^{10}
silver, hard drawn	7.75×10^{10}
steel, 0.38% C, annealed	20.0×10^{10}
tungsten, drawn	35.5×10^{10}

Table 9 SURFACE TENSION OF WATER AGAINST AIR

Temp. (°C)	*Surface tension* (n/m)
−8	7.70×10^{-2}
−5	7.64×10^{-2}
0	7.56×10^{-2}
+5	7.49×10^{-2}
10	7.42×10^{-2}
15	7.35×10^{-2}
18	7.30×10^{-2}
20	7.28×10^{-2}
25	7.20×10^{-2}
30	7.12×10^{-2}
40	6.96×10^{-2}
50	6.79×10^{-2}
60	6.62×10^{-2}
70	6.44×10^{-2}
80	6.26×10^{-2}
100	5.89×10^{-2}

Table 10 SURFACE TENSION OF VARIOUS LIQUIDS

Liquid	*In contact with*	*Temp.* (°C)	*Surface tension* (n/m)
carbon disulfide	vapor	20	3.233×10^{-2}
carbon tetrachloride	vapor	20	2.695×10^{-2}
ethyl alcohol	air	0	2.405×10^{-2}
ethyl alcohol	vapor	20	2.275×10^{-2}

Table 11 DENSITY AND SPECIFIC GRAVITY OF GASES

($T = 0°$C; $p = 760$ mm Hg) (Air standard for specific gravity)

Gas	*Formula*	*Density* (g/l)	*Sp. gr.*
acetylene	C_2H_2	1.17910	0.91202
air, dry, CO_2 free		1.29284	1.00000
ammonia	NH_3	0.77126	0.5965
argon	Ar	1.78364	1.3796
chlorine	Cl_2	3.214	2.486
carbon dioxide	CO_2	1.9769	1.529
carbon monoxide	CO	1.25004	0.9669
ethane	C_2H_6	1.3562	1.0490
helium	He	0.17846	0.138
hydrogen	H_2	0.08988	0.06952
hydrogen chloride	HCl	1.6392	1.268
methane	CH_4	0.7168	0.5544
neon	Ne	0.89990	0.6960
nitrogen	N_2	1.25036	0.9671
oxygen	O_2	1.42896	1.1053
sulfur dioxide	SO_2	2.9262	2.263

Table 12 EQUILIBRIUM VAPOR PRESSURE OF WATER

Temp. (°C)	*Pressure* (mm Hg)	*Temp.* (°C)	*Pressure* (mm Hg)	*Temp.* (°C)	*Pressure* (mm Hg)
0	4.6	25	23.8	90	525.8
5	6.5	26	25.2	95	633.9
10	9.2	27	26.7	96	657.6
15	12.8	28	28.3	97	682.1
16	13.6	29	30.0	98	707.3
17	14.5	30	31.8	99	733.2
18	15.5	35	42.2	100	760.0
19	16.5	40	55.3	101	787.5
20	17.5	50	92.5	103	845.1
21	18.7	60	149.4	105	906.1
22	19.8	70	233.7	110	1074.6
23	21.1	80	355.1	120	1489.1
24	22.4	85	433.6	150	3570.5

Table 13 HEAT CONSTANTS

Material	Specific heat (cal/g C°)	Melting point (°C)	Boiling point (°C)	Heat of fusion (cal/g)	Heat of vaporization (cal/g)
alcohol, ethyl	0.581 (25°)	−115	78.5	24.9	204
aluminum	0.214 (20°)	660.2	2467	94	2520
	0.217 (0–100°)				
	0.220 (20–100°)				
	0.225 (100°)				
ammonia, liquid	1.047 (−60°)	−77.7	−33.35	108.1	327.1
liquid	1.125 (20°)				
gas	0.523 (20°)				
brass (40% Zn)	0.0917	900			
copper	0.0924	1083	2595	49.0	1150
glass, crown	0.161				
iron	0.1075	1535	3000	7.89	1600
lead	0.0305	327.5	1744	5.47	207
mercury	0.0333	−38.87	356.58	2.82	70.613
platinum	0.0317	1769	3827 ± 100	27.2	
silver	0.0562	960.8	2212	26.0	565
tungsten	0.0322	3410 ± 10	5927	43	
water	1.00		100.00		538.7
ice	0.530	0.00		79.71	
steam	0.481				
zinc	0.0922	419.4	907	23.0	420

Table 14 COEFFICIENT OF LINEAR EXPANSION

(Increase in length per unit length per Celsius degree)

Material	Coefficient ($\Delta l/l$ C°)	Temperature (°C)	Material	Coefficient ($\Delta l/l$ C°)	Temperature (°C)
aluminum	23.8×10^{-6}	20–100	invar (nickel steel)	0.9×10^{-6}	20
brass	19.30×10^{-6}	0–100	lead	29.40×10^{-6}	18–100
copper	16.8×10^{-6}	25–100	magnesium	26.08×10^{-6}	18–100
glass, tube	8.33×10^{-6}	0–100	platinum	8.99×10^{-6}	40
crown	8.97×10^{-6}	0–100	rubber hard	$8\bar{0} \times 10^{-6}$	20–60
pyrex	3.3×10^{-6}	20–300	quartz, fused	0.546×10^{-6}	0–800
gold	14.3×10^{-6}	16–100	silver	18.8×10^{-6}	20
ice	50.7×10^{-6}	−10–0	tin	26.92×10^{-6}	18–100
iron, soft	12.10×10^{-6}	40	zinc	26.28×10^{-6}	10–100
steel	10.5×10^{-6}	0–100			

The coefficient of cubical expansion may be taken as three times the linear coefficient.

Table 15 COEFFICIENT OF VOLUME EXPANSION

(Increase in Unit Volume per Celsius Degree at 20° C)

Liquid	*Coefficient*
acetone	14.87×10^{-4}
alcohol, ethyl	11.2×10^{-4}
benzene	12.37×10^{-4}
carbon disulfide	12.18×10^{-4}
carbon tetrachloride	12.36×10^{-4}
chloroform	12.73×10^{-4}
ether	16.56×10^{-4}
glycerol	5.05×10^{-4}
mercury	1.82×10^{-4}
petroleum	9.55×10^{-4}
turpentine	9.73×10^{-4}
water	2.07×10^{-4}

Table 16 HEAT OF VAPORIZATION OF SATURATED STEAM

Temp. (°C)	*Pressure* (mm Hg)	*Heat of vaporization* (cal/g)
95	634.0	541.9
96	657.7	541.2
97	682.1	540.6
98	707.3	539.9
99	733.3	539.3
100	760.0	538.7
101	787.5	538.1
102	815.9	537.4
103	845.1	536.8
104	875.1	536.2
105	906.1	535.6
106	937.9	534.9

Table 17 SPEED OF SOUND

Substance	*Density* (g/l)	*Velocity* (m/sec)	$\Delta v/\Delta t$ (m/sec C°)
GASES (S.T.P.)			
air, dry	1.293	331.45	0.59
carbon dioxide	1.977	259	0.4
helium	0.178	965	0.8
hydrogen	0.0899	1284	2.2
nitrogen	1.250	334	0.6
oxygen	1.429	316	0.56
LIQUIDS (25°C)	(g/cm³)		
acetone	0.79	1174	
alcohol, ethyl	0.79	1207	
carbon tetrachloride	1.595	926	
glycerol	1.26	1904	
kerosene	0.81	1324	
water, distilled	0.998	1498	
water, sea	1.025	1531	
SOLIDS (thin rods)			
aluminum	2.7	5000	
brass	8.6	3480	
brick	1.8	3650	
copper	8.93	3810	
cork	0.25	500	
glass, crown	2.24	4540	
iron	7.85	5200	
lucite	1.18	1840	
maple (along grain)	0.69	4110	
pine (along grain)	0.43	3320	
steel	7.85	5200	

Table 18 INDEX OF REFRACTION

($\lambda = 5893$ Å; *Temperature* = 20°C *except as noted*)

Material	*Refractive index*
air, dry (S.T.P.)	1.00029
alcohol, ethyl	1.360
benzene	1.501
calcite	1.6583
	1.4864
canada balsam	1.530
carbon dioxide (S.T.P.)	1.00045
carbon disulfide	1.625
carbon tetrachloride	1.459
diamond	2.4195
glass, crown	1.5172
flint	1.6270
glycerol	1.475
ice	1.310
lucite	1.50
quartz	1.544
	1.553
quartz, fused	1.45845
sapphire (Al_2O_3)	1.7686
	1.7604
water, distilled	1.333
water vapor (S.T.P.)	1.00025

Table 19 ELECTROCHEMICAL EQUIVALENTS

Element	*g-at wt* (g)	*Valence* (+ ionic charge)	*z* (g/c)
aluminum	27.0	3	0.0000932
cadmium	112	2	0.0005824
calcium	40.1	2	0.0002077
chlorine	35.5	1	0.0003674
chromium	52.0	6	0.0000898
chromium	52.0	3	0.0001797
copper	63.5	2	0.0003294
copper	63.5	1	0.0006558
gold	197	3	0.0006812
gold	197	1	0.0020435
hydrogen	1.01	1	0.0000104
lead	207	4	0.0005368
lead	207	2	0.0006150
magnesium	24.3	2	0.0001260
nickel	58.7	2	0.0003041
oxygen	16.0	2	0.0000829
potassium	39.1	1	0.0004051
silver	108	1	0.0011179
sodium	23.0	1	0.0002383
tin	119	4	0.0003075
tin	119	2	0.0006150
zinc	65.4	2	0.0003388

Table 20 RESISTIVITY

(*Temperature* = 20° C)

Material	*Resistivity* (Ω cm)	*Melting point* (°C)
advance	48×10^{-6}	1190
aluminum	2.824×10^{-6}	660
brass	7.00×10^{-6}	900
climax	87×10^{-6}	1250
constantan (Cu 60, Ni 40)	49×10^{-6}	1190
copper	1.724×10^{-6}	1083
german silver (Cu 55, Zn 25, Ni 20)	33×10^{-6}	1100
gold	2.44×10^{-6}	1063
iron	$1\bar{0} \times 10^{-6}$	1535
magnesium	4.6×10^{-6}	651
manganin (Cu 84, Mn 12, Ni 4)	44×10^{-6}	910
mercury	95.783×10^{-6}	−39
monel metal	42×10^{-6}	1300
nichrome	$10\bar{0} \times 10^{-6}$	1500
nickel	7.8×10^{-6}	1452
nickel silver (Cu 57, Ni 43)	49×10^{-6}	1190
plantinum	$1\bar{0} \times 10^{-6}$	1769
silver	1.59×10^{-6}	961
tungsten	5.6×10^{-6}	3410

Table 21 PROPERTIES OF COPPER WIRE

(*Temperature*, 20°C)

American Wire Gauge (B&S)—for any metal			for copper only	
Gauge number	*Diameter* (mm)	*Cross section* (mm^2)	*Resistance* (Ω/km)	(m/Ω)
0000	11.68	107.2	0.1608	6219
000	10.40	85.03	0.2028	4932
00	9.266	67.43	0.2557	3911
0	8.252	53.48	0.3224	3102
1	7.348	42.41	0.4066	2460
2	6.544	33.63	0.5027	1951
3	5.827	26.67	0.6465	1547
4	5.189	21.15	0.8152	1227
5	4.621	16.77	1.028	9729
6	4.115	13.30	1.296	771.5
7	3.665	10.55	1.634	611.8
8	3.264	8.366	2.061	485.2
9	2.906	6.634	2.599	384.8
10	2.588	5.261	3.277	305.1
11	2.305	4.172	4.132	242.0
12	2.053	3.309	5.211	191.9
13	1.828	2.624	6.571	152.2
14	1.628	2.081	8.258	120.7
15	1.450	1.650	10.45	95.71
16	1.291	1.309	13.17	75.90
17	1.150	1.038	16.61	60.20
18	1.024	0.8231	20.95	47.74
19	0.9116	0.6527	26.42	37.86
20	0.8118	0.5176	33.31	30.02
21	0.7230	0.4105	42.00	23.81
22	0.6438	0.3255	52.96	18.88
23	0.5733	0.2582	66.79	14.97
24	0.5106	0.2047	84.21	11.87
25	0.4547	0.1624	106.2	9.415
26	0.4049	0.1288	133.9	7.486
27	0.3606	0.1021	168.9	5.922
28	0.3211	0.08098	212.9	4.697
29	0.2859	0.06422	268.5	3.725
30	0.2546	0.05093	338.6	2.954
31	0.2268	0.04039	426.9	2.342
32	0.2019	0.03203	538.3	1.858
33	0.1798	0.02540	678.8	1.473
34	0.1601	0.02014	856.0	1.168
35	0.1426	0.01597	1079	0.9265
36	0.1270	0.01267	1361	0.7347
37	0.1131	0.01005	1716	0.5827
38	0.1007	0.007967	2164	0.4621
39	0.08969	0.006318	2729	0.3664
40	0.07987	0.005010	3441	0.2906

Table 22 THE CHEMICAL ELEMENTS

(The more common elements are printed in color)

Name of element	*Symbol*	*Atomic number*	*Atomic weight*	*Name of element*	*Symbol*	*Atomic number*	*Atomic weight*
actinium	Ac	89	[227]	lawrencium	Lr	103	[257]
aluminum	Al	13	26.9815	lead	Pb	82	207.19
americium	Am	95	[243]	lithium	Li	3	6.939
antimony	Sb	51	121.75	lutetium	Lu	71	174.97
argon	Ar	18	39.948	magnesium	Mg	12	24.312
arsenic	As	33	74.9216	manganese	Mn	25	54.9380
astatine	At	85	[210]	mendelevium	Md	101	[256]
barium	Ba	56	137.34	mercury	Hg	80	200.59
berkelium	Bk	97	[249*]	molybdenum	Mo	42	95.94
beryllium	Be	4	9.0122	neodymium	Nd	60	144.24
bismuth	Bi	83	208.980	neon	Ne	10	20.183
boron	B	5	10.811	neptunium	Np	93	[237]
bromine	Br	35	79.904	nickel	Ni	28	58.71
cadmium	Cd	48	112.40	niobium	Nb	41	92.906
calcium	Ca	20	40.08	nitrogen	N	7	14.0067
californium	Cf	98	[251]	nobelium	No	102	[254]
carbon	C	6	12.01115	osmium	Os	76	190.2
cerium	Ce	58	140.12	oxygen	O	8	15.9994
cesium	Cs	55	132.905	palladium	Pd	46	106.4
chlorine	Cl	17	35.453	phosphorus	P	15	30.9738
chromium	Cr	24	51.996	platinum	Pt	78	195.09
cobalt	Co	27	58.9332	plutonium	Pu	94	[242]
copper	Cu	29	63.546	polonium	Po	84	[210*]
curium	Cm	96	[247]	potassium	K	19	39.102
dysprosium	Dy	66	162.50	praseodymium	Pr	59	140.097
einsteinium	Es	99	[254]	promethium	Pm	61	[147*]
erbium	Er	68	167.26	protactinium	Pa	91	[231]
europium	Eu	63	151.96	radium	Ra	88	[226]
fermium	Fm	100	[253]	radon	Rn	86	[222]
fluorine	F	9	18.9984	rhenium	Re	75	186.2
francium	Fr	87	[223]	rhodium	Rh	45	102.905
gadolinium	Gd	64	157.25	rubidium	Rb	37	85.47
gallium	Ga	31	69.72	ruthenium	Ru	44	101.07
germanium	Ge	32	72.59	samarium	Sm	62	150.35
gold	Au	79	196.967	scandium	Sc	21	44.956
hafnium	Hf	72	178.49	selenium	Se	34	78.96
helium	He	2	4.0026	silicon	Si	14	28.086
holmium	Ho	67	164.930	silver	Ag	47	107.868
hydrogen	H	1	1.00797	sodium	Na	11	22.9898
indium	In	49	114.82	strontium	Sr	38	87.62
iodine	I	53	126.9044	sulfur	S	16	32.064
iridium	Ir	77	192.2	tantalum	Ta	73	180.948
iron	Fe	26	55.847	technetium	Tc	43	[99*]
krypton	Kr	36	83.80	tellurium	Te	52	127.60
kurchatovium	Ku	104	[257]	terbium	Tb	65	158.924
lanthanum	La	57	138.91	thallium	Tl	81	204.37

A value given in brackets denotes the mass number of the isotope of longest known half-life, or for those marked with an asterisk, a better known one. The atomic weights of most of these elements are believed to have no error greater than ± 0.5 of the last digit given.

Table 22 THE CHEMICAL ELEMENTS (cont'd)

(The more common elements are printed in color)

Name of element	*Sym-bol*	*Atomic number*	*Atomic weight*	*Name of element*	*Sym-bol*	*Atomic number*	*Atomic weight*
thorium	Th	90	232.038	vanadium	V	23	50.942
thulium	Tm	69	168.934	xenon	Xe	54	131.30
tin	Sn	50	118.69	ytterbium	Yb	70	173.04
titanium	Ti	22	47.90	yttrium	Y	39	88.905
tungsten	W	74	183.85	zinc	Zn	30	65.37
uranium	U	92	238.03	zirconium	Zr	40	91.22

Table 23 MASSES OF SOME NUCLIDES

Element	*Symbol*	*Atomic mass** (u)**	*Element*	*Symbol*	*Atomic mass** (u)**
hydrogen	$^{1}_{1}H$	1.007825	sodium	$^{23}_{11}Na$	22.98977
deuterium	$^{2}_{1}H$	2.01410	magnesium	$^{24}_{12}Mg$	23.98504
helium	$^{3}_{2}He$	3.01603		$^{25}_{12}Mg$	24.98584
	$^{4}_{2}He$	4.00260		$^{26}_{12}Mg$	25.98259
lithium	$^{6}_{3}Li$	6.01513	chlorine	$^{35}_{17}Cl$	34.96885
	$^{7}_{3}Li$	7.01601		$^{37}_{17}Cl$	36.96590
beryllium	$^{6}_{4}Be$	6.0198	potassium	$^{39}_{19}K$	38.96371
	$^{8}_{4}Be$	8.00531		$^{41}_{19}K$	40.96184
	$^{9}_{4}Be$	9.01219	krypton	$^{95}_{36}Kr$	94.9
boron	$^{10}_{5}B$	10.01294	molybdenum	$^{100}_{42}Mo$	99.9076
	$^{11}_{5}B$	11.00931	silver	$^{107}_{47}Ag$	106.9041
carbon	$^{12}_{6}C$	12.00000		$^{109}_{47}Ag$	108.9047
	$^{13}_{6}C$	13.00335	technetium	$^{137}_{52}Te$	137.0000***
nitrogen	$^{12}_{7}N$	12.0188	barium	$^{138}_{56}Ba$	137.9050
	$^{14}_{7}N$	14.00307	lead	$^{214}_{82}Pb$	213.9982
	$^{15}_{7}N$	15.00011	bismuth	$^{214}_{83}Bi$	213.9972
oxygen	$^{16}_{8}O$	15.99491	polonium	$^{218}_{84}Po$	218.0089
	$^{17}_{8}O$	16.99914	radon	$^{222}_{86}Rn$	222.0175
	$^{18}_{8}O$	17.99916	radium	$^{228}_{88}Ra$	228.0303
fluorine	$^{19}_{9}F$	18.99840	uranium	$^{235}_{92}U$	235.0439
neon	$^{20}_{10}Ne$	19.99244		$^{238}_{92}U$	238.0508
	$^{22}_{10}Ne$	21.99138	plutonium	$^{239}_{94}Pu$	239.0522

*Atomic mass of neutral atom is given.
**1 atomic mass unit (u) = 1.66043×10^{-27} kg.
***Mass not accurately known. Maximum possible value is given.

Table 24 FOUR-PLACE LOGARITHMS

n	0	1	2	3	4	5	6	7	8	9
10	0000	0043	0086	0128	0170	0212	0253	0294	0334	0374
11	0414	0453	0492	0531	0569	0607	0645	0682	0719	0755
12	0792	0828	0864	0899	0934	0969	1004	1038	1072	1106
13	1139	1173	1206	1239	1271	1303	1335	1367	1399	1430
14	1461	1492	1523	1553	1584	1614	1644	1673	1703	1732
15	1761	1790	1818	1847	1875	1903	1931	1959	1987	2014
16	2041	2068	2095	2122	2148	2175	2201	2227	2253	2279
17	2304	2330	2355	2380	2405	2430	2455	2480	2504	2529
18	2553	2577	2601	2625	2648	2672	2695	2718	2742	2765
19	2788	2810	2833	2856	2878	2900	2923	2945	2967	2989
20	3010	3032	3054	3075	3096	3118	3139	3160	3181	3201
21	3222	3243	3263	3284	3304	3324	3345	3365	3385	3404
22	3424	3444	3464	3483	3502	3522	3541	3560	3579	3598
23	3617	3636	3655	3674	3692	3711	3729	3747	3766	3784
24	3802	3820	3838	3856	3874	3892	3909	3927	3945	3962
25	3979	3997	4014	4031	4048	4065	4082	4099	4116	4133
26	4150	4166	4183	4200	4216	4232	4249	4265	4281	4298
27	4314	4330	4346	4362	4378	4393	4409	4425	4440	4456
28	4472	4487	4502	4518	4533	4548	4564	4579	4594	4609
29	4624	4639	4654	4669	4683	4698	4713	4728	4742	4757
30	4771	4786	4800	4814	4829	4843	4857	4871	4886	4900
31	4914	4928	4942	4955	4969	4983	4997	5011	5024	5038
32	5051	5065	5079	5092	5105	5119	5132	5145	5159	5172
33	5185	5198	5211	5224	5237	5250	5263	5276	5289	5302
34	5315	5328	5340	5353	5366	5378	5391	5403	5416	5428
35	5441	5453	5465	5478	5490	5502	5514	5527	5539	5551
36	5563	5575	5587	5599	5611	5623	5635	5647	5658	5670
37	5682	5694	5705	5717	5729	5740	5752	5763	5775	5786
38	5798	5809	5821	5832	5843	5855	5866	5877	5888	5899
39	5911	5922	5933	5944	5955	5966	5977	5988	5999	6010
40	6021	6031	6042	6053	6064	6075	6085	6096	6107	6117
41	6128	6138	6149	6160	6170	6180	6191	6201	6212	6222
42	6232	6243	6253	6263	6274	6284	6294	6304	6314	6325
43	6335	6345	6355	6365	6375	6385	6395	6405	6415	6425
44	6435	6444	6454	6464	6474	6484	6493	6503	6513	6522
45	6532	6542	6551	6561	6571	6580	6590	6599	6609	6618
46	6628	6637	6646	6656	6665	6675	6684	6693	6702	6712
47	6721	6730	6739	6749	6758	6767	6776	6785	6794	6803
48	6812	6821	6830	6839	6848	6857	6866	6875	6884	6893
49	6902	6911	6920	6928	6937	6946	6955	6964	6972	6981
50	6990	6998	7007	7016	7024	7033	7042	7050	7059	7067
51	7076	7084	7093	7101	7110	7118	7126	7135	7143	7152
52	7160	7168	7177	7185	7193	7202	7210	7218	7226	7235
53	7243	7251	7259	7267	7275	7284	7292	7300	7308	7316
54	7324	7332	7340	7348	7356	7364	7372	7380	7388	7396

Table 24 FOUR-PLACE LOGARITHMS (cont'd)

n	0	1	2	3	4	5	6	7	8	9
55	7404	7412	7419	7427	7435	7443	7451	7459	7466	7474
56	7482	7490	7497	7505	7513	7520	7528	7536	7543	7551
57	7559	7566	7574	7582	7589	7597	7604	7612	7619	7627
58	7634	7642	7649	7657	7664	7672	7679	7686	7694	7701
59	7709	7716	7723	7731	7738	7745	7752	7760	7767	7774
60	7782	7789	7796	7803	7810	7818	7825	7832	7839	7846
61	7853	7860	7868	7875	7882	7889	7896	7903	7910	7917
62	7924	7931	7938	7945	7952	7959	7966	7973	7980	7987
63	7993	8000	8007	8014	8021	8028	8035	8041	8048	8055
64	8062	8069	8075	8082	8089	8096	8102	8109	8116	8122
65	8129	8136	8142	8149	8156	8162	8169	8176	8182	8189
66	8195	8202	8209	8215	8222	8228	8235	8241	8248	8254
67	8261	8267	8274	8280	8287	8293	8299	8306	8312	8319
68	8325	8331	8338	8344	8351	8357	8363	8370	8376	8382
69	8388	8395	8401	8407	8414	8420	8426	8432	8439	8445
70	8451	8457	8463	8470	8476	8482	8488	8494	8500	8506
71	8513	8519	8525	8531	8537	8543	8549	8555	8561	8567
72	8573	8579	8585	8591	8597	8603	8609	8615	8621	8627
73	8633	8639	8645	8651	8657	8663	8669	8675	8681	8686
74	8692	8698	8704	8710	8716	8722	8727	8733	8739	8745
75	8751	8756	8762	8768	8774	8779	8785	8791	8797	8802
76	8808	8814	8820	8825	8831	8837	8842	8848	8854	8859
77	8865	8871	8876	8882	8887	8893	8899	8904	8910	8915
78	8921	8927	8932	8938	8943	8949	8954	8960	8965	8971
79	8976	8982	8987	8993	8998	9004	9009	9015	9020	9025
80	9031	9036	9042	9047	9053	9058	9063	9069	9074	9079
81	9085	9090	9096	9101	9106	9112	9117	9122	9128	9133
82	9138	9143	9149	9154	9159	9165	9170	9175	9180	9186
83	9191	9196	9201	9206	9212	9217	9222	9227	9232	9238
84	9243	9248	9253	9258	9263	9269	9274	9279	9284	9289
85	9294	9299	9304	9309	9315	9320	9325	9330	9335	9340
86	9345	9350	9355	9360	9365	9370	9375	9380	9385	9390
87	9395	9400	9405	9410	9415	9420	9425	9430	9435	9440
88	9445	9450	9455	9460	9465	9469	9474	9479	9484	9489
89	9494	9499	9504	9509	9513	9518	9523	9528	9533	9538
90	9542	9547	9552	9557	9562	9566	9571	9576	9581	9586
91	9590	9595	9600	9605	9609	9614	9619	9624	9628	9633
92	9638	9643	9647	9652	9657	9661	9666	9671	9675	9680
93	9685	9689	9694	9699	9703	9708	9713	9717	9722	9727
94	9731	9736	9741	9745	9750	9754	9759	9763	9768	9773
95	9777	9782	9786	9791	9795	9800	9805	9809	9814	9818
96	9823	9827	9832	9836	9841	9845	9850	9854	9859	9863
97	9868	9872	9877	9881	9886	9890	9894	9899	9903	9908
98	9912	9917	9921	9926	9930	9934	9939	9943	9948	9952
99	9956	9961	9965	9969	9974	9978	9983	9987	9991	9996

Appendix C —Equipment and Supplies

Quantities listed are for *12 stations.* For a smaller number of stations the quantities should be reduced accordingly. The *Equipment* lists are subdivided for Mechanics, Heat, Sound, Light, Electricity and Electronics, and Atomic and Nuclear Physics. *Supplies* listed cover the complete range of experiments. Equipment needed in more than one section is listed in each section in which it is used. *Note to Teacher:* if this list is used for drawing up an order for equipment, please bear in mind that there is some duplication of standard items such as balances, beakers, meter sticks, etc.

EQUIPMENT LIST
(Mechanics: Exps. 1–19)

12	Acceleration board, Packard's
6	Air table
12	Balance, 0.01-g sensitivity
12	Balance, inertia
12	Balance, platform, with set of masses, 1-1000 g
12	Balance, triple beam
36	Balance, spring, 20-n capacity (2000-g capacity may be used)
24	Beaker, 100 ml
12	Block, wood, rectangular
12	Block, wood, rectangular, paraffin coated
12	Buret, 50 ml
24	Car, dynamics
12	Car, metal (Hall's car)
12	Clamp, buret
48	Clamp, C, 100-g to 150-g mass
12	Compass, pencil
12	Composition of forces apparatus
12	Crane boom, simple
12	Cylinder, numbered, of brass or aluminum
12	Dry cell, No. 6
12 yds	Fishline, nylon, 27 lb test
12	Glass tube, 15 cm × 1.0 cm O.D.
12	Graduated cylinder, 100-ml or 250-ml capacity
1	Hard-rubber rod
12	Hooke's law apparatus
12	Inclined plane board and table support
12	J-tube apparatus
12	Level, spirit, for Packard's acceleration board
1 roll	Masking tape
12	Medicine dropper
12	Meter stick
60	Meter stick clamp
12	Meter stick support, knife edge
12	Micrometer caliper, metric
1 lb	Nails, small
1 box	Paper clips
12	Pendulum bob, metal
12	Pendulum bob, wood
12	Pendulum clamp
12	Petri dish
6	Polaroid camera and stand
12	Protractor
6 sets	Pucks, for air table
24	Pulley, single sheave
24	Pulley, 2 or 3 sheaves
12	Pulley, single sheave, with table clamp
12	Ring, iron, 1 in.
36	Ring, iron, 3 in.
12	Ring stand
12	Rubber stopper, No. 6, 1-hole
12	Ruler, metric, 15-cm or 30-cm length
12	Solid, irregular, lumps of coal, rock, or metal
24	Solids, various, with masses less than 1 kg
1	Stop watch, 10 sec sweep
1 ball	String
6	Strobe light, white, for air table
12	Switch, knife, SPST
12	Thermometer, Celsius − 10° to 110°
1 spool	Thread, No: 8
12	Timer, recording with accessories
1 ball	Twine
12	Vernier caliper, metric
60	Weight hanger
12	Weights, slotted, set
2 lb	Wire, annunciator, No. 18
¼ lb	Wire, copper, bare, No. 18
¼ lb	Wire, copper, bare, No. 22
¼ lb	Wire, copper, bare, No. 30

EQUIPMENT LIST (Heat: Exps. 20–25)

12	Asbestos board, 30 cm × 30 cm
12	Balance, platform, with masses, or triple beam
12	Beaker, 100 ml
12	Beaker, 600 ml
24	Block, metal, assortment of lead, aluminum, brass, and copper
12	Burner and tubing

12	Calorimeter
12	Charles' law tube
24	Clamp, buret
12	Coefficient of linear expansion apparatus
8 ft	Glass tubing, 6 mm
12	Hydrometer jar
12	Magnifier
12	Meter stick
4	Pan, for ice (pneumatic trough may be used)
12	Ring, iron
12	Ring stand, tall, or table support
24	Rod for expansion apparatus, assortment of aluminum, brass, copper, and steel
12 ft	Rubber tubing, 3/16 in.
30 ft	Rubber tubing, 1/4 in.
12	Steam boiler
12	Stirring rod
12	Test tube, 125 mm × 15 mm
12	Thermometer, Celsius -1° to 101° in 0.2° div.
12	Tripod for steam boiler
12	Towel
1 ball	Twine
12	Water trap
12	Wire gauze

EQUIPMENT LIST (Sound: Exps. 26-29)

6 sets	Coil spring: Slinky and heavy metal coil matched to give approximately equal amplitudes of reflected and transmitted pulses
2 pkg	Corks, student assortment of 100
6	Glass, flat, shaped like a prism or lens, for ripple tank
12	Glass tube, 2.5 cm-4.0 cm in diameter, 40 cm long
12	Hydrometer jar
12	Meter stick
4	Paint brush, 1 in.
18	Paraffin blocks or separators
6	Protractor
6	Ripple tank, with attachments to produce linear waves and waves from a point source
12 ft	Rubber hose, 2.0 cm O.D.
1	Sheet of metal, thin, bent in the shape of concave and convex mirrors, and other irregular shapes, for ripple tank
1	Stop watch
6	Stroboscope, hand
12	Thermometer, Celsius -10° to 110°
24	Tuning fork, assortment of C, E, G, and C′ forks, concert pitch. (Physical pitch forks may be used.)
12	Tuning fork hammer
1	Vibrograph
12	Vibrograph glass plate or strip of paper
6	Wooden rod, 2.0 cm O.D.

EQUIPMENT LIST (Light: Exps. 30-43)

12	Bristol board, white, at least 12.5 × 15 cm
12	Bunsen burner, with hose
12	Calcite (Iceland spar) crystal, approx. 2 cm long
12	Candle, or small illuminated object
12 pkg	Cellophane sheets, for thickness plates
12	China-marking pencil
12	Cobalt glass filter
12 sets	Color filters, various colors
1 pkg	Construction paper, containing red, orange, yellow, green, blue, and violet
12	Diffraction grating, transmission replica, about 4×10^3 lines/cm
12	Diffraction slits, single and double, of various widths and spacings
12	Evaporating dish, size 00
12	Ferrotype plate, 5 × 5 cm
12	Glass bottle, molded
12	Glass plate, 7 × 7 × 0.9 cm
1 gro	Glass plate, 5 × 5 cm
12	Glass prism, equilateral faces, 7.5 cm long and 8 mm thick
1 set	Glass slides, including red, yellow, green and blue, to fit color apparatus
30	Glass stirring rod
12	Grating holder and support for meter stick

12	Image screen, Bristol board with metric scale
6	Image screen, 25-cm square
1 pkg	Index cards, white
12	Lamp, clear glass, straight filament
12	Lamp, new 40 watt
36	Lamps, assorted incandescent, including 25 w, 40 w, 60 w
12	Lamp socket, standard base, with extension cord and plug
24	Lens, converging, 5-cm focal length, 3.75-cm diam
12	Lens, converging, 10-cm focal length, 3.75-cm diam
12	Lens, converging, 15-cm focal length, 3.75-cm diam
12	Lens, converging, 25 to 35-cm focal length, 3.75-cm diam
12	Lens, diverging, 10-cm focal length, 3.75-cm diam
12	Lens, diverging, 15-cm focal length, 3.75-cm diam
24	Lens holder, for 3.75-cm diam lens
12	Magnifier
1	Mercury vapor discharge tube and power supply
12	Meter, galvanometer, zero-center
12	Meter stick
24	Meter stick supports
12	Metric scale and slit, for meter stick mounting
36	Metric paper scale, 20 cm, for horizontal use
12	Metric rule, steel, graduated in 0.5 mm
12	Microscope, 16 mm (10×) and 32 mm (4×) objectives. (Borrow from Biology Dept.)
12	Microscope objective lens, 32 mm (4×) (Needed only if 16 mm (10×) objective is not divisible)
½ gro	Microscope slides 2.5 × 7.5 cm
12	Mirror, plane, rectangular
12	Mirror, spherical, 4.0 cm, concave and convex, 25-cm focal length
12	Object box, with electric lamp, extension cord and plug
12	Object screen
12	Object screen, black Bristol board with triangular wire-gauze aperture
1 ream	Paper, drawing
12	Photometer, Joly or Bunsen, or *photoelectric
12	Photoelastic specimens, U-shaped transparent plastic
1 box	Pins, large, straight
12 pr	Polaroid disks, 4-cm diam, or 5-cm square
1 set	Poster paints, containing red, yellow, green, and blue
12	Protractor
12	Ring stand, rectangular base, with 1-1 in ring and 1-2 in ring
1 box	Rubber bands
12	Ruler, metric
36	Screen holder
1	Tape measure, metric
12	Vernier caliper, metric
2 lb	Wire, copper, annunciator, No. 18
12	Wooden block, rectangular, to support plane mirror

EQUIPMENT LIST
(Electricity and Electronics: Exps. 44–63)

12	Ampere's law stand
12	Balance, platform, accurate to 0.1 g
12	Balance, triple beam, 0.01 g sensitivity
12 sets	Balance masses, brass, in wood block, 1 g to 1 kg
12	Ball, brass, drilled, 1 in. diam
36	Battery, "B", 22.5-v and 45-v terminals, Burgess 5308 or equivalent
12	Battery, "C", 9 v
12	Battery, 12-v storage or dry (Burgess 732 or equivalent)
12	Boiler, or large beaker
12	Burner, with tripod
12	Calorimeter, electric, 2-5 ohm heating coil
12	Capacitor, 25 microfarads, 25 v
24	Capacitor, 50 microfarads, 25 v
1	Capacitor, demonstration, for gold-leaf electroscope

1	Capacitor decade box, 0.01 microfarad to 1.1 microfarad in 0.01 microfarad steps
24	Coil, for induction, on brass spool to accommodate 19 × 6 mm bar magnet
12	Coil set, primary and secondary, student type
48	Compass, 1-cm diam
12	Compass, 5-cm diam
6 doz	Connector, brass, double
12	Contact key, or push-button switch
4 ft^2	Copper sheet
6 pkg	Cross-section paper, cm-square
12	Demonstration cell, student form
12	Diode, type 6H6
48	Dry cell, standard No. 6
12	Electric bell, 1-3 v
12	Electrode, carbon, flat, 12.5 × 2 cm
12	Electrode, copper, flat, 12.5 × 2 cm
24	Electrode, lead, flat 12.5 × 2 cm
24	Electrode, zinc, flat, 12.5 × 2 cm
1	Electroscope, gold leaf, demonstration type
12	Electroscope, leaf, flask type
12	Electroscope, pith-ball
12	Enameled pan
12	Exciting pad, fur or wool
12	Exciting pad, silk
12	Extension cord, split line, and plug
3	Filter chokes ranging from 8 h to 30 h
24	Flexible leads, short, with alligator clips
12	Friction rod, ebonite or hard rubber
12	Friction rod, glass
24	Fuse, 6 a, standard screw base
12	Fuse block, double, for standard screw base
1	Galvanometer, lecture form, zero center
12	Galvanoscope, 3 windings of 1, 25, and 100 times
12	Lamp, tungsten, 40 watt
48	Lamp, tungsten, 60 watt
12	Lamp, 6 v, No. 40 miniature screw base, 150 ma
12	Lamp, 6 v, carbon, standard base
1	Lamp, 117 v, clear glass, 7-25 watts
12 pr	Magnets, bar, 19 × 6 mm cross section
12	Magnet, horseshoe, with keeper
12	Magnet core, iron, 10 cm long, 8 mm diam
12	Magnifier, 3 to 10×
12	Meter, a-c ammeter, 0-3 a
12	Meter, a-c milliammeter, 0-50/200/500 ma
12	Meter, a-c voltmeter, 0-7.5 v
12	Meter, a-c voltmeter, 0-150/300/600 v
1	Meter, a-c/d-c VTVM, multirange
12	Meter, d-c ammeter, 0-1/3/30 a
12	Meter, galvanometer, zero-center
12	Meter, d-c milliammeter, 0-15/50/150 ma
12	Meter, d-c voltmeter, 0-3/7.5/15 v
12	Meter, d-c voltmeter, 0-150 v, 20,000 ohms/volt
12	Motor, St. Louis, with 2 bar magnets and electromagnet
12	Octal socket, mounted
12	Potentiometer, 1000 ohms, 4 watts
12	Potentiometer, 4000 ohms, 10 watts
12	Power supply, variable d-c
1	Power supply, 6.3 vac at 0.3 a, 180 vdc at 10 ma
12	Resistance box, plug type, of the order of 100 ohms
2	Resistance box, plug type, of the order of 1000 ohms
1	Resistance box, plug type, of the order of 10,000 ohms
12 sets	Resistance spools, each set consisting of the following spools: 30 ga - 200 cm - nickel silver 28 ga - 200 cm - nickel silver 30 ga - 160 cm - nickel silver 30 ga - 2000 cm - copper (Constantan or German silver may be used instead of nickel silver)
12	Resistor, 10,000 ohms, 5 watts
12	Resistor, 20,000 ohms, 5 watts
12	Resistor, 40,000 ohms, 5 watts
12	Resistor, 80,000 ohms, 5 watts
12	Rheostat, 10 ohms, 25 watts
12	Rheostat, 50 ohms, 25 watts
1	Rheostat, 200 ohms (for galvanometers of the order of 100-ohms resistance), or 500 ohms (for galvanometers of the order of 50-ohms resistance)

12	Rheostat, tubular, 25 ohms
1 pkg	Sandpaper, fine grit
12	Stop watch, 10-second sweep
12	Storage battery, 12 v, tapped for 6-v service
12	Switch, knife, DPDT
12	Switch, knife, DPST
24	Switch, knife, SPST
12	Temperature coil, copper, approx. 3 ohms
1 pr	Test leads, 1000-v insulation
24	Test lead, with alligator clips
12 pr	Test lead, with pointed probes
12	Test tube, 200 mm × 25 mm
12	Test tube, 150 mm × 18 mm
12	Thermometer, Celsius, $-10°$ to $110°$
1 spool	Thread, silk
12	Transformer, 6-v filament
12	Triode, type 6J5
2 ft	Tubing, rubber, 6 mm I.D.
1 lb	Twine, cotton
12	Wheatstone bridge, slide-wire form
2 lb	Wire, annunciator, No. 18
1 spool	Wire, copper, bare, No. 14
1 spool	Wire, copper, bare, No. 30
2 lb	Wire, copper, insulated, No. 28

EQUIPMENT LIST
(Atomic and Nuclear Physics: Exps. 64–67)

1	Cat's fur
12	Dish, large, such as ripple tank
1	Electroscope
1	Geiger tube and associated counting apparatus
1	Hard-rubber rod
1 roll	Masking tape
12	Manila folder
12	Masonite or plywood, about 30 cm × 40 cm
12	Medicine dropper
1	Radioactive mineral sample, or radioactive isotope
1	Radioactive sample
1	Radioactivity sample holder
12	Ring stand, tall, or table support
12	Ruler, metric, 15 cm or 30 cm
1	Stop watch
20	Sheet, aluminum, 10 cm square, 1.0 mm thick
15	Sheet, cardboard, 10 cm square, 1.0 mm thick
15	Sheet, lead, 10 cm square, 1.0 mm thick
1	Uranium, metal or compound sample

SUPPLIES

1 lb	Acetamide crystals
6 lb	Acid, hydrochloric, tech
7 lb	Acid, nitric, tech
18 lb	Acid, sulfuric, tech
4 gal	Alcohol, ethyl, denatured
	Chalk dust, as needed
3 lb	Copper sulfate, tech
5 gal	Distilled water
	Ice cubes, as needed
1 oz	Lycopodium powder
16 lb	Mercury, tech
1 lb	Naphthalene crystals
1 oz	Oleic acid
1 pt	Rubber cement
1 pt	Whiting suspended in alcohol

Appendix D

Mathematics Refresher

1. Introduction.

The topics selected for this Mathematics Refresher are those which sometimes trouble physics students. You may want to study this section before working on certain types of physics problems. The presentation here is not so detailed as that given in mathematics textbooks, but there is sufficient review to help you perform certain types of mathematical operations.

The following references are given for your convenience in reviewing these topics also:

Section 2.6 defines *significant figures* and gives rules for their notation.

Section 2.7 explains the *exponential notation* system.

Section 2.12 explains the *orderly procedure* you should use *in problem solving*, and the proper method of handling units in computations.

2. The slide rule.

With the slide rule, you can perform many mathematical operations in a fraction of the time that it would take you to do them on paper.

1. *Locating numbers.* The numbers on a slide rule are not evenly spaced. The upper figure (below) shows the location of the number 246 on the ***C*** and ***D*** scales. The lower figure shows the location of 1865.

2. *Multiplication.* The scales on a slide rule have a number 1 at each end. These end numbers are called *indices*. To multiply, set an index of the ***C*** scale above one of the factors on the ***D*** scale. Then slide the cursor (the transparent window containing a hairline) to the other factor on the ***C*** scale. The product will be under the hairline on the ***D*** scale. The figure on the next page shows the setting for the problem $18 \times 26 = 468$.

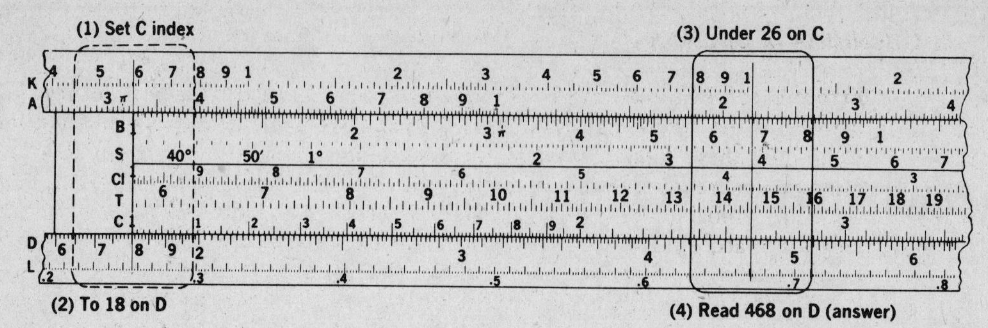

3. *Division.* The process for division is the reverse of the multiplication process. Hence, the figure above also shows the setting for the problem $468 \div 26 = 18$. When problems involve numbers with decimal points, disregard the decimal points when making the settings and place the decimal point in the answer by inspection.

4. *Squares and square roots.* Only the ***A*** and ***D*** scales are needed for these problems. To find the square of a number, move the hairline to that number on the ***D*** scale and read the answer under the hairline on the ***A*** scale. To find the square root of a number, locate the number on the ***A*** scale and read the square root on the ***D*** scale. If the number is greater than 1 and contains an odd number of digits to the left of the decimal point, use the left half of the ***A*** scale. If it has an even number of digits to the left of the decimal point, use the right half of the ***A*** scale. If the number is less than 1, count off the digits to the right of the decimal point in groups of two. If the first group with significant figures contains only 1 significant figure, use the left half ot the ***A*** scale. If it contains two significant figures, use the right half ot the ***A*** scale.

5. *Porportions.* With any setting of the ***C*** and ***D*** scales, all readings opposite each other on the two scales are in the same ratio. For example, in the previous figure, $1:18 = 26:468 = 30:540$, etc.

3. *Conversion of a fraction to a decimal.*

Divide the numerator by the denominator, carrying the answer to the required number of significant figures.

Example

Convert $\frac{45}{85}$ to a decimal.

Solution

Dividing 45 by 85, we find the equivalent decimal to be 0.53, rounded to two significant figures.

```
      0.529
85)45.000
   42 5
    2 50
    1 70
      800
      765
```

4. *Calculation of percentage.*

Errors occur in measurements of physical phenomena. In physics, an error is defined as the difference between experimentally obtained data and the accepted values for these data. Errors in measurement always exist because measuring instruments are inaccurate to some degree. The instruments you use in laboratory work are probably much less accurate than those used by the physicists who obtained the accepted values for the data you find in tables. Errors may also result from your own inaccuracies in measurement. In most experimental work in physics, the percentage error (relative error) is of much greater significance than the actual difference between the observed value and its accepted value (absolute error). Hence, as a physics student you must be able to calculate percentages and percentage error, and convert fractions to percentages and percentages to decimals.

Example

What percentage of 19 is 13?

Solution

We divide 13 by 19, and multiply the quotient by 100%. A multiplication by 100% does not change the value of the quotient, since 100% actually has a value of 1. When we multiply by 100%, we multiply the number by 100, and affix the % sign to the product: $\frac{13}{19} \times 100\% = 68\%$ rounded to two significant figures.

5. *Calculation of percentage error.*

$$\text{Percentage Error} = \frac{\text{Absolute Error}}{\text{Accepted Value}} \times 100\%$$

where the *Absolute Error* is the difference between the *Observed Value* and the *Accepted Value.*

Example

In a laboratory experiment carried out at 20.0° C, a student found the speed of sound in air to be 329.8 m/sec. The accepted value at this temperature is 343.5 m/sec. What was his percentage error?

Solution

$$\begin{aligned}\text{Absolute Error} &= 343.5\ \text{m/sec} - 329.5\ \text{m/sec}\\ &= 13.7\ \text{m/sec}\end{aligned}$$

$$\text{Percentage Error} = \frac{13.7\ \text{m/sec}}{343.5\ \text{m/sec}} \times 100\% = 3.99\%$$

6. *Conversion of fractions to percentages.*

To convert a fraction to a percentage, divide the numerator by the denominator, and multiply the quotient by 100%.

Example

Express $\frac{11}{13}$ as a percentage.

Solution

$\frac{11}{13} \times 100\% = 85\%$, rounded to two significant figures.

7. *Conversion of percentages to decimals.*

To convert a percentage to a decimal, move the decimal point two places to the left, and remove the percent sign.

Example

Convert 62.5% to a decimal.

Solution

If we express 62.5% as a fraction, it becomes 62.5/100. If we actually perform the indicated division, we obtain 0.625 as the decimal equivalent. Observe that the only changes have been to move the decimal point two places to the left, and remove the percent sign.

8. *Proportions.*

Many physics problems involving the temperature, pressure, and volume relationships of gases may be solved by using proportions.

Example

Solve the proportion $\frac{x}{225\text{ ml}} = \frac{273°}{298°}$ for x.

Solution

Multiply both sides of the equation by 225 ml, the denominator of x, giving $x = \frac{225\text{ ml} \times 273°}{298°}$. Solving, $x = 206$ ml.

9. *Fractional Equations.*

The equations for certain problems involving lenses, mirrors, and electric resistances produce fractional equations where the unknown is in the denominator. To solve such equations, clear them of fractions by multiplying each term by the lowest common denominator. Then isolate the unknown, and complete the solution.

Example

Solve $\frac{1}{s_o} + \frac{1}{s_i} = \frac{1}{f}$ for f.

Solution

The lowest common denominator of s_o, s_i, and f is $s_o s_i f$. Multiplying the fractional equation by this product, we obtain

$$\frac{s_o s_i f}{s_o} + \frac{s_o s_i f}{s_i} = \frac{s_o s_i f}{f} \quad \text{or} \quad s_i f + s_o f = s_o s_i$$

Thus, $f(s_o + s_i) = s_o s_i$ and $f = \frac{s_o s_i}{s_o + s_i}$

10. *Equations.*

When a known equation is employed in solving a problem and the unknown quantity is not the one usually isolated, the equation should be solved algebraically to isolate the unknown quantity of the particular problem. Then the values (with units) of the known quantities can be substituted and the indicated operations performed.

Example

From the equation for potential energy, $E_p = mgh$, we are to calculate h.

Solution

Before substituting known values for E_p, m, and g, the unknown term h is isolated and expressed in terms of E_p, m, and g. This is accomplished by dividing both sides of the basic equation by mg.

$$\frac{E_p}{mg} = \frac{mgh}{mg} \quad \text{or} \quad h = \frac{E_p}{mg}$$

If E_p is given in joules, m in kilograms, and g in meters/second², the dimension of h is

$$h = \frac{\text{j}}{\text{kg m/sec}^2} = \frac{\text{kg m}^2\text{/sec}^2}{\text{kg m/sec}^2} = \text{m}$$ Thus, h is expressed in meters.

11. Laws of exponents.

Operations involving exponents may be expressed in general fashion as $a^m \times a^n = a^{m+n}$; $a^m \div a^n = a^{m-n}$; $(a^m)^n = a^{mn}$; $\sqrt[n]{a^m} = a^{m/n}$. For example:

$$x^2 \times x^3 = x^5 \quad 10^5 \div 10^{-3} = 10^8 \qquad (t^2)^3 = t^6 \quad \sqrt[4]{4^2} = 4^{2/4} = 4^{1/2} = 2$$

12. Quadratic formula.

The two roots of the quadratic equation $ax^2 + bx + c = 0$ in which a does not equal zero, are given by the quadratic formula

$$x = \frac{-b \pm \sqrt{b^2 - 4ac}}{2a}$$

13. Right triangles.

In physics some facts about triangles are used to solve problems about forces and velocities.

1. *30°—60°—90° right triangle.* It is useful to remember that the hypotenuse of such a triangle is twice as long as the side opposite the 30° angle. The length of the side opposite the 60° angle is $\mathbf{s}\sqrt{3}$, where **s** is the length of the side opposite the 30° angle.

2. *45°—45°—90° right triangle.* The sides opposite the 45° angles are equal. The length of the hypotenuse is $\mathbf{s}\sqrt{2}$, where **s** is the length of a side.

3. *Trigonometric functions.* Trigonometric functions are ratios of the lengths of sides of a right triangle and depend on the magnitude of one of its acute angles. Using right triangle **ABC**, below, we define the following trigonometric functions of $\angle \mathbf{A}$:

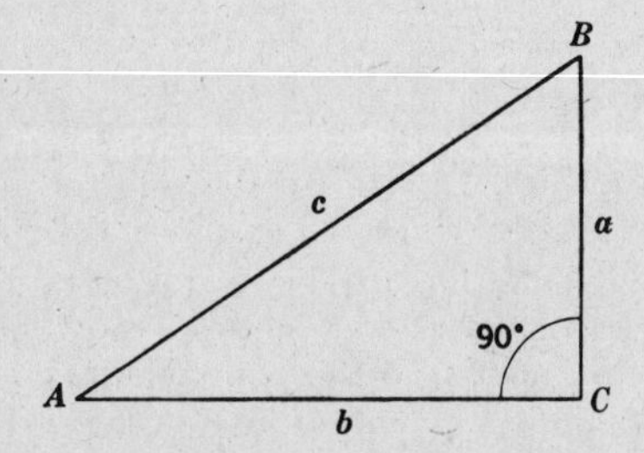

$$\sin \angle \mathbf{A} = \frac{\mathbf{a}}{\mathbf{c}}$$

$$\text{cosine} \angle \mathbf{A} = \frac{\mathbf{b}}{\mathbf{c}}$$

$$\text{tangent} \angle \mathbf{A} = \frac{\mathbf{a}}{\mathbf{b}}$$

These functions are usually abbreviated as sin **A**, cos **A**, and tan **A**. Table 6, Appendix B, gives values of trigonometric functions.

Example

In a right triangle, one side is 25.3 and the hypotenuse is 37.6. Find the angle between these two sides.

Solution

Using the designations in the triangle shown above, $b = 25.3$ and $c = 37.6$. You are to find $\angle A$. Hence you will use the trigonometric function $\cos A = b/c$, or $\cos A = 25.3/37.6 = 0.673$. From Table 6, Appendix B, the cosine of 47.5° is 0.676 and the cosine of 48.0° is 0.669. The required angle is $3/7 \times 0.5°$, or 0.2°, greater than 47.5°. Thus, $\angle A = 47.7°$.

Example

In a right triangle, one angle is 23.8° and the adjacent side (not the hypotenuse) is 43.2. Find the other side.

Solution

Again using the designations in the above triangle, $\angle A = 23.8°$ and $b = 43.2$. The function involving these values and the unknown side, a, is $\tan A = a/b$. Solving for a: $a = b \tan A$. From Table 6, Appendix B, $\tan 23.8° = 0.438$. Thus, $a = 43.2 \times 0.438 = 18.9$.

4. *Sine law and cosine law.* For any triangle **ABC**, below, the sine law and the cosine law enable us to calculate the magnitudes of the remaining sides and angles if the magnitudes of one side and of any other two parts are given.

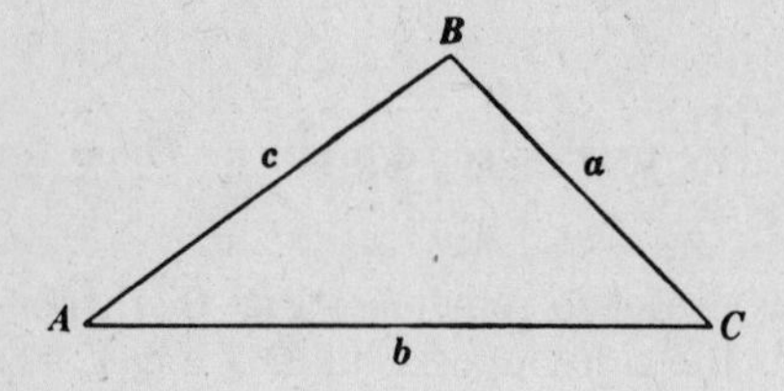

Sine law: $\dfrac{\mathbf{a}}{\sin \mathbf{A}} = \dfrac{\mathbf{b}}{\sin \mathbf{B}} = \dfrac{\mathbf{c}}{\sin \mathbf{C}}$

Cosine law: $\mathbf{a} = \sqrt{\mathbf{b}^2 + \mathbf{c}^2 - 2\mathbf{bc} \cos \mathbf{A}}$

$\mathbf{b} = \sqrt{\mathbf{c}^2 + \mathbf{a}^2 - 2\mathbf{ca} \cos \mathbf{B}}$

$\mathbf{c} = \sqrt{\mathbf{a}^2 + \mathbf{b}^2 - 2\mathbf{ab} \cos \mathbf{C}}$

14. Circles.

The circumference of a circle is $c = \pi d$, where d is the diameter.
The area of a circle is $A = \pi r^2$, where r is the radius.

15. Cylinders.

The volume of a right circular cylinder is $V = \pi r^2 h$, where r is the radius of the base and h is the height.

16. Spheres.

The surface area of a sphere is $A = 4\pi r^2$, where r is the radius.
The volume of a sphere is $V = 4/3\pi r^3$, where r is the radius.

17. Logarithms.

The common logarithm of a number is the exponent or the power to which 10 must be raised in order to obtain the given number. A logarithm is composed of two parts: the *characteristic*, or integral part; and the *mantissa*, or decimal part. The characteristic of the logarithm *of any whole or mixed number* is one less than the number of digits to the left of its decimal point. The characteristic of the logarithm *of a decimal fraction* is always negative and is numerically one greater than the number of zeros immediately to the right of the decimal point. Mantissas are always positive and are read from tables such as Table 24 in Appendix B. Proportional parts are used when numbers having four significant figures are involved. In determining the mantissa the decimal point in the original number is ignored since its position is indicated by the characteristic.

Logarithms are exponents and follow the laws of exponents:

Logarithm of a product = sum of the logarithms of the factors

Logarithm of a quotient = logarithm of the dividend minus logarithm of the divisor

To find the number whose logarithm is given, determine the digits in the number from the table of mantissas. The characteristic indicates the position of the decimal point.

Example

Find the logarithm of 35.76.

Solution

There are two digits to the left of the decimal point. Therefore the characteristic of the logarithm of 35.76 is one less than two, or 1. To find the mantissa, ignore the decimal point and look up 3576 in Table 24, Appendix B. Read down the *n* column to 35. Follow this row across to the 7 column, where you will find 5527. This is the mantissa of 3570. Similarly, the mantissa for 3580 is 5539. To calculate the mantissa for 3576, take 6/10 of the difference between 5527 and 5539, or 7, and add it to 5527. Thus, the required mantissa is 5534, and the complete logarithm of 35.76 is 1.5534.

Example

Find the logarithm of 0.4692.

Solution

There are no zeros immediately to the right of the decimal point, hence the characteristic of the logarithm of 0.4692 is -1. However, only the characteristic is negative; the mantissa is positive and is found from Table 24, Appendix B, to be 6714. The complete logarithm of 0.4692 may be written as $\bar{1}.6714$, or $0.6714 - 1$, or $9.6714 - 10$.

Example

Find the logarithm of (1) 357,600,000 (2) 0.0000003576

Solution

(1) $357{,}600{,}000 = 3.576 \times 10^8$ log = 8.5534

(2) $0.0000003576 = 3.576 \times 10^{-7}$ log = $\bar{7}.5534$, or $0.5534 - 7$, or $3.5534 - 10$